Wireless
Crash
Course

Second Edition

Paul Bedell

McGraw-Hill/Osborne
New York   Chicago   San Francisco   Lisbon
London   Madrid   Mexico City   Milan
New Delhi   San Juan   Seoul   Singapore
Sydney   Toronto

McGraw-Hill

A Division of The McGraw-Hill Companies

Copyright © 2005 by The McGraw-Hill Companies, Inc. All rights reserved. Printed in the United States of America. Except as permitted under the United States Copyright Act of 1976, no part of this publication may be reproduced or distributed in any form or by any means, or stored in a data base or retrieval system, without the prior written permission of the publisher.

1 2 3 4 5 6 7 8 9 0 DOC/DOC 0 1 9 8 7 6 5

ISBN 0-07-145280-X

The sponsoring editors for this book were Steve Chapman and Jane Brownlow and the production supervisor was David Zielonka. It was set in Century Schoolbook by MacAllister Publishing Services, LLC.

Printed and bound by R. R. Donnelley & Sons.

To Margot, Steve, Aimee, and Mark

To Ric, for always being there

And to a budding 8-year-old writer,
my son Bob

ABOUT THE AUTHOR

Paul Bedell obtained a master of science degree in telecommunications management, with distinction, from Chicago's DePaul University in 1994. In 1995, Bedell designed and has since taught a survey course on cellular and wireless telecommunications in the Graduate School of Computer Science, Telecommunications, and Information Systems at DePaul University. Bedell also developed and launched an advanced wireless engineering and deployment course in spring 2004.

He began his telecommunications career in the United States Army Signal Corps, serving in (West) Germany from 1985 to 1988, where he worked as a multichannel communications equipment operator at a remote signal site. Upon discharge in 1988, Bedell spent 5 years working for several Fortune 500 companies as a telecommunications analyst.

From there, he moved to the wireless industry, where he spent 5 years working for three leading wireless carriers in both the cellular and PCS industries. In those positions, Bedell worked in network planning, fixed network engineering, information technology network planning, and corporate LAN operations.

Bedell moved to Ameritech's long-distance business unit, Ameritech Communications, Inc. (or ACI), in 1998. He managed the implementation of ACI's 42-node data network, supporting its new 2,500-mile Synchronous Optical Network (SONET) system, which spanned five midwestern states. He also managed the installation of two voice over IP (VoIP) networks at ACI, the first-ever VoIP networks installed Ameritech-wide. After the merger with SBC, Bedell moved into SBC product marketing in August 2000, managing a metro area Ethernet product known as Giga-MAN®. Bedell managed the expansion of the product into seven additional states (along with the midwestern states) and oversaw the development of multiple enhancements to the product. In May 2004, Bedell began product management work supporting SBC's Hosted IP Communication Service, a hosted VoIP service. He can be reached at bedell14@sbcglobal.net.

CONTENTS

Contents

Contents

Contents

Contents

ACKNOWLEDGMENTS

As usual, I have to thank my wife and boys for being patient with me and tolerating my absences as I secluded myself in my office for several months—especially during the 2004 holiday season. Their support has no price tag, and I appreciate it more than they know. Paula, Aaron, Ryan, Bob: I salute you!

My old, reliable friend and wireless guru Ric Biederwolf, as always, was there for me when I turned to him for help on verifying the information in many chapters of this book. I asked him many questions out of the blue, and he always, always took the time to answer my questions with clarity, good information, and in a timely manner. Ric has about 20 years of wireless industry experience in designing and engineering just about every type of wireless network technology that exists. His treasure trove of experience and knowledge has helped this book become an accurate reflection of real-world wireless. Ric—you never disappoint, and I appreciate your help and friendship.

Clint Smith, another McGraw-Hill author and wireless guru, also helped me out by sanity checking certain chapters of this new edition. Thanks for your help, Clint.

I would like to thank the following people from Andrew Corporation, a world leader in the manufacture of cell base station and microwave radio equipment, as they supplied me with many of the photos seen in this new edition: Rick Aspan, Bernie Surtz, and Chris McFarland.

Thanks to Rob Strauch of Cingular Wireless for his updates on intercarrier networking and interconnection to the PSTN. Steve Shepard of Shepard Communications gave me valuable insight on some topics. Thanks to Julie Brandt of the International Engineering Consortium, and Luis Rodriguez of the Cellular Telecommunications and Internet Association. I appreciate the help of Mike Catalanotti, Web site proprietor of NewEnglandCellularsites.net, for allowing the use of many base station and tower photos.

Thank you to Mike Pratt and Jonathan Kramer for the use of photos from their awesome web pages (http://kramerfirm.com, and www.prattfamily.demon.co.uk/mikep/gsmnet.html). Anyone looking for excellent photos of base stations and towers is encouraged to go to these sites.

And finally, thanks to the following people for their contributions large and small to the success of this book: David Velazquez, Jessica Becker, Bill Locke, Caprice DeLorm, Melissa Rolnicki, fellow author Steven Shepard, and my friend Alyssa Williams.

PREFACE

Wow! So much has happened to this industry since I finished writing *Wireless Crash Course* in late 2000; it is truly mind boggling. Aside from the Internet explosion, no service technology has changed more than wireless in the last four years. This fact is underscored by witnessing how often many major publications run articles on wireless technology. A January 2005 *Life Magazine* insert in *The Chicago Tribune* ran a cover story on camera phones. *The Chicago Tribune* has featured three editorials on various aspects of wireless technology from 2003 to 2004. *Business Week* ran a cover story on Wi-Fi technology in 2004. *Newsweek Magazine* published a cover story titled "Way Cool Cell Phones" in June 2004. It is plain to see that wireless technology has had a profound, lasting impact on world culture.

It is not necessary to go into the changes to this second edition in this preface—they are amply noted throughout the book and in the table of contents. Even more than in the past, I have strived as much as possible to present the material in this book in a clean, simple manner.

Like the first edition, I try to maintain a real world focus to the wireless industry. Like its predecessor, this book is largely devoid of a heavy-duty, theoretical approach, and that is the way I like it. There is nothing wrong with the theoretical side of wireless if you are a researcher, an antenna or radio frequency engineer, a base station equipment engineer, or a wireless software engineer. But that is not the target audience for this book. Readers of this book should have a basic understanding of telecommunications—although that is not completely necessary—and a desire to understand the underpinnings of how those wonderful cell phones of theirs work. I have found that some of the readers of the first edition were people working in the finance industry who wanted, and needed, to become acclimated to wireless technology. Other readers included people new to the industry who wanted a crash course in its technology.

Like the first edition, I provide overviews of wireless concepts, dive down a few feet, then stop. There are plenty of technical books on wireless available; that is not the aim of this book. This book can be a prerequisite to more technical books, if the reader is looking to start with the basics and move on from there. Reading and comprehending this entire book will present the reader with a great foundation—a summation—of how the wireless world works.

So, what is new with this edition? *Many* things! As with the first edition, I do my best to deliver information in a logical, building-block format to ensure the book blends everything together. You will see how I have largely removed any semblance of analog technology—that has been confined to a brief chapter on the history of the industry. You will also notice how I now refer to existing wireless carriers in two categories: traditional 850 MHz carriers (the original cellular carriers) and 1,900 MHz carriers (the PCS carriers). Please remember that the term "cellular" will always refer to wireless networks because the term reflects their core design: the use of hexagon cells that resemble the honeycombs of bees. I have completely redrawn about 90 percent of the diagrams in the book to make them look more clean and professional, using a more modern version of Microsoft Visio. Another differentiating factor about this edition is the widespread use of actual photos to illustrate concepts, topics, towers, equipment, and cell base stations. Like they say, a picture is worth a thousand words (I know, I know . . . cliché). I have added many sections to the existing chapters from the first edition—for example, a section that explains how spread spectrum radio technology works in the Digital Wireless Technologies chapter. . This chapter has also been completely rewritten. The wireless data chapter has been rewritten from scratch. There is a new chapter devoted entirely to third-generation technology. Many chapters from the first edition have been completely eliminated and several new chapters have been added, namely on Wi-Fi, WiMAX, home networking, and the "new age of cell phones." These are just some of the major changes.

Please remember as you are reading that there are *entire books* out there on many of the topics discussed here. Some of the topics in this book also have *many* books written about them—Wi-Fi and wireless data, for example. If you are wondering why I do not cover some specific topics very extensively, it is because I am covering many, many topics in a survey form. These days, any Internet search using the word "wireless" or most of the topics in this book would result in hundreds or thousands of hits. I have selected what I believe are the best, high-level concepts to cover.

In late summer 2005, the following Web site will go live: www.wireless crashcourse.com. I will periodically post updates or corrections to the book on the site: If errors are spotted in the book, I would certainly appreciate being notified of them there as well. Like the first edition, what sets this book apart from all others that I am aware of is the fact that it covers every conceivable aspect of wireless system operations. It covers many topics that are simply not covered in any other survey books on wireless. To that end, it is my sincere hope that the reader walks away from reading this book feeling like a wireless guru. If you have any questions, my e-mail address is bedell14@sbcglobal.net. Read, enjoy, and thanks.

Regards,
Paul Bedell

Cellular Radio History and Development

1.1 Definition of Cellular Radio: The Cellular Concept

There are two different ways to view the definition of cellular systems from a high-level perspective:

Federal Communications Commission (FCC) definition: A high-capacity land mobile system in which the assigned radio spectrum is divided into discrete channels that are assigned in groups to geographic cells covering a cellular geographic service area (CGSA). The discrete channels are capable of being reused in different cells within the service area through a process known as frequency reuse.

Layman's definition: A system that uses radio transmission rather than physical wirelines to provide telephone service comparable to that of regular business or residential telephone service.

 Key: The cellular concept can be explained as follows: Instead of having just a few radio channels that everyone must share (i.e., citizens' band [CB] radio), cellular radio channels are reused simultaneously in nearby geographic areas, yet customers do not interfere with each other's calls. This is the key concept underlying cellular communications.

The cellular system is similar in functional design to the legacy, circuit-switched public-switched telephone network (PSTN), or *landline* network. It contains subscribers (customers), transmission systems, and switches. The existence and control of the radio function in a wireless network is what differentiates cellular technology from landline telephone service (or the PSTN). When launched in 1983, the cellular radio telephone system was the culmination of all prior mobile communication systems from a development and technology standpoint.

▦ 1.2   Cellular System Objectives

When Bell Labs developed the original cellular concept known as Advanced Mobile Phone Service (AMPS), the major system objectives were the efficient use of radio spectrum and widespread availability. While the cellular concept had been developed earlier, several critical technologies came together simultaneously in the late 1970s to propel the cellular industry forward. These new technologies enabled small, relatively lightweight subscriber equipment to be manufactured cheaply. Vastly improved integrated circuit-manufacturing techniques allowed for major advances in computer technology, as well as the miniaturization of critical equipment elements, especially within portable mobile telephones. Functionally, several fundamental attributes were necessary to make cellular service a reality for the masses:

- ▦ Frequency agility in the radiotelephone, which allows the mobile phone to operate on any given number of frequencies in FCC-assigned radio spectrum (which subsequently allows for call handoff)

- ▦ Call-handoff capability—the act of handing off a call in progress seamlessly throughout a wireless network

- ▦ A contiguous arrangement of cell base stations so that the mobile phone can always operate at acceptable radio signal levels

- ▦ A fully integrated, transparent "fixed' network (backhaul network) to manage these operations

    Key: There are five main components to a cellular (wireless) telephone system: the mobile telephone, the cell base station, the mobile switching center (MSC), the fixed network (transmission systems), and connectivity to the PSTN. Since around 2000, wireless *data* network elements have also become standard fixtures in wireless carrier architectures; this includes connections to the Internet. See Figure 1-1.

Figure 1-1
Five main
components of
wireless
networks

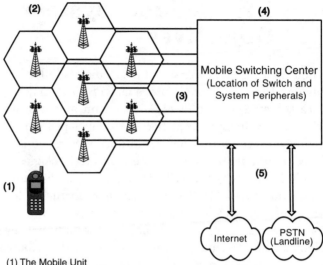

Figure 1-1
Five main
components of
wireless
networks

(1) The Mobile Unit
(2) The Cell Base Station
(3) The Backhaul Network (aka "Fixed Network")
(4) The Mobile Switching Center
(5) Interconnection to PSTN and Internet ("other networks")

 ## 1.3   AMPS: The American Cellular Standard

In an effort to use the airwaves more efficiently, AT&T Bell Labs engineers decided to stretch the limited number of radio frequencies available for mobile service by scattering multiple low-power transmitters, or base stations, throughout a metropolitan area, and *handing off* calls from transmitter to transmitter as customers moved from place to place.

Key: This new technique would allow many more customers to access the systems *simultaneously*. When more capacity was needed, the area served by each transmitter could be "divided" again to create a new base station. The development of this concept was the birth of wireless technology as we know it today.

AMPS is the American analog cellular standard. In 1970, several key developments occurred:

- The FCC set aside new radio frequencies for land-mobile communications. These frequencies were ultrahigh frequency (UHF) television channels in the 800 MHz band that had never been used.
- AT&T proposed to build the first high-capacity cellular telephone system. It dubbed the system AMPS and selected Chicago as the first test city.

At the inception of the cellular industry, the FCC initially granted a total of 666 channels in each market. At first, AT&T thought it would get national rights to *all* cellular frequencies, thereby making AT&T the only national cellular carrier. This would also have made the cellular industry a monopoly. At that time, AT&T never anticipated the growth potential of this pent-up demand by the general public for widespread availability of mobile communication services. They estimated only one million cellular customers would exist by the year 2000. In reality, as of 2005, the wireless subscriber base in the United States alone is over 175 million, while worldwide there are over two billion people using wireless service.

However, bowing to intense pressure from radio common carriers (RCCs), the FCC determined that the cellular industry should have two carriers per market, and 333 channels were allocated per carrier per market. This marked the birth of the A band/B band carrier concept (see Section 1.6). The number of channels was later increased to a total of 832 total cellular channels, or 416 channels per carrier per market. This change was brought about by cellular industry pressure on the FCC to relinquish reserve spectrum to relieve capacity and congestion problems. A detailed review of channelization will be delivered in Chapter 4.

In 1977, although the FCC realized it had to create a regulatory scheme for the new service, the Commission also decided to authorize construction of two developmental cellular systems. One system, located in Chicago, was licensed to Illinois Bell, and a second system, serving Baltimore and Washington, DC, was licensed to a nonwireline company: American Radio Telephone Service (an RCC).

Once the regulatory framework was decided upon by the FCC, the first commercial cellular network began operation in Chicago on October 13, 1983. The very first commercial cellular telephone call was made at

Soldier Field in Chicago to a descendant of Alexander Graham Bell in West Germany. The second system was activated a short time later in the Baltimore/Washington, DC corridor in December 1983. It was these systems that gave rise to the fastest-growing consumer technology in history, an industry that adds tens of thousands of customers per day.

1.4   AMPS Technical Specifications

AMPS will be described in this section in the past tense because it is an analog system whose time has come and gone. Analog systems do not deliver enough capacity or capability to warrant their use in today's wireless networks. AMPS cellular systems used FM radio transmission where available spectral bandwidth was divided into 30 kHz channels, and each channel was capable of carrying one-half of a conversation or serial data stream at a time. Frequency-division multiple access (FDMA) was the analog cellular (AMPS) modulation standard. Two 30 kHz channels were required for every AMPS conversation: one channel for the uplink (mobile to base station) and one channel for the downlink (base station to mobile). FDMA describes the process of subdividing a large block of radio spectrum into many smaller, discrete blocks of spectrum. These smaller blocks of spectrum were then divided into thousands of usable channels. These channels (frequencies) were reused over and over again throughout a wireless system. The 50 MHz of FCC-assigned radio spectrum for cellular licenses in the United States was parceled into channels, and each channel was used for either the transmit or receive portion of a cellular telephone call (uplink or downlink).

   Key: Cellular telephony is a duplex mode of communications and two channels are needed for each call: one channel for transmit and one channel for receive.

AMPS frequencies allocated by the FCC in the United States are as follows:

- Mobile transmit (base receive): 824—849 MHz
- Base transmit (mobile receive): 869–894 MHz

It should be noted that these frequencies are still allocated and used by the 850 MHz carriers operating in the United States today. Due to the evolution of the industry, they are now using these frequencies for digital wireless systems, not analog AMPS. See Figure 1-2 for a listing of AMPS cellular frequency allocations.

1.5 The Cellular Market Structure: MSAs and RSAs

The general concept of a metropolitan statistical area (MSA) is one of a large population nucleus with adjacent communities that have a high degree of economic and social integration within that nucleus.

Each MSA must contain either a place with a minimum population of 50,000 or a Census Bureau-defined urbanized area and a total MSA population of at least 100,000. An MSA comprises one or more counties and may also include one or more outlying counties that have close economic and social relationships with the central county. An outlying county must have a specified level of commute to the central counties and must also meet certain standards regarding metropolitan character, such as population density, urban population, and population growth.

The FCC divided the United States into 734 cellular markets. The structure of personal communication services (PCS) market sizes is different and will be reviewed later in Chapter 8. There are 306 MSAs in the United States, and they are labeled according to the largest city within their boundaries. The smaller, rural markets are known as rural service areas (RSAs); there are 428 RSAs in the United States. RSAs are labeled according to their state, beginning with the number 1, from the north or west side of the state, progressing in sequence to the south or

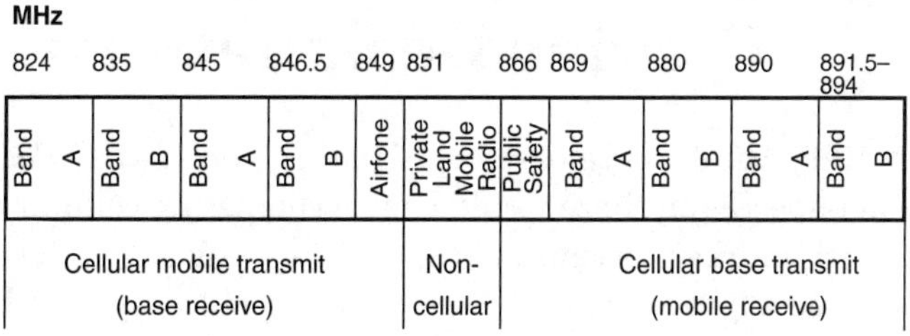

Figure 1-2
AMPS cellular frequency allocations

MHz

	824	835	845	846.5	849	851	866	869	880	890	891.5–894

Band A	Band B	Band A	Band B	Airfone	Private Land Mobile Radio	Public Safety	Band A	Band B	Band A	Band B

| Cellular mobile transmit (base receive) | | | | Non-cellular | | Cellular base transmit (mobile receive) | | | | |

east end of the state. For example, DeKalb, Illinois, is located in Illinois RSA 1 at the north end of the state, and Kentucky RSA 1 is located in the western end of the state of Kentucky.

The FCC developed and demarcated MSAs first, and then, a few years later, it developed and demarcated RSAs. This prioritization underscored the FCC's emphasis on developing markets with higher populations first so cellular carriers could serve higher subscriber densities first. MSA and RSA borders run along county lines, as defined by state governments. Most cellular markets encompass more than one county. See Figure 1-3 for an illustration of MSAs and RSAs in the United States.

Key: MSAs cover approximately 75 percent of the population and 20 percent of the landmass of the United States, while RSAs cover approximately 25 percent of the population and 80 percent of the landmass. In 1983, the FCC decreed that cellular carriers had to provide coverage to 75 percent of the population of a given market or 75 percent of the geographic area of a given market. This rule gave carriers the freedom to build out their markets more efficiently, as they saw fit.

For example, they could concentrate their system buildouts in MSAs to just the urban area itself, in order to serve as many customers as possible as quickly as possible. Conversely, in RSAs, they could spread the buildout throughout the market's key population areas to serve as many people as possible in as short a time frame as possible. For instance, in the states of Wyoming or Arizona, a wireless carrier could build out the market along interstate highways versus desert areas.

1.6   The "A" Carrier and "B" Carrier Designations

In 1981, the FCC released its *Final Report and Order* on cellular systems, specifying there would be two competing cellular carriers licensed in each market. Because of this FCC specification, the cellular industry is a duopoly by decree. A *duopoly* is an industry that by definition has only two competitors. Once PCS licenses were issued by the FCC in 1995, this increased the number of wireless competitors in each market. Once

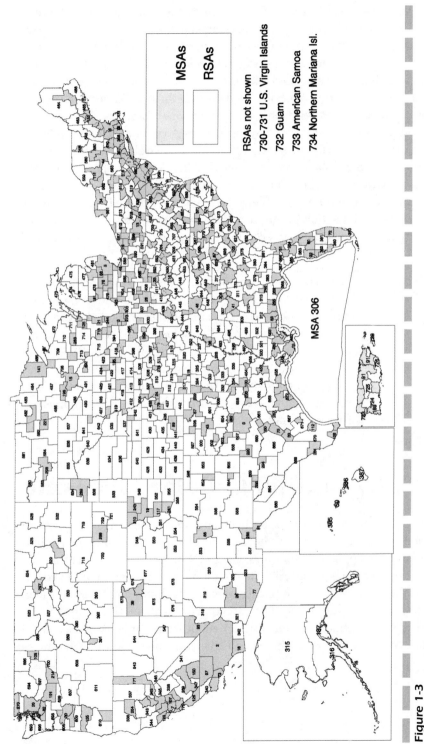

Figure 1-3
Cellular market areas: MSAs and RSAs. These are not the same as PCS markets. (Figure courtesy www.fcc.gov)

its rules were in place, the FCC accepted applications for the 60 largest cities during 1982. The Commission decided to license cellular systems in the 306 MSAs first and then move on to the 428 RSAs. The FCC decreed that one license in each market would be reserved for the local telephone company. This is the so-called wireline license, and the licensee is known as the B carrier in that market. The B stands for Bell, indicating the (B) carrier's association with the local telephone company. The second licensee in cellular markets is known as the A carrier. The A stands for alternative and marks the (A) carrier as the nonwireline carrier, indicating that this carrier has no association with the local telephone company in that particular market. The FCC set up the cellular regulatory framework in this way for a reason: It ensured that at least one licensee in each market would have telecommunications industry experience and would already have some type of telecommunications infrastructure in place.

Key: The distinction between landline telephone companies as the B carrier in cellular markets and nonwireline companies as the A carrier in cellular markets has become irrelevant over the years. Today, any qualified company can obtain either the A or B license in any MSA/RSA. Now, wireless markets in general are traded amongst carriers, the trades being subject to FCC approval. In many cases, entire market clusters are even traded between competing wireless carriers.

1.7   Initial Cellular Licensing

The FCC decided initially to conduct comparative hearings to select the most qualified applicants for the A and B block licenses when more than one company applied for the same license. "Comparative" means that the FCC examined the prospective licensee's proposed coverage and growth plans. In early 1983, 567 applications were filed just for the 30 markets ranked 61st to 90th in size. The FCC quickly realized that comparative hearings would take too long, and a new licensing process needed to be found. At the time, it was taking 10 to 18 months of deliberation and more than $1 million in costs to award a single license. So in October 1983, the FCC amended its rules and specified that lotteries would be used to select the A and B carriers in cellular markets among competing applicants in all but the top 30 markets. The lottery winners faced a detailed legal, technical, and financial review before the FCC would issue a license to construct a cellular system. A key purpose in conducting this review was to ensure that the prospective licensees had a documented network build plan that made sense. By 1990, construction permits had been issued for at least one system in every cellular market in the United States.

It is important to note that the terms "cellular market," "cellular system," and "MSA/RSA" are all synonymous and interchangeable.

 Key: When a wireless carrier has multiple markets that abut each other in a geographically contiguous area, this is known as a *market cluster.*

Test Questions

True or False?

1. _____ MSAs are cellular markets containing populations of 1,000,000 or less.

2. _____ Cellular market (RSA/MSA) boundaries run along county lines, as depicted by Rand McNally.

3. _____ The distinction between landline telephone companies as the B carrier in cellular markets and nonwireline companies (radio common carriers) as the A carrier in cellular markets has become blurred and unimportant in recent years.

4. _____ The "A" in "A Carrier" stands for the word "access."

5. _____ Cellular RSAs are labeled according to the largest city within their boundaries.

Multiple Choice

1. The FCC created the cellular industry as a duopoly with an A carrier and a B carrier existing in each cellular market. At the inception of the cellular industry, the FCC decreed that the B carrier would be associated with the local telephone company because

   a. AT&T wanted it that way.

   b. The A carriers insisted that a local telephone company (telco) should be their chief competition.

   c. This structure ensured that at least one cellular licensee in each market would have telecommunications industry experience and would already have some type of telecommunications infrastructure in place.

   d. A similar market structure was already in place in Japan, and the FCC was following that country's lead.

   e. Motorola recommended this structure to the FCC.

2. MSAs cover approximately 75 percent of the population and ____ percent of the landmass of the United States?

   a. 50%

   b. 25%

   c. 20%

   d. 75%

   e. 10%

CHAPTER 2

Basic Wireless Network Design and Operation

2.1 Frequency Reuse and Planning

The term "cellular" can apply both to descriptions of cellular carriers and their networks (i.e., systems that began as AMPS networks), as well as to PCS carriers. The cellular concept and its design implications—the term "cellular" itself—are equivalent in both worlds. This concept will be addressed in Section 2.5. To that end, the terms "cellular" and "wireless" are also interchangeable for purposes of this book.

Cellular technology enables mobile communications through the use of a sophisticated two-way radio link that is maintained between the user's mobile phone and the wireless network. The underlying concept behind the two-way radio link is ingenious. It involves using individual radio frequencies (or radio channels) over and over again throughout a market with minimal interference in order to serve a large number of simultaneous conversations. This concept is the central tenet of cellular system design and is known as *frequency reuse*. The frequency-reuse concept is what distinguishes cellular systems from all preceding mobile radio systems.

The major drawback with previous mobile communication systems was the inefficient use of allocated radio spectrum. Repeatedly *reusing* radio frequencies over a given geographic area provides the means for supporting a number of *concurrent* conversations far in excess of the number of voice channels derived from simply parceling out available spectrum.

Most frequency-reuse plans are produced in groups of seven cells, where seven is the number of cells in the frequency-reuse plan. There are dozens or hundreds of seven-cell frequency-reuse groups in each cellular carrier's MSA or RSA, depending on the size of the market. This means that all available 416 channels (conversation frequencies) in a cellular market are plotted throughout the seven-cell frequency-reuse plan over and over again. Higher-traffic cells will have more radio channels assigned to them in order to optimize the capacity of the system according to customer usage (also known as "subscriber density"). See Figure 2-1 for an illustration of the N = 7 frequency-reuse plan.

A frequency-reuse plan is defined as how radio frequency (RF) engineers (working for wireless carriers) subdivide and assign the FCC-allocated radio spectrum throughout the carrier's market. The proximity of existing cell base stations and the existing frequency-reuse plan must be taken into account when examining where to build new cells into an existing system.

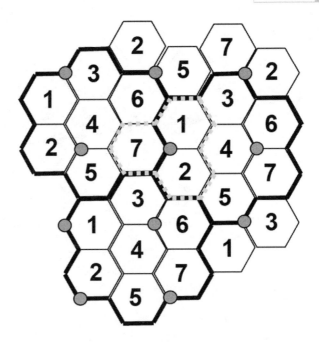

Figure 2-1
N = 7
frequency-reuse cluster:
Note how identical frequency sets are laid out symmetrically throughout a network.

Key: The main challenge of wireless network engineering is to grow a system in capacity by adding cell base stations and radio channels (to existing cells), so that the base stations and channels throughout a system do not interfere with each other (interference will be covered in another chapter). Note the hexagon grid design (see Section 2.5).

2.2 Distance-to-Reuse Ratio

The *distance-to-reuse ratio* (D/R) defines the geographic distance that is required between cells using identical frequencies in order to avoid interference between the radio/channel/frequency transmissions at these cells. The D/R ratio is actually a complex mathematical formula that was derived from extensive field research and testing. The D/R ratio is a critical cellular design principle and a core element of cellular system development. The overall geographic coverage size of cell base stations throughout a system determines the actual D/R ratio, along with the radio power level transmitted from cell base stations transmitting the

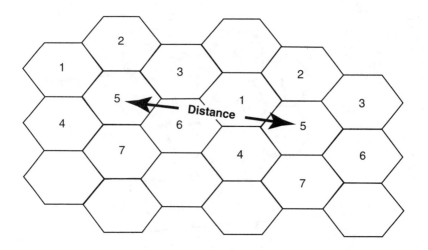

Figure 2-2
The size of the cell base station's coverage area and radio power level transmitted from cell base stations transmitting the same set of (N 5 7) frequencies dictates the D/R ratio in a wireless system. Note the hexagon grid design (see Section 2.5).

same set of seven frequencies. See Figure 2-2 to see how the D/R ratio is viewed from an engineering perspective.

The *frequency agility* concept goes hand in hand with the D/R frequency-reuse concept.

    Key: Frequency agility describes the capability of mobile phones to operate on any given frequency in their assigned radio spectrum. The spectrum referred to here is that which is assigned by the FCC to a particular carrier in conjunction with how mobile phone manufacturers define the specifications for their equipment in the manufacturing process. The mobile telephone has the ability to change frequencies instantly upon command from the wireless system.

In other words, frequency agility defines the ability of the mobile telephone to perform a seamless transfer from one radio frequency (channel) to another. Frequency agility allows for reuse on a massive scale of the relatively small amount of radio spectrum that has been assigned to wireless carriers by the FCC. This capability also allows for another key cellular system concept, known as call-handoff.

The D/R ratio applies in the digital wireless world, as well as to analog wireless systems.

▓▓▓ 2.3　Call-Handoff

What makes cellular technology work at all is its ability to have the subscriber unit (the mobile phone) change frequency as the unit moves throughout a market. This is known as *call-handoff*.

　　Key: Call-handoff is the process where a *call-in-progress* is seamlessly transferred from one cell base station to another (via a channel change) while maintaining the call's connection to the cellular system via the mobile switching center (MSC).

Along with frequency reuse and frequency agility in the mobile phone, the call-handoff capability is the driving force behind wireless technology. The call-handoff process goes hand in hand with the concept of frequency reuse. Without call-handoff, frequency reuse would not be technically feasible, and vice versa.

The rationale behind the handoff process is the need to keep the subscriber unit receiving a usable signal. The wireless switch (the MSC) tracks the radio power levels of all wireless phone calls. When the system sees a mobile's signal level going down, it seeks a neighboring cell base station that can hear the subscriber better via a stronger received signal level from these neighboring cells. When a call-handoff is required, the switch knows which surrounding site is the best candidate to receive the call-handoff. The call is then handed off to that neighboring cell base station.

Base station, or radio, coverage from each cell overlaps with all the cells that are adjacent to that cell. This overlapping radio coverage is what actually allows for call-handoff to occur. Without seamless coverage, wireless calls would be dropped during the handoff process. The call-handoff process is microprocessor controlled, and the call-handoff process is supposed to be transparent to the user. The call-handoff capability accounts for some of the complexity within a wireless handset (i.e., the frequency synthesizer, the control, and memory functions). Call-handoff demonstrates why all mobile telephones must have frequency agility or the ability to change from one channel to another. Call-handoff is known as *call-handover* in other parts of the world, especially in Europe. See Figure 2-3 for an illustration of how call-handoff operates.

2.4   Maps Used in Wireless System Design

The types of maps that are used in wireless system design are United States Geological Survey (USGS) maps. It is an FCC requirement that USGS maps be used for depicting service contours.

All cell site locations are denoted using polar coordinates (latitude and longitude). Wireless system engineers need to learn how to read latitude and longitude on maps regardless of the scale of the map. They need to learn how to read ground elevations on maps, and the overall map symbology. For example, they must be able to determine where ridges, mountains, forests, bodies of water, and airports are located.

Most USGS maps are 1:500,000 scale, where 1 inch equals approximately 8 miles. In addition to large USGS maps, special maps known as *7.5 minute maps* are used for detailed analysis of topography, population

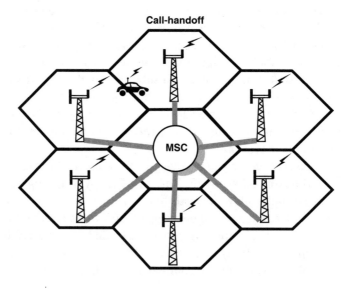

Figure 2-3
Wireless calls in progress are transferred from one cell to another cell while maintaining the call's connection to the wireless network.

Call-handoff

centers, and roads. These maps are 1:24,000 scale, where 1 inch equals 0.4 miles. These maps are used to determine terrain features in relation to where a wireless carrier wants to install a cell base station. The 7.5 minute map is also the foundation for cell location searches. USGS maps can be obtained through Rand McNally, the U.S. government, the International Map Service, and similar companies.

2.5   The Hexagon Grid

To facilitate the design of a wireless network with frequency reuse in mind, cells are laid out as hexagons in design documents. This is known as the *hex metaphor*. The hexagon, or *hexagon grid*, is the predominant engineering-design tool in the wireless industry. Hex grids are usually printed on a large, clear plastic sheet and allow engineers to visualize a wireless system for design purposes, regardless of the actual terrain in a given market. Hex grids are laid over USGS maps when engineers design wireless systems. The hexagon is used because it best simulates a grid of overlapping circles. The overlapping circles represent neighboring cell base stations in a wireless system (see Figure 2-4).

Key: The hex grid can contain hexes of many different sizes, depending on the density of a given wireless market. Urban markets, such as Chicago or Los Angeles, will have smaller hex grids because there will be more base stations in these urban markets. They have a more dense population base and, therefore, more installed cell base stations overall.

Base stations in rural areas cover a larger geographic area than base stations in urban areas because the populations in rural areas are slimmer and more dispersed. Therefore, the size of each hex in a rural grid will be larger to represent the less dense subscriber base in these areas. Call handoff *cannot* occur unless there is some overlapping of radio coverage between any given cell base station and all its neighboring cell base stations. The key benefit of using the hex grid is that it allows a wireless carrier to grow in an orderly fashion in terms of coverage and the frequency-reuse plan.

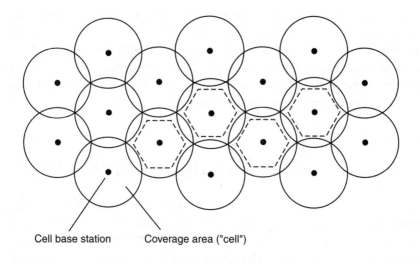

Cell base station   Coverage area ("cell")

Figure 2-4
Hexagon grids are ideal cellular system design and engineering tools because they graphically and functionally depict overlapping radio coverage between, and among, adjacent cell base stations.

For design purposes, hexes can be laid out on paper with the points facing up, or any flat side facing up. This depends on the preference of cellular engineers or the actual configuration of the wireless system being addressed. Hex layouts can all be modified according to the needs of the particular cellular system. The hex grid allows a wireless carrier to design a system with an eye toward reducing or eliminating signal interference and it facilitates long-term frequency planning. Failure to use a grid system for planned growth would make it harder to add additional cells into a wireless market. The hex grid also facilitates cellular design in terms of identifying call-handoff candidate cells for every given cell in the system. The hex grid can also be used for traffic analysis purposes (see Figure 2-4).

 Key: The term "cellular" is derived from the fact that the hexagons used in wireless system design resemble the cells in a bee's honeycomb. The reader should understand that the term cellular was derived from this meaning and usually refers to the earliest commercial mobile systems. But for all practical purposes today, "cellular" still means any commercial wireless service available to the public.

2.6 Fundamental Wireless System Components

There are five key components to any wireless network. They are (1) the mobile phone, (2) the cell base station, (3) the backhaul network (also known as the fixed network), (4) MSC, and (5) interconnection to the public switched telephone network (PSTN) and the Internet.

2.6.1 The Mobile Phone

There are many types of wireless telephones in use today. The functionality of today's wireless phones comes in many, many flavors. From a high-level perspective, though, all wireless phones fall into one of the following two categories:

- *Portable telephones* are the type of handset that users are able to carry with them anywhere. They are the most popular form of wireless phone. They are small and relatively lightweight, and today many people wear them on their belts for easy access and to easily hear them ring. Since the mid-1990s, wireless handset manufacturers have made great breakthroughs in producing wireless phones that are very small and lightweight. Most wireless subscribers today seek the convenience of a phone they can take with them in their pockets or on their belts in order to truly achieve "anytime, anywhere" communication ability. Chapter 17 will review the design and functionality of today's cell phones in greater detail.

- *Mobile telephones* are stationary, nonremovable phones that are mounted in automobiles. These phones are not produced or sold on a mass-market basis anymore. They have become very rare due to the appeal, size, and ease of use of the modern portable cell phone. The only mobile phones that are still being produced are for automobile manufacturers as add-ons to some car models. One such application is the OnStar system used by some automobile manufacturers. The mobile phone is built right into the vehicle, and the only part of the phone that is seen by the user is the send button.

In the mid-1990s, wireless carriers migrated to digital wireless technologies, and digital wireless phones are now ubiquitous in the marketplace. These phones are smaller, more streamlined, and more functional than their analog predecessors. All wireless phones sold today are digital phones. This means that although the traffic (voice, data, and/or video) that is sent through the air still uses an analog sine wave to be transmitted, the information transmitted "on top of" the sine wave is all digitized—it's all ones and zeroes.

These digital phones come in several types. First, there are *multimode* digital phones. These are phones that operate in two or more digital wireless technology modes—for example, a phone that operates using the Code Division Multiple Access (CDMA) digital wireless standard as well as the Global System for Mobile Communication (GSM) digital wireless standard. Due to research and development by Motorola and other handset manufacturers, there are now mobile phones that operate in three and four standard wireless modes. For example, these are mobile phones that can operate in CDMAOne mode, CDMA 1X mode (see Chapter 7), GSM mode, IS-136 mode, and possibly even Wi-Fi mode. These types of phones will exist in the commercial marketplace by the end of 2005. Cell phones that can operate in three modes are known as trimode phones. Cell phones that can operate in four modes are known as quadmode phones.

Along with multimode digital wireless phones, there are also dual or multiband phones. Multiband phones are phones that operate in two or more frequency ranges: for example, a wireless phone that operates in the 850 MHz cellular frequency band (in the United States) as well as the 1,900 MHz PCS band. See Figure 2-5 for a picture of a typical modern cell phone.

 Key: Today's digital wireless phones can operate using both multimode *and* multiband functionality at the same time.

Figure 2-5
Typical cell
phone

2.6.2   The Cell Base Station

The *cell base station* is the physical location of the radios (also known as *transceivers*), adjunct equipment, and antennas that serve wireless subscribers. Base stations can be installed just about anywhere, especially because the components themselves have shrunken so much in the early twenty-first century. This capability has allowed manufacturers to produce smaller and smaller base station components and equipment.

2.6.3   The Backhaul Network

The *backhaul network*, also known as the *fixed network*, is the web of transmission systems that connect all cell base stations to the base station controller (BSC), which then connects to the mobile switching center. This network is what effectively connects the mobile subscriber to the outside world, and vice versa.

2.6.4   The Mobile Switching Center (MSC)

The MSC is the brain of a wireless carrier's network. A mobile switching center is responsible for connecting calls together by switching digital voice data packets from one network path to another—a process known as *call routing*. MSCs also provide additional information to support mobile service subscribers, including user registration, authentication, and location updating.

The MSC monitors all active mobile phones and all mobile calls throughout a system and handles the switching of all calls. In most systems today, cell base stations are connected to a base station controller, which is then connected to the MSC.

2.6.5   Interconnection to the Public Switched Telephone Network (PSTN) and the Internet

For wireless calls to connect to the landline public network, and for landline callers to reach a mobile subscriber, some type of linkage between the wireless network and the landline network must exist. These links are in the form of DS-n- or OC-n-type circuits (transmission systems), TCP/IP connections, or ATM connections between the wireless and land-line networks. Today, there is a subnetwork that has managed in this vein known as the *PDSN*, or the *public data service node*. The PDSN is what connects wireless data and/or Internet traffic from the wireless network to all other data networks.

▓▓▓ 2.7   POP Counts

POP counts, also known as POPs, is a term that means population counts, or potential customers. *One person equals one POP count.*

▬ ▬ ▬ ▬ ▬ ▬ ▬ ▬ ▬ ▬ ▬ ▬ ▬ ▬ ▬ ▬ ▬
Key: The density and value of wireless markets is mea-sured in terms of POP counts and POPs covered. Wireless markets are bought, sold, and traded on the basis of POP counts.
▬ ▬ ▬ ▬ ▬ ▬ ▬ ▬ ▬ ▬ ▬ ▬ ▬ ▬ ▬

Wireless carriers try to place the majority of their cell sites in areas where the population is dense. The carriers can thus maximize the potential for generating revenue and obtaining return on the investment of building their systems.

▓▓▓ Test Questions

True or False?

1. _____ The hexagon grid does not allow a cellular carrier to design its system to reduce or eliminate signal interference and facilitate long-term frequency planning.

2. _____ The frequency-agility concept goes hand in hand with the D/R concept.

Multiple Choice

1. Which design and technical concept sets cellular communications apart from all preceding mobile radio systems?
   a. TDMA technology
   b. Duplex functionality
   c. Frequency reuse
   d. None of the above

2. The frequency reuse plan is usually divided into cell groupings where the number of cells equals _____.
   a. 3
   b. 10
   c. 7
   d. 416
   e. 21

3. Which cellular engineering technical construct requires determining the geographic distance that is required between cells using identical frequencies to avoid interference between the radio transmissions at these cells?
   a. The Hertz theory
   b. Call handoff
   c. The D/R ratio
   d. The hexagon metaphor

4. What is the name of the activity that describes how wireless calls in progress are transferred from one cell base station to another cell base station while maintaining the call's connection to the wireless network?
   a. Automatic call delivery
   b. Cell split
   c. Call handoff
   d. None of the above

The Cell Base Station

▨▨▨ 3.1 Overview

A cell base station is a physical area where a particular set of frequencies serves users, with adjacent cells using different frequencies to avoid interference. The cell base station is also known as a base transceiver station (BTS). The cell base station serves as the air interface between the mobile phone and the wireless system. It is the first or last transmission leg of every wireless phone call, whether that call is originated or received by the wireless subscriber. Cell base stations are capable of handling many simultaneous conversations and serve as the initial access point to the wireless network. There are many low-power radio transmitter/receivers (called transceivers) located within each base station to support customers' wireless transmissions—whether they be conversations, accessing the Internet, or sending a picture to a friend or relative. Each mobile phone is, effectively, also a stand-alone transceiver. The radio frequencies emitted by a cell base station cover a given geographic area, and cell base stations are created by carving up counties, cities, and even buildings into small areas called *cells*, hence the name *cellular*. The terms "cell," "cell site," "cell base station," and "base station" are all synonymous and interchangeable.

 Key: Cell sizes are not the same throughout a given market. Cells can range in size from a few hundred meters (referred to as a "microcell") to a mile across in urban areas and from 10 to 15 miles across in some rural areas. The actual size of a cell depends upon terrain, system capacity needs, and geographic location (urban or rural). Generally, cell sizes are very small in urban areas and larger in rural areas, as a result of subscriber (population) density and the overall amount of traffic on the wireless network. See Figure 3-1 for an illustration of how cell sizes vary when moving from a densely populated urban area to a less populated rural area.

As the popularity of wireless service exploded in the 1990s, cell sizes had to shrink and simultaneously multiply in order to support the drastically increased number of subscribers. For example, in Chicago in 1998, Cingular Wireless's average cell size was 2 miles. In downtown Chicago,

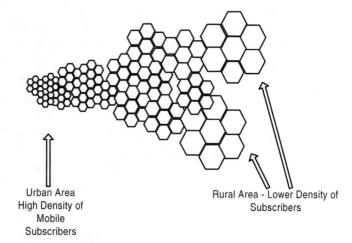

Figure 3-1
Cell size
change from
urban to rural
areas

Urban Area
High Density of
Mobile
Subscribers

Rural Area - Lower Density of
Subscribers

the cell sizes are now half a mile or less. When the Chicago system was first built, the average size of a cell was 5 to 10 miles. Cell sizes became smaller as traffic on the system grew and more cells were added to accommodate the increased capacity required.

■ 3.2 Criteria and Methods for Cell Placement

Cell base stations can be deployed in multiple venues. They include the following:

- ■ *Raw land sites:* Wireless carriers purchase or lease raw land, build a tower and install equipment, and then turn up the base station.

- ■ *Rooftop sites:* Wireless carriers lease space on the rooftop of an existing building. They mount base station antennas on the rooftop and place the base station equipment either on the rooftop or in leased space somewhere in the building itself. The roof can be the roof of an office building, a church, an apartment building, or even a skyscraper.

- ■ *Water tank sites:* Wireless carriers lease space on an existing water tank, mount base station antennas on the tank, and run coaxial cables from the antennas down to the ground where the base station equipment is located.

■ *Colocated sites*: Wireless carriers lease space on an existing tower where other wireless carriers are located and operate their own base stations. They mount their own base station antennas on the tower and run the coax cables down the tower to their own equipment shelters. Collocation can include mounting base station antennas on electric utility towers, as shown in Figure 3-2.

■ *Stealth sites*: These sites are base stations that are hidden from noticeable view because they are concealed within other structures such as gas stations or hotel signs. These sites can also be hidden from noticeable view because they are using special towers or other structures that allow them to blend seamlessly into their immediate surroundings. These stealth towers will be reviewed at length in Chapter 9. Also see Figure 3-3, which shows a photo of a cactus and boulder that are a disguised cell base station.

Refer to Figures 3-4 and 3-5 to see how the base stations described in the preceding passage would look in a real-world environment. Similar photos are shown in Chapter 9.

The main criterion for determining where a cell base station will be located is where the customers are or where a high concentration of wireless transmission exists according to traffic studies conducted by wireless carriers. This focus was initially on mobile subscribers (i.e., car phone users), and thus sites were usually placed along major roads and highways. Information on vehicular traffic was found from Rand McNally interstate road maps and traffic studies by state and local departments of transportation. This information was used to determine the optimal placement for new base stations. The focus on where to install cell sites has shifted since the early 1990s to include areas of high pedestrian traffic such as convention centers and arenas, stadiums, shopping malls, downtown areas, and nightlife areas. This shift is largely due

Figure 3-2
BTS colocated
on electric
utility tower

Figure 3-3
Cactus (tower)
and boulder
(BTS) stealth
base station
(Photo
courtesy
Shepard
Communica-
tions)

to the immense popularity of the portable cellular telephone and wireless service itself.

Overall, site selection today is based on population densities, traffic densities on highways and major roads, the proximity of existing cells, and the existing frequency-reuse plan.

Key: Maintaining interference-free service while growing a wireless network is the most challenging task faced by wireless engineers.

Figure 3-4
Raw land site for monopole tower base station and shelter

Figure 3-5
Note dual use of tower as antenna mount and light pole.

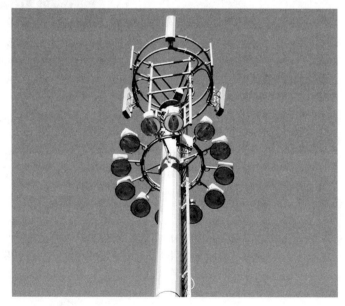

▨▨▨ 3.3   Selecting Cell Base Station Locations

Several key parameters must be examined to determine an ideal location for a cell base station. First and foremost, what area does the wireless carrier want to cover? Where is there an obvious need for new system capacity, based on system traffic studies? What is the height of the terrain above mean sea level (AMSL)? Is the area in a flood plain? Ideal urban cell site locations are in business areas such as industrial parks and strip malls. Residential areas are usually avoided, when possible, for aesthetic reasons. In the heart of urban areas, such as downtown centers in major cities, base station antennas are often placed at street level to accommodate the large volume of pedestrian traffic, which translates into a large volume of wireless traffic. Today, major advances have been made in the development and deployment of camouflaged base stations.

The geography of certain areas sometimes makes it impossible to place cell base stations in convenient locations. For example, in mountainous regions in the western United States, some cells are placed right at the peak of a mountain in order to maximize the radio coverage area for that cell. These locations may be accessible only by helicopter and can be snowed in for months at a time. In order to reach some of these locations, a cell site technician might have to use a four-wheel-drive vehicle for part of the run up a hill or mountain, then a snowmobile, and then walk the remaining distance to the site. These cells can run on solar power or huge propane tanks that are filled before the winter weather sets in. The tanks can hold enough fuel to operate the cell nonstop until spring. In areas like those described in the preceding passage, some cells are even constructed on stilts above the projected snow line so that technicians can enter the cell unimpeded, when necessary.

In the mid-1980s, Ameritech Mobile Communications actually used a miniature blimp to simulate signal characteristics (and, hence, quality of coverage) at potential base station locations. The blimp was about 14 feet tall and 40 feet long. It would be tethered at a potential cell location about 300 feet above ground level. An antenna was inside the blimp with a coax cable running down to an amplifier located at ground level. By using the blimp to simulate a cell site, Ameritech radio frequency (RF) engineers could determine whether the location would be good or less than ideal for deployment of an actual cell base station. The blimp was a cost-effective alternative to renting a crane, which could cost up to $10,000 per day.

▓▓▓▓ 3.4   Cell Base Station Deployment

Today, the locations for constructing new cell sites in existing wireless markets can be determined by joint decisions between wireless carrier marketing departments and RF design engineers.

Key: The proximity of existing cells and the existing frequency-reuse plan must be taken into account when examining potential sites for new base stations.

Wireless system engineers use search rings to designate areas where site acquisition specialists should seek a location for a new cell site. Remember: This location can be a rooftop, an existing tower, or raw land, to name a few options. Search rings describe a three-ringed geographic area designated on a USGS map that is deemed optimal for a new base station. The site acquisition specialists should try to obtain a new cell location within the center of the search rings' geographic area. If they are not successful in finding a location in the center of the area, then they should seek a location within the second ring in the target area, and possibly into the third ring, until they are successful. See Figure 3-6 for an illustration of the search ring concept.

Key: Where a new base station is ultimately placed will affect the frequency-reuse plan and RF power levels at nearby base stations.

Cell base stations may now be located at just about any land area or structure: on farms, in church steeples, on water towers, in hotel and motel signs, in apartments, and even on light towers at athletic fields. Today base station antennas are even mounted on bridge overpasses on interstate highways. They are usually painted to blend into the color of the bridge itself. Some wireless carriers have even mounted their antennas on the top of grain elevators. When referring to cell site locations in this manner, we are actually referring to places where the carrier's antennas may be mounted. Regardless of where the antennas are

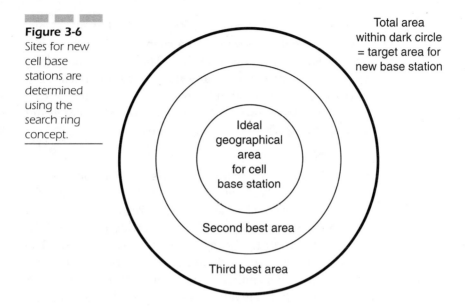

Figure 3-6
Sites for new cell base stations are determined using the search ring concept.

Total area within dark circle = target area for new base station

Ideal geographical area for cell base station

Second best area

Third best area

mounted, coax cable will run from the antenna's location back to where the BTS equipment actually exists.

Local municipalities are becoming much more stringent when approving new base station installations, especially in urban areas. This is due to community resistance to unsightly towers and is sometimes known as the *not in my back yard* (NIMBY) phenomenon in municipal political circles. To that end, tower vendors have undertaken initiatives since the mid-1990s to create "stealth" towers that are camouflaged (see Chapter 9). Many times wireless carriers have to dispatch special teams to testify before municipal zoning boards to get approval to install base station towers. These teams emphasize the benefits of wireless service to the community and try to come to an agreement with the zoning boards as to what it will take to get the approval to install a new base station, or tower, in a specific area. Sometimes zoning boards will even extract certain concessions from wireless carriers, such as mandating that the carriers paint the towers a certain color, restricting the towers to certain heights, or mandating that the carrier share its tower with other wireless carriers. Some zoning boards will even request that the wireless carrier repave village parking lots or build new community parks to demonstrate its commitment to the local community.

Key: The irony of the NIMBY issue is that all people, even elected officials and zoning boards, desire great wireless coverage by service providers. They want to be able to make a call anywhere, at any time. But sometimes these very same people stand in the way of allowing wireless carriers to construct towers to provide additional coverage and capacity to a given village, city, or area.

Today, many wireless carriers deploy their base stations on community water tanks. This type of deployment can be a win-win situation for all parties involved: the wireless carrier is spared from constructing an expensive tower, and a municipality receives a monthly payment from the wireless carrier for renting the space on the water tower. The monthly charge for renting space on top of a water tower can be anywhere from $500 to $3,000, depending on the location of the water tower and the local government that is involved. See Figure 3-7 for a picture of a water tank base station.

Sometimes wireless carriers get very creative in their efforts to mount a base station on top of a water tower. One community in a suburb of Chicago allowed a wireless carrier to mount its base station antennas on the top of the water tower. However, the carrier had to repaint the water tower (which was in dire need of repainting) with the name of the village on it as a condition of being allowed to mount its site on the tower. On the plus side, the carrier also was allowed to paint *its* name and logo onto the

Figure 3-7
Cell base station on top of a water tank: Note use of tank midsection for microwave antennas. (Photo courtesy kramerfirm. com)

tower as well, which resulted in free advertising. This is a classic win-win situation. Many times community water tanks will even support multiple carriers' antennas. One water tower in west suburban Chicago has a base station (set of antennas) on top of the water tower and three additional base stations (sets of antennas) from three different carriers on the neck, or *stem*, of the water tower. Sometimes antennas that are installed on a water tower have a collar. A *collar* is a nylon or fiberglass wrap that is wound around the antenna mounting to make the antenna structure invisible to the general public. This is done to hide the antennas from public view.

In urban areas, wireless carriers also deploy base stations on rooftops of multistory buildings. Rooftops are ideal places for base stations as they represent a replacement for the construction of a tower; thus wireless carriers are spared from the capital cost of tower construction. The building owners also win in this situation as they are generating revenue in the form of monthly payments for space that might otherwise go unused. One scenario where this has become very commonplace is on the rooftops of school buildings. As school districts across the country are financially squeezed, they are seeking new sources of revenue. Many schools gladly accept proposals by wireless carriers to mount base station antennas on their rooftops in order to get the additional monthly income that such a service provides.

Monthly rent for a rooftop base station can run anywhere from $500 to $3,000, but the average cost is about $1,000 per month. Similar to water tower rent, rooftop rental fees are dependent on the building owner and possibly the height of the building itself.

Key: A trend that has become commonplace with base station deployment is for municipalities to concentrate all wireless carrier cell sites within a small geographic area, around a half-mile to a mile in diameter. These cell site clusters minimize the negative environmental impact of unsightly towers by concentrating all the sites to one specific area. See Figure 3-8, which shows several clustered sites.

In some cases, base stations are deployed in areas known as *antenna farms*. An antenna farm is a plot of land that has been dedicated by the owner to the placement of towers for all different types of wireless

Figure 3-8
Clustered cell sites (Photo courtesy P's UK GSM and UMTS Pages)

services offered by all types of wireless carriers and even public service agencies. This could include 850 MHz wireless operators, 1,900 MHz wireless operators, paging carriers (rapidly becoming extinct), Nextel (the E/SMR carrier), and even state or local police and fire agencies.

Cell sites are usually named after the area where they are located—for example, a town, a highway, or a mountain. Sometimes cell sites are even named after company executives or employees. In many urban areas, they are simply numbered, as that is the easiest way to designate them because they are so numerous. Things like base station naming conventions are usually company specific.

The industry average cost to build an entire base station today is approximately $350,000, depending on the manufacturer, type, and amount of equipment used. This includes the costs for the land (lease or buy), the tower (if one is necessary), the shelter that houses the equipment, *all* the base station equipment, and antennas and coaxial cable. Circa 1995, this cost was as high as $1,000,000. Several factors contributed to the decrease in the cost of base stations:

- A reduction in cell site equipment costs due to larger volumes being ordered by wireless carriers, specifically by the new PCS carriers. The manufacturers gained economies of scale due to huge orders, and they passed on these savings to the wireless carriers.

- Carrier migrations to all-digital systems, which means all-digital base stations. Digital base stations are less expensive than their analog predecessors: fewer moving parts, more integrated circuitry (IC) in the equipment, and fewer overall parts required to make the equipment work.

The average lead time to install a new base station is approximately 6 months to a year. This reflects the time period from having no lease to having a fully operational cell site.

▬▬ 3.5 Base Station Shelters

Older cell base stations and some newer PCS base stations deploy small, self-contained shelters at the bottom of the towers at cell sites. The shelters look like small huts and house all the base station equipment from standard 19- or 23-inch racks that is necessary to make a cell site run. These shelters are also sometimes known by the acronym CEV, which stands for *controlled environmental vault*. See Figure 3-9 for a photo of a typical base station shelter.

Newer, all-digital base stations are completely self-contained cabinets that house all the equipment required to make a cell site function. These cabinets are stand-alone objects and are about 4 to 5 feet long, approximately 5 feet high, and about 2 feet in depth. In many cases, these cabinets have taken the place of the costlier shelter (the CEV) for carriers that choose to deploy them. The main purpose of shelters and cabinets is to keep equipment safe from the elements—that is, keep them cool and dry. Easy access to equipment for maintenance purposes is also a factor for selecting this type of base station. See Figures 3-10 and 3-11, which show photos of digital base stations.

With the immense popularity of wireless service and the accompanying proliferation of base stations, more communities are cracking down on building and *cabinet* "pollution." They want shelters to blend into the environment better. Shelters that are aesthetically pleasing are increasingly requested by community zoning boards, and CEV manufacturers have responded by developing entire product lines of aesthetically pleasing and disguisable huts. Just as disguised antennas and towers have become popular with zoning boards, attractive and less obtrusive shelters are on their way to setting a new trend in cell-site planning. There are now shelter structures that have brick or stone appearances on the

▬▬ ▬▬ ▬▬
Figure 3-9
Aluminum CEV
(Photo
courtesy Nortel
Networks)

Figure 3-10
Typical modern
digital base
station

Figure 3-11
Typical modern
digital base
station

outside. Brick shelters blend in to residential neighborhoods better than
steel or fiberglass shelters. Sometimes a carrier will just paint a shelter,
or add shrubbery to the outside, to have it match the surrounding buildings. This has proven to be an effective and low-cost method some shelter vendors, and hence wireless carriers, have employed.

One option for keeping shelters from being an eyesore is to keep them
to a minimum. Today, a single tower can have several providers leasing
space on it, but the drawback is that each of those providers will also be
deploying their own shelter at the base of the tower. With the variety of
shelter designs, colors, and fence heights, the base of a tower can begin to
resemble a haphazard shantytown. Modular shelters can alleviate this
situation. These shelters are one physical unit that allows multiple carriers to be corralled together within a singular structure. Each carrier gets
its separate space, separated by a chain-link fence within the structure.

Some manufacturers even make bulletproof shelters because in some areas the shelters are used for target practice. Some shelter manufacturers have even developed shelters that look like boulders or monuments so that the shelter will completely blend into its desert landscape. Earlier in this chapter, Figure 3-3 showed a picture of a base station shelter that is actually a boulder, where the cactus shown is a "stealth" miniature tower.

Security is also an issue to consider when installing a shelter. Keeping the equipment safe is the most important goal when selecting a shelter design, according to industry engineers. To that end, most shelters have 8- to 10-foot fences erected around them, topped with barbed wire or razor wire. Shelter doors are made of steel and are outfitted with steel pry-resistant latch guards and tamper-resistant hinge pins. Some carriers also use motion detectors and alarms on some of their remote shelters to dissuade trespassers.

3.6   Cells on Wheels (COWs)

A COW is essentially a cell site that is custom designed, engineered, and manufactured for a particular wireless carrier. The customization is required to fit the needs of a given wireless carrier and to include the particular digital wireless technology that the carrier uses in its network. It is an enclosure that is mounted on a large truck or flatbed tractor-trailer. It contains all the equipment that is normally used at a regular cell site. They are fully functioning base stations, including the ability to provide call-handoff. COWs have retractable antennas (or minitowers); the makeshift tower on a COW is usually a telescopic monopole or a minilattice tower that is erected from the truck itself.

In the telecommunication world today, immediacy and mobility are critical. COWs are used for emergency purposes in any number of situations. For instance, if a tower fell down or an entire base station became inoperable, a COW could be deployed to that location. Other examples of where a COW deployment could come into play to provide urgently needed capacity in a wireless system include the following:

- Political conventions
- PGA golf tournaments
- Major, ongoing news events
- Professional sports playoff or championship games

In the scenarios listed here, more than one COW could be deployed as well.

The design of many COWs is based on a tandem trailer for over-the-road operation. The unique construction of some COWs also offers the option for deployment to remote locations via helicopter lifting or transportation by C-130 aircraft. All controls are remotely actuated, and two men can set up a mobile cell site in four hours. Mobile cell units are well insulated with efficient lighting, making it possible to operate under virtually any weather conditions, domestically or internationally. MSCs can also be housed on tractor-trailers for emergency purposes. Figure 3-12 shows a picture of a COW.

3.7 Enhancers

Enhancers, sometimes known as repeaters, are special radios (transceivers) that are used to boost RF signals in areas where RF coverage is inadequate. The types of areas that are typically good candidates for enhancer deployment—where enhancers improve coverage—are tun-

Figure 3-12
COW (Photo
courtesy
Operation
Gadget.com)

nels, malls, large *dead spots* (also known as coverage shadows), and geographic areas between cell sites. Areas between cell sites could be candidates in a situation where, for example, a road is wedged between two ridgelines in a rural area. Enhancers do not contribute any capacity to a wireless system; they improve *coverage*. Due to advances in technology today, enhancers can be placed in small cabinets and mounted almost anywhere. Buildings to house equipment are not necessary; the main components of an enhancer are radio equipment and an antenna. Enhancers use RF channels from donor cells to increase the coverage in a particular area that is incapable of being supported by the donor cell.

Today's enhancers are known as *pass-through enhancers*. These enhancers use channels that are fed from a donor cell. A highly directional antenna known as a *pickup antenna* is used at the enhancer location to receive the signals propagated from the antenna at the donor cell. This highly directional antenna ensures that the pass-through enhancer's pickup antenna receives only the assigned channels from the donor cell. Pass-through enhancers can enhance RF channels from any range of frequencies supported by equipment manufacturers, which are usually all FCC-allocated frequencies (i.e., 850 MHz band and 1,900 MHz band). Subscriber traffic that is picked up and carried within the enhancer's coverage area via call originations or call handoffs is also routed to the donor cell through these directional antennas for call processing. A directional antenna is then also used at the pass-through enhancer location to provide RF coverage to very specific geographic areas around the enhancer's location where dead spots may exist. Examples of dead spots include a specific highway corridor, a small town nestled between two high ridges, tunnels, and even subways. If desired, a wireless carrier could place more than one directional antenna at an enhancer location to provide coverage in several directions (i.e., both up *and* down a particular highway corridor).

The power used at a pass-through enhancer location could be high or low, depending on the type of amplifier used. Fiber optic technology is now being deployed to feed the donor signal to the repeater or enhancer from the donor cell location. The donor base station can be housed in a central location called a hub or *hotel*. From this location, fiber optics is used to transmit the radio signal from the hub (donor) radios to the enhancers. Enhancers can be deployed without concern over interference since the RF signal is transmitted via fiber optic cable instead of utilizing donor and pickup antennas. The fiber can be buried under streets and sidewalks, or strung from utility poles. See Figure 3-13 for an illustration of pass-through enhancer operation.

▓▓▓ 3.8 Microcells

The *macrocell* network is the base station infrastructure in the outdoors —rooftop cell sites, tower-based cell sites, and water-tank cell sites, for example.

Like the trend toward smaller handsets, the wireless industry has seen a trend toward smaller infrastructure equipment. This trend has been spurred by several things:

- The migration to all-digital wireless systems that began in the mid-1990s. Digital base station equipment is smaller than analog equipment by definition; it is completely microcircuitry based.

- The NIMBY phenomenon has fostered the development of smaller, easier-to-hide equipment.

- The move toward more collocation amongst carriers and tower sharing.

- Microcells are less expensive than full-fledged base stations and can deliver similar results in terms of coverage and capacity.

Figure 3-13
Sample pass-through enhancer deployment

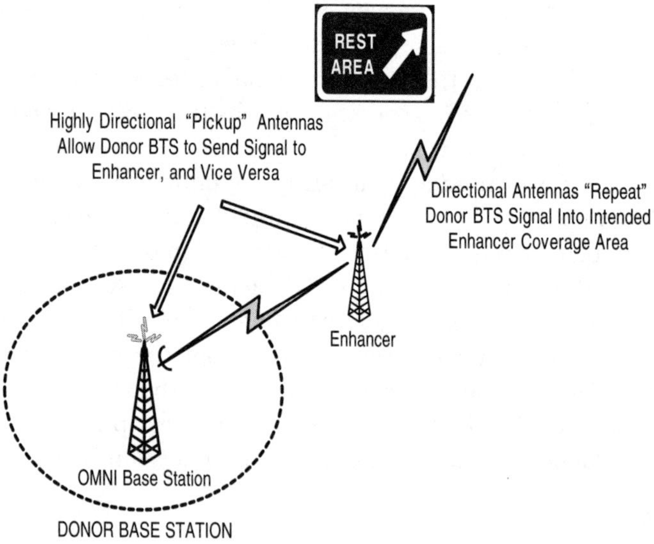

Highly Directional "Pickup" Antennas Allow Donor BTS to Send Signal to Enhancer, and Vice Versa

REST AREA

Directional Antennas "Repeat" Donor BTS Signal Into Intended Enhancer Coverage Area

Enhancer

OMNI Base Station

DONOR BASE STATION

Microcell technology is fostering this trend with new technology patents that provide more features, greater range, and expanded capacity to carriers. Once seen as a niche product to fill gaps in coverage areas and as a replacement for enhancers, modern microcell technology gives carriers a way to deploy base stations quickly and cost-effectively, while minimizing impact on the surrounding area. Small microcellular base stations can provide the same capacity and coverage area as that of their traditional-size counterparts, *macrocellular* base stations. Carriers can plug-and-play microcells, installing them to fit their coverage and capacity needs while simultaneously reducing deployment costs.

Physical size and weight are the primary distinguishing features between microcells and traditional base stations. Traditional base stations are very large and take up a lot of space. They require weather-proofed, air-conditioned enclosures and environmental aesthetics. They are difficult to disguise, particularly in residential areas where they are usually least welcome.

The new generation of microcell base stations are ruggedized boxes that generally measure about 6 cubic feet and weigh less than 100 pounds. Microcell base stations are designed to withstand temperatures ranging from −40° to 158° F. They require no additional heating or cooling systems like a traditional base station requires. Because of the small form factor, microcells are readily installed wherever they can be easily and imaginatively concealed: on the side of a building, in a tunnel, on top of a utility pole, or even under a bridge. A microcell base station's unobtrusive size is ideal for suburban residential areas where zoning restrictions tend to be the most stringent. Outside of residential areas, microcell base stations can be used to improve or expand coverage in office buildings, shopping malls, sport stadiums, airport terminals, and other indoor locations where people are likely to engage in heavy wireless phone use. Because no buildings or special power sources are necessary, microcell installation time is reduced from days to hours. Outdoor, directional antennas are usually used for microcells. Indoor, omni-antennas, directional antennas, or *leaky* coax cable may be used for microcell coverage.

Key: Leaky coax is a coaxial cable that has notches cut into the copper conductor at regular intervals so that radio waves can enter and exit the cable to provide wireless service indoors. Radiax is a popular brand of leaky coax. Leaky coax is very expensive compared to antennas. See Figure 3-14 for a picture of Radiax.

Figure 3-14
Radiax "leaky"
coax cable
(Photo
courtesy
Andrew Corp.)

Note notches
in cable

Wireless carriers can always weigh several options when deciding what type of radio equipment they deploy in any given area. One of these options is price. The cost of installing a microcell can be about one-fifth the cost of installing a large, traditional macrocellular base station.

3.9   Picocells and Nanocells

A *picocell* is the smallest of the cells in a wireless system, normally covering an office area to support an in-building wireless system. The term picocell is most commonly used in connection with third-generation (3G) personal communications systems, especially GSM. Picocells are almost always placed indoors.

A *nanocell* is a miniature, portable radio unit that establishes cell phone service wherever coverage or capacity is needed. Installation is fast and easy. Nanocells require only a power source and a means for connecting the base station to a laptop PC, and then the mini-cell is ready to operate. Nanocells are the most cost-effective cell node available today. The design of a nanocell combines a small form factor with low power consumption to enable installation in a wide range of environments. Like picocells, a nanocell can be mounted on the wall, in a vehicle, in a briefcase, or in a weatherproof outdoor enclosure. While there is no limit to the number of users that can be allowed on a nanocell, up to seven handsets may access the network at one time in a typical installation. For greater coverage or capacity, multiple distributed nanocell base stations can be linked together using standard Ethernet cables. Up to four nanocell units can be combined and configured as a single network managed by one computer.

The actual range of a nanocell or picocell will be determined by antenna gain, terrestrial obstructions, and building materials among other things.

Key: The main difference between macrocells, microcells, picocells, and nanocells is the radius of coverage provided by each. This coverage is defined by the power output and antenna gain used in each system. Typically, the following cell radius ranges apply:

- Macrocells are base stations that supply coverage to areas that are greater than 1,000 meters.
- Microcells supply coverage to areas where the base station radius is between 100 meters and 1,000 meters. It may take three to five microcells to provide coverage that is equivalent to that of one macrocell.
- Picocells supply coverage in areas where the base station radius is less than 100 meters.

Some industry scientists envision a future where nanocells are located every few hundred feet, providing wireless access anywhere, anytime. These people predict a day when we may see nanocells perched on top of utility poles, street lights, traffic lights, and so on.

In summary, microcell costs range from $60,000 to $100,000 to install. The average traditional base station can cost between $125,000 and $350,000. The average enhancer costs between $15,000 and $50,000.

Test Questions

True or False?

1. _____ Picocells are mostly used in wireless in-building systems.

2. _____ Pass-through enhancers can exist without a signal from a donor cell.

3. _____ School districts throughout the United States will gladly accept proposals from wireless carriers to mount base station antennas on the rooftops of school buildings.

4. _____ Cell sites (base station antennas) can be located only on towers.

Multiple Choice

1. Microcells:
   a. Have all the functionality of a regular cell site, but with lower power output
   b. Can sometimes be used to cover large geographic areas
   c. Are used to fill small coverage holes such as buildings, shopping malls, and convention centers
   d. Have a range of 10 miles or greater
   e. a and c only
   f. All of the above

2. The areas that system engineers designate for site acquisition specialists to seek land for a new cell site are known as:
   a. Hex rings
   b. O rings
   c. Search rings
   d. D-R rings

3. A nylon or fiberglass wrap that is wound around the antenna mounting to make the antenna structure invisible to the general public is:
   a. A hex grid
   b. A collar
   c. A search ring
   d. A flexometer
   e. None of the above

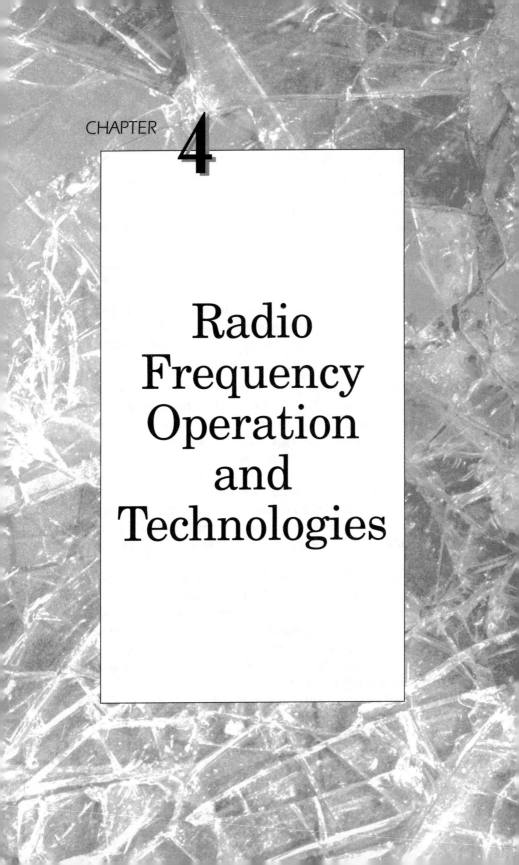

CHAPTER

4

Radio
Frequency
Operation
and
Technologies

4.1   Introduction

Radio propagation is an electromagnetic phenomenon discovered by Heinrich Hertz in the late 1800s in which energy travels in waves through a given medium. This medium could be air, water, a wired cable, or optical fiber. In the air, radio signals propagate at the speed of light: 186,282 miles per second. This equates to nearly a billion feet per second. The only significant functional difference between cellular systems and the conventional landline telephone system is the radio link that connects the subscriber unit to the wireless network via the cell base station. Radio frequencies differ in energy and their ability to propagate media.

Key: In many respects, radio wave propagation is very similar to the propagation of white light, and thus the comparison between the two is often made in technical venues.

Different radio frequencies have different physical properties. In other words, lower frequencies behave differently when propagated than higher frequencies do. This is why certain radio frequencies are relegated to certain functions while others are not. Radio signals can bend through the atmosphere (*refraction*), bend around obstructions (*diffraction*), and bounce off obstructions and solid objects (*reflection*). An example of refraction is how light bends through a prism. An example of diffraction would be the way in which wind careens over a moving vehicle and flows over the back end of the car after flowing over the front hood and the roof. A reflected radio signal is roughly equivalent to how light is reflected off a mirror.

Radio-frequency (RF) *propagation* refers to how well a radio signal radiates, or travels, into free space. In the wireless industry, it refers to how well RF radiates into the area where RF coverage must be provided, which is determined by RF engineers working for wireless carriers.

Omnidirectional (all directions) radio-frequency propagation can be compared to the waves created by throwing a pebble into a pond. The waves made by the pebble emanate in all directions equally. The waves are proportional to the size and weight of the pebble and the force with which it was thrown into the pond. This pebble-in-a-pond analogy applies to the design of RF coverage for a cell.

Key: Actual RF coverage from any given base station is mainly determined by three key factors: the height of the antennas at the base station, the type of antenna used, and the RF power level emitted. These three factors, taken together, equate to the size and weight of the pebble thrown in the pond and the force with which the pebble was thrown. It should be noted that the height of the antennas could apply to a tower, a rooftop, a water tank, or any other structure.

4.2   Signal Fading and Degradation

The following sections review ways that RF propagation can be impaired and how wireless carriers can compensate for these impairments. These phenomena are inherent by-products of operating a wireless system and must be dealt with on a constant and ongoing basis.

4.2.1   Ducting

Ducting is defined as the atmospheric trapping of a cell base station's RF signal in the boundary area between two air masses known as a *duct channel*. The ducting channel size is directly related to the wavelength of radio propagation and ducting can be caused by a channel of hot air over cold air, or vice versa. Ducting of a base station RF signal is caused by an atmospheric anomaly known as *temperature inversion*. It is an anomaly of nature that can affect how well RF propagates through a given coverage area. Ducting is not a constant occurrence, and hence this will affect wireless system design due to the unpredictability of ducting. This phenomenon can occur in microwave radio systems as well as base station transmissions and is especially problematic when microwave radio is used for backhaul purposes over the wireless fixed network (see Chapter 12). This potential to impact microwave backhaul transmissions is usually the larger problem because the outage of microwave links will result in dozens or even hundreds of dropped wireless calls.

If the ground temperature is 30° F up to 300 feet above the earth, and then there is a layer of air that is colder or warmer than 30 degrees, this could create a duct. The duct is effectively an airway that can trap a base station RF signal between these two air masses, and the base station signal could propagate for as long as the duct exists. Ducts have been known to go on for hundreds or even thousands of miles. This creates a problem because instead of being absorbed by ground clutter (which is anticipated to a degree) the ducted signal, once trapped, could cause interference in distant wireless systems. Ducting is undesirable in RF propagation design but also unavoidable. It is a natural meteorological phenomenon that has to be addressed and managed.

Key: The downtilting of base station antennas may compensate for ducting. *Downtilting* is the act of electrically or mechanically directing the RF emitted by a base station antenna toward the ground. See Chapter 5 for an explanation of antenna downtilting.

The most common place for ducting to occur is across large bodies of water like the Great Lakes. For example, Michigan base station transmissions are picked up routinely along the Illinois and Wisconsin shore lines, and Illinois stations are routinely picked up in Michigan.

There are other physical anomalies that may affect propagation of a base station signal. What could happen to the signal must be examined and hopefully compensated for by RF engineers when designing cell sites. Problems may not always be foreseen and resolved, though, as the anomalies mentioned can have unpredictable consequences on RF propagation.

There are three basic types of fading of wireless signals, aside from ducting: *absorption, free-space loss,* and *multipath fading,* also known as *Rayleigh* fading. These are the types of fading that occur from every base station and impact propagation into the intended coverage area, whereas a phenomenon like ducting usually occurs only near major bodies of water.

4.2.2   Absorption

Absorption describes how a radio signal is absorbed by objects. When a radio wave strikes materials, it can be absorbed. RF can be absorbed by buildings, trees, or hills. The greater the amount of absorption of an RF signal, the less geographic area covered.

> *Key:* Organic materials tend to absorb more wireless signals than inorganic materials. Pine needles are noted for absorbing a great deal of RF transmissions because their needle length is close to 1/4 the wavelength of base station RF signals. *Phasing* of RF signals in the 800 MHz range means the signals are more apt to being absorbed by pine needles because the length of the pine needle is equivalent to 1/4 the wavelength of a base station signal in the 800 MHz range. Phasing describes how the wavelengths of radio signals can mix together to either complement the signal or detract from it. It will be reviewed in Section 4.2.4.

Absorption can be compensated for in the following ways:

- By using higher-gain antennas
- By using higher RF power levels in order to cover the same geographic area
- By decreasing the distances between base stations

These concepts will be described in detail in Chapter 5.

4.2.3   Free-Space Loss

Free-space loss, also known as *path loss*, describes the attenuation of a radio signal over a given distance, or the *path length* of the signal. In other words, free-space loss defines how a base station signal can fade over distance. As a radio signal propagates through the air, it is continually fading and losing its strength.

Key: The higher the frequency of a radio signal, the greater the free-space loss. This means that higher frequency signals (emitted by base stations) will fade much faster than those from a base station that is using a lower frequency. To put this in context, this means that a base station signal from a PCS carrier operating in the 1,900 MHz range will fade much more rapidly than a signal from a wireless carrier operating in the 850 MHz range. This phenomenon requires PCS carriers to install many more base stations throughout any given area to provide coverage equivalent to that of carriers operating in the 850 MHz range.

The power level of a base station signal (and antenna gain) will determine at what point a signal will fade to the point that it is unable to sustain a wireless call.

4.2.4   Multipath (Rayleigh) Fading

Multipath, or Rayleigh, fading describes a condition where the transmitted base station signal is reflected by physical features or structures, creating multiple signal paths between the base station and the user terminal (mobile phone). Rayleigh fading can and will occur in both the mobile-to-base station path as well as in the base station-to-mobile signal path.

Key: There will always be a direct signal from the mobile phone to the base station, which will be the *strongest* signal. There will also be multiple reflected, or indirect, signals from the mobile to the base station.

One signal will always arrive at an antenna (either mobile or base station) as the strongest, most direct signal while other reflected signals are weaker, multipath, indirect signals. Indirect signals will reflect off any and all objects in the path between the transmitting antenna and the receiving antenna. These indirect signals arrive at receiving antennas via reflection paths. The indirect, reflected signals will arrive at an antenna and be either in phase or out of phase with the strongest, direct signal. The indirect signals combine with the direct signal to either complement or detract from the direct signal, and this depends on whether or not the indirect signals are in phase or out of phase with the direct signal. If the indirect signals are in phase with the direct signal (i.e., symmetric wavelengths), then the indirect signals will complement the direct signal when they combine with it as both signals enter the receiver. If the indirect signals are out of phase with the direct signal (that is, direct and indirect wavelengths are offset from each other by 180 degrees), then the indirect signals will detract from or weaken the combined signal entering a receiver. See Figure 4-1 for an illustration of two signals that are 180 degrees out of phase with each other.

Multipath signals can reflect off bodies of water, vehicles, buildings, or really anything on their way to a receiving antenna. See Figure 4-2 for an illustration of Rayleigh fading.

4.3 Wireless Frequency Bands

The radio-frequency spectrum that was assigned to the cellular industry in the late 1970s was mostly unused UHF television spectrum. The 850 MHz spectrum had been unused for years, and the television networks vigorously fought the FCC against reallocating their spectrum, even though it was grossly underutilized. The TV industry did not use the spectrum but also did not want to give it up. The FCC won.

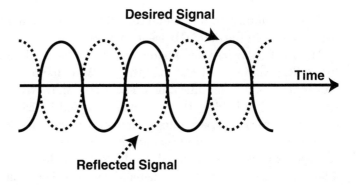

Figure 4-1
Two signals 180 degrees out of phase with each other. Dotted line indicates signal 180 degrees out of phase with main signal.

Desired Signal

Time

Reflected Signal

Figure 4-2
Rayleigh fading

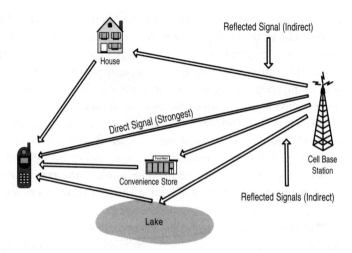

The following describes the nature of radio propagation at 850 MHz and explains why the 800 MHz frequency range is mostly beneficial:

- It has a very short signal wavelength (about 12 inches).

- It tends to be line of sight, similar to light itself. It is not really subject to skipping or bouncing off the ionosphere, as with short-wave radio.

- The 850 MHz signal is easily reflected off buildings, cars, and trucks.

- The 850 MHz signal is easily absorbed by foliage (i.e., trees and the forest). This can be both good and bad for base station coverage. It's good because it allows for efficient frequency reuse. It's bad because it can cause problems when major coverage points of the base stations (e.g., highways and population centers) are in heavily wooded areas.

These combined factors make the 800 MHz spectrum ideal for radio transmission in the 850 MHz frequency band.

The 1,900 MHz frequency band (or 1.9 GHz) was allocated by the FCC to the PCS carriers in 1991. Like the 850 MHz band, it is well-suited to personal mobile communications. The 1,900 MHz frequency band that was assigned to PCS by the FCC was occupied by part of the common carrier microwave spectrum. Incumbent microwave operators were ordered by the FCC to find *comparable facilities* replacements for their transport needs (this issue will be explored further in Chapter 8). The 1,900 MHz band is good for wideband services. However, its very short wavelength may cause building penetration problems. This is partially due to the fact that it does not propagate as far as lower frequencies (i.e., 850 MHz frequency).

4.4   In-Building Coverage

Through the 1980s and 1990s, all wireless carriers (both cellular and PCS) feverishly expanded coverage in their markets by building thousands of base stations. At the same time, wireless subscribership grew at an average rate of 45 to 60 percent annually. As the carriers expanded their coverage by building out these base stations, the focus was primarily on expanding their macrocell footprint. From the carrier's perspective, any RF coverage that was afforded to customers in large (office) buildings has been, to some degree, incidental. Of course, wireless carriers would like all customers to be able to make calls regardless of where they are physically located, and this includes large buildings. But the objective up until around 2001 was to concentrate base station construction in the macrocell environment.

But the evolution of the wireless industry since the late 1990s—the explosion in the popularity of its service—has required that wireless carriers now focus more intently on providing coverage indoors in major buildings in urban areas and inner cities. In many large buildings (i.e., convention centers, hotels, sports stadiums, airports), wireless carriers install full-fledged base stations or microcells. Increasingly, the focus could be on 802.11 Wireless Fidelity (Wi-Fi) hot spots that are effectively dual-mode systems, serving both voice and data needs (see Chapter 20).

There are and will be certain impediments to expanding coverage into large buildings though. Some large office buildings have a metallic coating added to their window glass to reduce the amount of heat taken through the window from direct sunlight. These are usually newer buildings, and their windows give them an appearance that resembles a green or gold mirror. Not only are the infrared rays of the sun blocked from these buildings to reduce heating or air conditioning requirements, but radio waves are also severely attenuated and reflected in this scenario. This is because the metallic nature of the window coatings reflects base station radio waves, making it difficult for RF to penetrate these types of buildings. This can severely impact in-building RF coverage. In these cases, a wireless carrier could install an enhancer or a microcell in the building to compensate for the lack of coverage from base stations in the macrocellular wireless network. One means of doing so could be to install a base station in a building (either in the basement or in a leased space anywhere else in the building) and run leaky coaxial cable through the building's riser system to provide coverage inside the structure. Figure 3-14 showed a picture of leaky coax cable. Figure 4-3 illustrates how leaky coax could be used to provide in-building coverage in a multistory building.

4.5   Frequency Coordination

Frequency coordination defines the effort to assign frequencies to base stations in a logical, orderly way to minimize interference *within* a carrier's own system as well as with neighboring systems of different wireless carriers. There are two kinds of frequency coordination: *intramarket* and *intermarket*.

Intramarket frequency coordination is the effort, by RF engineers, to carefully assign and manage base station frequency assignments throughout a carrier's market that are internal to the system. It is the effort to keep frequency assignments as far apart as practical and is based on a frequency-reuse growth plan using the hex grid. Internal frequency coordination is done to minimize interference and optimize system operations.

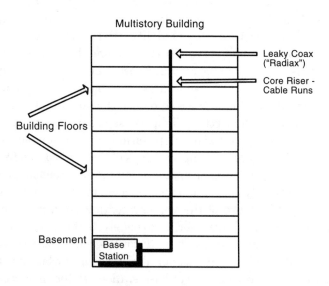

Figure 4-3
In-building
coverage using
Radiax®

Multistory Building

Leaky Coax
("Radiax")

Core Riser -
Cable Runs

Building Floors

Basement

Base
Station

Intermarket frequency coordination is the effort by wireless carriers to manage the effect(s) of their frequency assignments that are at market borders, which can impact other wireless carrier's systems in their own markets. It involves coordinating frequency assignments with neighboring systems' border cells within 70 miles of the common border, according to FCC rule. The FCC dictates that "all reasonable actions must be taken to limit and/or reduce interference between two different wireless systems." There are at least two ways to approach intermarket frequency coordination: (1) An RF engineer could simply call the engineering department of a neighboring wireless carrier. Engineers from both companies would exchange information regarding frequency assignments at border cells for their respective markets and work together to resolve potential interference problems. (2) An outside vendor could be hired to handle the situation. This is not advisable, because it costs extra money, and the results may not be as accurate and reliable as approaching the other carrier directly.

4.6   System Interference

Interference is defined as adverse interaction between two or more radio signals that causes noise or effectively cancels out *both* signals. Interference usually occurs between two transmitting radio signals whose frequencies are too close together, or even identical.

Base transmit and mobile transmit frequencies in wireless networks are normally assigned with a given separation between them to avoid interference. This required separation is a function of the frequencies being used. However, most interference experienced in wireless systems is still internally generated. In most cases, this interference is a by-product of frequency reuse.

Higher frequencies may require less of a separation between transmit and receive channels than lower frequencies require to reduce or eliminate interference.

4.6.1   Cochannel Interference

Cochannel interference occurs when the same base station carrier frequency reaches the same receiver (mobile phone) from two separate transmitters (base stations).

Key: This type of interference is usually caused when channels have been assigned to two cells that are not far enough apart geographically and their signals are strong enough to overlap with each other, thereby causing interference.

When cochannel interference occurs, it is because it is a by-product of the basic tenet of wireless system design: frequency reuse. Though the basic principle of wireless system design is to reuse FCC-assigned frequencies over and over again throughout a system, it is also very important to ensure that the frequencies are reused far enough apart, *geographically*, to ensure that no interference occurs between identical frequencies/channels. Therefore, there must be enough base stations installed between cochannel cells to provide a level of protection to ensure that interference is thwarted or eliminated. In conjunction with ensuring that the cell sites are physically placed far enough apart, the appropriate power levels must be maintained at cell base stations throughout a system as well to avoid cochannel interference. If power levels at one cell are too high, the RF coverage could overlap and hence possibly interfere with a cochannel cell. Figure 4-4 depicts how cochannel interference can occur.

So the following factors must be carefully assessed to reduce the potential for cochannel interference to occur:

- RF power levels at cochannel cells (most important factor)
- Geographic distances between cochannel cells
- Types of antennas used at cochannel cells

As more cells are added to a system operating on cochannel frequencies, it becomes more difficult to keep cochannel interference at a level that is not noticeable to subscribers. Cochannel interference could manifest itself in the form of cross talk, static, intermittent speech (choppy conversations), or simply dropped calls.

Cochannel cells must *never* be direct neighbors of each other in a wireless system. Cells with different frequency assignments must be placed in between cochannel cells to provide a level of protection against signal interference. *Management of cochannel interference is the number one limiting factor in maximizing the capacity of a wireless system.*

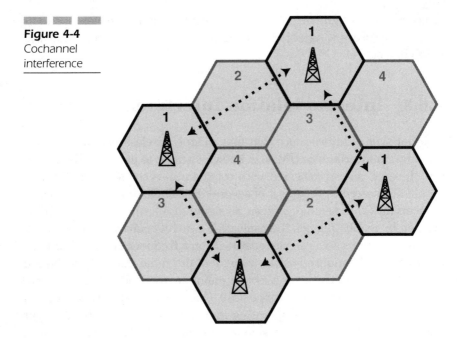

Figure 4-4
Cochannel
interference

The following options are available to wireless carriers to reduce or eliminate cochannel interference:

- Use downtilt antennas when and where appropriate.
- Use reduced-gain antennas.
- Decrease RF power output at base stations.
- Reduce the height of towers.

These topics are reviewed extensively in Chapter 5.

4.6.2   Adjacent-Channel Interference

Adjacent channel interference is caused by the inability of a mobile phone to filter out the signals (frequencies) of adjacent channels assigned to nearby cell sites (e.g., channel 361 in cell A, channel 362 in cell D, where cells A and D are in the same N=7 frequency reuse cluster). Adjacent channel interference occurs more frequently in heavily used cells.

Good system design can minimize adjacent channel interference temporarily by preventing adjacent channel assignments in cells that are near each other.

4.6.3   Intermodulation Interference (IM)

Another type of interference that may plague wireless systems is intermodulation interference (IM, also known as *imod* to people in the industry). If a cell site is colocated with other radio-based services, IM may result. IM describes the effect of several signals mixing together to produce an unwanted signal, or even no signal at all. Another type of IM is created by mobile phones themselves. If a mobile subscriber is on a call in close proximity to a cell site operating on a frequency band other than their assigned band, the power from all of the radio channels in the cell can cause the receiver in the mobile phone to overload. When this happens, the result could be a dropped call. One way to resolve this situation is for all wireless carriers operating in a market to place cell sites near each other so that a stronger signal is maintained in the mobile phone. In the early twenty-first century, this is not a problem as carriers are often forced to colocate together, usually on towers. Competent engineering practices should overcome intermodulation interference.

4.7   Radio Frequency Channelization

This section explains how radio spectrum is carved up to allow for the transmission of wireless phone calls and the management of the wireless network. All wireless carriers, regardless of what air interface technology they are using, have to break up their FCC-assigned radio spectrum into usable, discrete, manageable portions. How this spectrum is broken down is defined by the digital wireless standards that are being used by any given wireless carrier. Every wireless air interface technology (AMPS, GSM, CDMA) has its own standardized method of approaching channelization.

4.7.1   Paired Channels

Regardless of the air interface technology that is used, all wireless conversations require a *paired channel* to function. Because mobile service is a full-duplex system (simultaneous two-way transmission), two radio channels are required for every wireless conversation. Depending on the technology used, these channels may or may not use the same frequency. When the channels use different frequencies, this is known as *frequency-division duplexing*. When the channels use the same frequency, it is known as *time-division duplexing*. The only time two channels will not be required in a wireless environment is obviously in the case of one-way transmissions, such as short message service (SMS, also known as text messaging), or in the case of streaming media. Streaming media would involve the one-way transmission of music, video clips, or other media—directly to the handset from the wireless carrier or a content provider (see Chapter 16).

Key: The channel transmitted from the base station to the subscriber's mobile phone is known as the downlink, or the *forward channel*. The channel transmitted from the mobile phone to the base station is known as the uplink, or the *reverse channel*. See Figure 4-5.

A paired channel is the combination of the forward channel and the reverse channel that is necessary for every wireless call to take place.

In the late 1970s, there were a total of 832 channel pairs allocated per cellular market by the FCC (per MSA/RSA). Because the wireless industry was a duopoly in 1983—only two competitors per market—each carrier in every cellular market was allocated 416 channel pairs. Four hundred sixteen channels were used for the base transmit/mobile receive side, and 416 channels were used for the mobile transmit/base receive side.

Most cellular carriers partitioned their 416 channel pairs into the [N=7] frequency-reuse format, the de facto industry standard. Base station transmit and receive bands in the 850 MHz spectrum (traditional cellular spectrum) are separated by 45 MHz to avoid interference between the uplink and the downlink channels.

Figure 4-5
Paired
channels

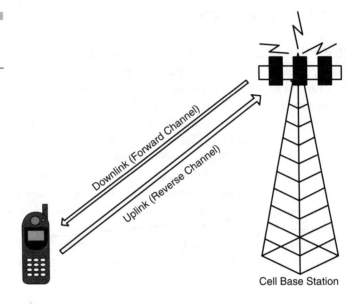

Cell Base Station

4.7.2 Channel Spacing

Channel spacing refers to the actual bandwidth space that is allocated for every wireless channel out of the total amount of spectrum allocated to a wireless carrier by the FCC. This concept applies to both the uplink and the downlink for every paired channel. The same amount of channel space (bandwidth) is used on both the uplink and downlink.

The Advanced Mobile Phone System (AMPS) is the oldest wireless air interface standard in the United States. Because it is the oldest standard analog, its approach to channel spacing was applied to some extent to subsequent digital wireless technologies that were developed and launched in the early to mid-1990s.

In AMPS, the channel spacing is 30 KHz. Each uplink and downlink channel occupies 30 KHz of bandwidth. This means that every AMPS cellular call actually occupied a total of 60 KHz of FCC-allotted cellular spectrum: 30 KHz for the forward channel and 30 KHz for the reverse channel. In other words, 30 KHz for the uplink, 30 KHz for the downlink. Each designated pair of frequencies that will be reused has been assigned a specific channel number under FCC guidelines and standards. This channel number equates directly to one specific paired channel and its associated frequencies.

Global System for Mobile Communications (GSM) systems allot their radio spectrum in 200 KHz carriers, where each carrier allocates 25 KHz to the uplink or downlink.

Code Division Multiple Access (CDMA), by definition, is unique when it comes to channel spacing. In the most technical, literal sense, channel spacing in a CDMA system is 1.25 MHz because all calls that are carried on a 1.25 MHz CDMA carrier are spread out over the entire swath of that carrier. That is why CDMA is known as a spread spectrum technology.

Key: The terms "frequency reuse" and "channel reuse" are synonymous in wireless networks. "Frequency reuse" is synonymous with the term "channel reuse" because all frequencies in FCC-assigned spectrum have been divided into discrete channels.

In the real world, different reuse patterns have been used, as well as different channel plans. Twenty-four channel reuse plans have been actively deployed, as well as N=4 reuse patterns and N=12 reuse patterns for omni cells. But N=7 is still the most-used reuse plan.

4.7.3   Control Channels

The control channel is a data signaling channel that handles the administrative overhead of the wireless system via signaling between mobile phones, cell base stations, base station controllers (BSCs), and the mobile switching center (MSC). It is used to administer the following tasks:

- Setup of wireless calls, both mobile-originated and mobile-terminated, and locating (paging) mobile phones in order to connect mobile-terminated calls
- Collecting call information to support billing operations
- Collecting traffic measurements on base stations
- Autonomous mobile registration, such as registering phones on the system (both home and roaming phones)
- Initiating or assisting in mobile call-handoffs

Each channel set in a wireless system has at least one control channel assigned to it. In sectorized base stations, each sector will have its own channel set. In these cases, that means that, by extension, each sector will also have its own control channel.

All mobile phones, once they are powered on and throughout the time they are powered on, tune to the control channel in their assigned band that they receive the strongest signal from. This frequency is one of a set of frequencies that are labeled *control channel* frequencies when a wireless carrier develops and launches their system. This frequency is programmed into their phones either at a fulfillment vendor's warehouse or at the wireless business store (a fulfillment vendor is a third-party company that can manage the programming and stocking of any company's merchandise before it is sent to a store or another venue where it can be sold to the public). Each subscriber unit automatically retunes the control channels in its band at predetermined intervals, based on system and carrier parameters. This interval can range from every 2 minutes to every 60 minutes.

Key: When a subscriber pushes the send button when placing a call, the phone again rescans for the strongest control channel signal in its assigned frequency band.

CDMA networks use what is known as a *pilot channel*, which functions very much like a control channel.

4.8 Bluetooth

Bluetooth wireless technology is a de facto standard as well as a specification for small form factor, low-cost, short-range radio links between laptops, mobile phones, and other portable devices, including home appliances. Bluetooth enables any electrical device to wirelessly communicate in the 2.5 GHz ISM (license-free) frequency band. It enables users to connect a wide range of computing and telecom devices easily and simply, without the need to buy, carry, or connect cables. It delivers opportunities for rapid ad hoc connections and the possibility of automatic connections between devices. It will virtually eliminate the need to purchase additional or proprietary cabling to connect devices. Because Bluetooth can be used for a variety of purposes, it will also potentially replace multiple cable connections via a single radio link.

Bluetooth technology is essentially an enabling technology. It is designed to be incorporated in a very wide range of products and to allow them to intercommunicate rather than being a product in its own right. Having said that, some early products include things like Compact Flash of Dongle devices to add Bluetooth functionality into existing legacy products, such as laptops and PDAs. Today, most wireless headsets are Bluetooth-enabled, using this technology to transmit conversations from base units to headsets. If you look closely, you can even see the word "Bluetooth" printed in very small font on many of today's wireless headsets.

Bluetooth was named after Harald Blatand (nicknamed Bluetooth), a tenth-century Danish Viking king who had united and controlled large parts of Scandinavia, which are today Denmark and Norway. The name was chosen to highlight the potential of the technology to unify the telecommunications and computing industries. It was chosen as an internal codename at Ericsson and was never at the time expected to survive as the name used in the commercial arena.

The main features of Bluetooth are as follows:

- Bluetooth operates in the 2.4 GHz frequency band with no license requirement for wireless communications (2.4 GHz is an unlicensed spectrum).

- Bluetooth's real-time data transfer is usually possible between 10 and 100 meters.

- With Bluetooth, close proximity is not required as with infrared data (IrDA) communication devices because Bluetooth does not experience interference from obstacles such as walls.

- Bluetooth supports both point-to-point wireless connections without cables between mobile phones and personal computers, as well as point-to-multipoint connections to enable ad hoc local wireless networks.

- Bluetooth requires a low-cost transceiver chip be included in each device. It is designed and meant to be used worldwide because it operates in a spectrum that is available worldwide. The Bluetooth air interface is based on nominal antenna power of 0 dBm. The air interface complies with FCC rules for the Industrial, Scientific, and Medical (ISM) band at power levels up to 0 dBm. Spectrum spreading has been added to Bluetooth to facilitate options' operation at power levels up to 100 milliwatts worldwide. Spectrum spreading is accomplished by frequency hopping in 79 hops displaced by 1 MHz of spectrum each, starting at 2.40 GHz and stopping at 2.48 GHz (see Chapter 6).

Key: The nominal link range using Bluetooth is 10 centimeters to 10 meters but can be extended by more than 100 meters by increasing the transmit power. The maximum frequency hopping rate with Bluetooth is 1,600 hops per second.

In addition to data, up to three voice channels are available. Each device has a unique 48-bit address from the Institute of Electrical and Electronics Engineers (IEEE) 802 (LAN) standard. Connections can be point-to-point or multipoint. Data can be exchanged at a rate of 1 megabit per second (Mbps) and up to 2 Mbps in the second generation of the technology. The frequency hop scheme allows devices to communicate even in areas with a great deal of electromagnetic interference. Built-in encryption and verification are also provided.

Before any piconet connections are made using Bluetooth, all devices are in standby mode. In this mode, an unconnected unit periodically listens for messages every 1.28 seconds. Each time a device wakes up, it listens on a set of 32 hop frequencies defined for that particular unit. The number of hop frequencies varies in different geographic regions, but 32 is the number for most countries (except Japan, Spain, and France).

Bluetooth was originally conceived by Ericsson in 1994 when they began a study to examine alternatives to cables that linked mobile phone accessories. Ericsson already had a strong capability in short-range wireless, having been a key pioneer of the European Digital Enhanced Cordless Telecommunications (DECT) standard. DECT is a digital wireless telephone technology that is expected to make cordless phones much more common in both businesses and homes. Formerly called the digital European cordless telecommunications standard because it was developed by European companies, DECT's present name reflects its global acceptance. Out of their study was born the specification for Bluetooth wireless.

In February 1998, the Bluetooth Special Interest Group (SIG) was founded by a small core of major companies that included IBM, Intel, Nokia, Toshiba, and Ericsson. The objective of the SIG was to work together to develop the technology and to subsequently promote its widespread commercial acceptance. Six months later, the core Promoter Members publicly announced the global SIG and invited other companies to join, offering free access to the technology in return for commitment to support the Bluetooth specification. Adoption was rapid and 1998–1999 saw a boom in the market for Bluetooth conference organizers and vast amounts of hype regarding the potential of the technology.

In December 1999 it was announced that four more major companies had joined the SIG as Promoter Members: Microsoft, Agere Systems (then Lucent), 3Com, and Motorola. A wide range of Bluetooth technology companies are also marketing modules, software, subsystems, and chips among other equipment. There is and will continue to be competition between short-range wireless technologies, Bluetooth being one of them. Wi-Fi, Bluetooth, and Ultra-Wideband (UWB) wireless are all competing for a place at the wireless table of the future.

4.9   Ultra-Wideband Wireless (UWB)

UWB or *digital pulse wireless* is a wireless technology used to transmit large amounts of digital data over a wide spectrum of frequency bands (theoretically, *all* frequency bands) over short distances with very low power consumption. UWB radio not only can carry a huge amount of data over a distance up to 230 feet at very low power (less than 0.5 milliwatts), but it also has the ability to carry signals through doors and other obstacles that tend to reflect signals at more limited bandwidths and higher power.

UWB broadcasts *digital pulses* that are timed very precisely on a carrier signal across a very wide spectrum at the same time. UWB is also known as *nonsinusoidal communication technology* because it does not rely on a narrowband carrier frequency to transmit information. UWB transmitters and receivers must be coordinated to send and receive pulses with an accuracy of trillionths of a second. The most common technique for generating a UWB signal is to transmit pulses with durations of less than 1 nanosecond. On any given frequency band that may already be in use, an UWB signal has less power than the normal and anticipated background noise, so theoretically no interference is possible. Time Domain, a company applying to use the technology, uses a microchip manufactured by IBM to transmit 1.25 million bits per second but says there is the potential for a data rate in the billions of bits per second.

UWB technology has been around since the 1980s, but it has been mainly used for radar-based applications until recently because of the wideband nature of the signal that results in very accurate timing information. Due to recent developments in high-speed switching technology, UWB is becoming more attractive for low-cost consumer communications applications. UWB wireless technology can be defined as "any wireless transmission scheme that occupies a bandwidth of more than 25 percent of a center frequency, or more than 1.5 GHz."

The term UWB is not very descriptive, but it does help to separate this technology from older narrowband as well as newer wideband systems. It can also be defined as wireless communications technology that can transmit data at speeds between 40 megabits and 60 megabits per second, and eventually this rate will rise to 1 gigabit per second.

Key: Devices using UWB technology transmit ultra-low power radio signals with short electrical pulses in the picosecond (1/1000 th of a nanosecond) range across all frequencies simultaneously.

Two main differences exist between UWB and other narrowband or wideband systems:

- The bandwidth of UWB is much greater than the bandwidth used by any current technology for communication.

- UWB is typically implemented in a *carrierless* fashion. Conventional narrowband and wideband systems use RF carriers to move the signal in the assigned frequency domain from baseband to the actual carrier frequency. Conversely, UWB transmissions directly modulate an impulse that has a very sharp rise and fall time, thus resulting in a waveform that occupies several GHz of bandwidth

4.9.1   UWB Drivers

Many expect the combination of short-range wireless and wired Internet to become a fast-growing complement to next-generation cellular systems for data, voice, audio, and video. Four major trends are driving short-range wireless in general and UWB in particular:

- The growing demand for wireless data capability in portable devices at higher bandwidth rates but lower in cost and power consumption than the currently available means.

- Congested spectrum that is segmented and licensed by regulatory authorities in traditional ways.

- The growth of high-speed wired access to the Internet in businesses, homes, and public spaces.

- Shrinking semiconductor cost and power consumption for signal processing. This last trend makes possible the use of signal-processing techniques that would have been impractical only 10 years ago.

To understand the capabilities and benefits of UWB, a comparison of spatial capacity among IEEE 802.11b/a, Bluetooth, and UWB is helpful, where *spatial capacity* is defined as bits per second per square meter.

To increase performance, standards now under development in the Bluetooth SIG and IEEE 802 working groups would boost the peak speeds and spatial capacities of their respective systems still further, but none appear capable of reaching the speed potential of UWB. The reason is that all systems are bound by the channel capacity theorem: The upper bound on the capacity of a channel grows linearly with total available bandwidth. UWB systems, occupying 2 GHz or more of bandwidth, have greater room for expansion than do systems that are more constrained by bandwidth.

Transceiver architecture has great impact on wireless telecom devices. The following comparison between Bluetooth and UWB transceivers gives an idea of the advantages of UWB over current technologies.

Bluetooth uses a form of Frequency Shift Keying (FSK) where information is sent by shifting the carrier frequency high or low. UWB transceivers could be used for the same applications targeted for use with Bluetooth, but at higher data rates and lower emitted RF power. During continuous transmission, the Bluetooth transmitter is rated to deliver about 1 Mbps at an average of 1 microwatt of RF power to the antenna, and it provides an average operating range of about 10 meters. A 2.5 GHz wide UWB transmitter operating at less than 10 microwatts of average power could provide the same throughput and estimated coverage range. (In computer technology, throughput is the amount of work that a computer can do in a given time period). This could translate into a significant battery life extension for portable devices. Alternately, more UWB signal power could be used to increase range or data rate.

UWB technology offers many advantages:

- Has high throughput
- Is effective in environments with a high level of multipath fading (UWB is relatively immune to multipath cancellation and Rayleigh fading)
- Has a short-range high data rate
- Carries a large amount of data with low power consumption
- Boasts impressive accuracy of measuring positions of objects and people and is excellent in radar and imaging systems
- Has ability to carry signals through doors and other obstacles that tend to reflect signals at more limited bandwidths

- Can directly modulate a baseband pulse
- Can be made nearly all digital
- Provides low probability of intercept/interference
- Offers a low cost and simple transceiver
- Provides flexibility in the form of a dynamic tradeoff: throughput for distance

UWB can be used to provide connectivity in the following applications:

- WPAN or wireless personal access network
- Desktop and laptop PCs
- Printers, scanners, and storage devices among others
- Mobile devices (i.e., Blackberry)
- Mobile phones, MP3, PDAs, games, video
- Handheld PCs, 3G handset
- Wireless DVD player, wireless television, and HDTV
- Cameras players, camcorders
- Personal connectivity
- Wireless infrastructure for a *smart home* environment
- Control center for automating every thing from security, heating, and lighting systems to remote control appliances and home entertainment centers

One specific area where UWB's pulse technology has proven especially effective is in the areas of radar/sensor technology. In this context, UWB can be used for the following applications:

- Motion detector or range finder
- Radar imaging of objects buried under the ground or behind walls
- Radar imaging in medical field
- Military and commercial asset protection
- Antiterrorist/law enforcement
- Rescue applications (i.e., detecting people buried in debris caused by earthquakes or other disasters)

On February 14, 2002, the FCC established regulations that permitted the marketing and operation of certain types of new products incorporating UWB technology. In general, the FCC did not make any significant changes to existing UWB technical parameters, indicating that it is reluctant to do so until it has gained more experience with UWB devices. The FCC indicated that it intends to investigate the potential impact of UWB devices on various radio services. In response to petitions, the FCC amended its rules to facilitate the operation of through-wall imaging systems by law enforcement, emergency rescue, and firefighter personnel in emergency situations. Also, many wireless carriers have opposed the FCC sanctioning of UWB technology based on the argument that UWB effectively can and will use *all frequencies* to transmit. Wireless carriers have contended that this may cause interference within their networks. UWB proponents argue that UWB transmits at such very low power levels, spread across huge swaths of radio spectrum, that it is nearly impossible to cause interference in any waveband-based RF systems.

UWB receivers can be susceptible to being unintentionally jammed by traditional narrowband transmitters that operate within the UWB receiver's pass-band. There is also the potential for interference between UWB emissions and a global positioning system (GPS). Serious concerns exist regarding the interference the technology would create for broadcasters, air-traffic controllers, and mobile phone operators. And great efforts have been made to address these problems and find remedies for them.

UWB has the potential to displace not only Wi-Fi, but also the technologies that are used to build large-scale wireless voice and data networks, such as CDMA, General Packet Radio Service (GPRS)/Global System for Mobile Communication (GSM), Enhanced Data GSM Environment (EDGE), and Wideband CDMA (W-CDMA).

In office buildings, UWB is expected to replace 802.11b networking protocols because of its penetration abilities. Walls, cubicles, and people can all interfere with 802.11 technologies, yet UWB simply does not have this problem.

According to the research firm In-Stat/MDR, the industry expects a standard for UWB to be ratified by the first half of 2005 with the first UWB products becoming commercially available later in 2005.

According to West Technology Research Solutions, with a 4 percent global gross domestic product (GDP) growth rate, annual shipments for UWB chipsets into the communications industry will exceed 63 million units by 2007. This forecast is dependent on factors such as global economic growth and the strength of UWB in certain key markets.

4.10 Software-Defined Radio (SDR)

Software-defined radio (SDR), sometimes shortened to software radio (SR), defines wireless communication in which the transmitter modulation is generated or defined by a computer, and the receiver uses a computer to recover the signal intelligence, or transmission. To select the desired modulation type, the proper programs must be run by microcomputers that control the transmitter and receiver.

As communications technology has rapidly evolved to all-digital systems, more functions of contemporary radio systems are implemented in software, leading toward development of the software radio. A software radio is a radio whose channel modulation waveforms are defined in software. That is, waveforms are generated as sampled digital signals, converted from digital to analog via a wideband digital-to-analog converter (DAC). The receiver similarly employs a wideband analog-to-digital converter (ADC) that captures all of the channels of the software radio node. The receiver then extracts, downconverts, and demodulates the channel waveform using software built into a general-purpose processor.

Software radios employ a combination of techniques that include multiband antennas and RF conversion; wideband ADC and DAC conversion; and the implementation of IF, baseband, and bitstream-processing functions in general-purpose, programmable processors. The resulting software-defined radio in part extends the evolution of programmable hardware, increasing flexibility via increased programmability. It represents an ideal that may never be fully implemented but that nevertheless simplifies and illuminates tradeoffs in radio architectures that seek to balance standards' compatibility, technology insertion, and the compelling economics of today's highly competitive marketplaces.

A typical voice SDR transmitter, like one that might be used in mobile two-way radio or cellular telephone communication, consists of the following elements (note: items with asterisks represent computer-controlled circuits whose parameters are determined by the software programmed into the radio):

- A microphone
- An audio amplifier
- An ADC that converts the voice audio to American Standard Code for Information Interchange (ASCII) data* (a coder/decoder, codec of sorts), which is the most common format for text files in computers

and on the Internet. In an ASCII file, each alphabetic, numeric, or special character is represented with a seven-bit binary number (a string of seven 0s or 1s). One hundred twenty-eight possible characters are defined.

- A modulator that impresses the intelligence onto a radio-frequency carrier*

- A series of amplifiers that boosts the RF carrier to the power level necessary for transmission

- A transmitting antenna

A typical SDR receiver designed to intercept the described voice SDR signal would employ the following steps, essentially reversing the activity of the transmitter (again, items followed by asterisks represent programmable circuits):

- A receiving antenna

- A superheterodyne system that boosts incoming RF signal strength and converts it to a constant frequency (the term superheterodyne refers to a method of designing and building wireless communications or broadcast equipment, particularly radio receivers; sometimes a receiver employing this technology is called a "superheterodyne" or "superhet")

- A demodulator that separates the ASCII intelligence from the RF carrier*

- A DAC that generates a voice waveform from the ASCII data*

- An audio amplifier

- A speaker, earphone, or headset

Many things have driven the development of SDRs. First, wireless hardware usually requires major upgrades (even forklift-type upgrades) every 10 years. The existence of a programmable radio would mitigate this issue completely. Also, application-specific integrated processors (ASICs) have become very sophisticated since the late 1990s, and very cheap to manufacture. Moore's Law comes into play as well; it says that computing power doubles roughly every 12 months without an appreciable increase in cost. Another key driver of SDRs is the fact that multiple wireless technologies are in use across the world. Radio network architectures have evolved from early point-to-point and relatively chaotic peer networks (e.g., citizens band and push-to-talk mobile military radio networks). Channel data rates continue to increase through multiplexing and spectrum spreading.

Multiband multimode military radios and future commercial wireless systems that seamlessly integrate multimedia services across four or more such access modes will represent the high end of network evolution.

The complexity of functions, components, and design rules of radio systems continues to increase with each subsequent generation of radio network technology. Future seamless multimode networks will require radio terminals and base stations with flexible

- RF bands,
- Channel access modes,
- Data rates,
- Equalization,
- Bit error rates (BERs),
- Radiated power, and
- Application functionality.

Software radios have emerged to increase quality of service through such agility. At the same time, SR architectures simplify hardware component tradeoffs and provide new ways of managing the complexity of rapidly emerging standards. Therefore, software radios are needed by carriers who must have future-proof infrastructure, such as radio access points, cell sites, and wireless data network hubs, that can be reprogrammed to meet changing standards or maintained in parallel with newer infrastructure. This is the key benefit and driving force behind the development and deployment of software-defined radios.

Key: A single radio unit, typically a mobile terminal, participates in more than one network hierarchy. This becomes evident when mobile subscribers roam into other, nonhome markets. A software radio terminal could, therefore, operate in a GSM-based PCS network, an AMPS network, and a future satellite mobile network. This is really *the* key driver behind the development of SDR.

The most significant benefit of SDR is versatility. Wireless systems use protocols that vary from one service provider to another. Even when offering the same type of service, the wireless technologies and services often differ from country to country. A single SDR set with an all-inclusive software repertoire can be used in any mode, anywhere in the world. Changing the service type, the mode, and/or the modulation protocol involves simply selecting and launching the requisite computer program and making sure the batteries are adequately charged if portable operation is contemplated. SDR also enables multifunction devices to seamlessly roam in a shared infrastructure and supports software-only upgrades for the faster and cheaper rollout of new technologies or service capabilities. An SDR also reduces the risk of settling on a specific standard that may change dramatically in a relatively short period of time.

In general, the more diverse the network infrastructure, the more potential there is to glean the benefits and flexibility offered by SRs. As a result, the military has historically been the developer of this technology and some key U.S. wireless service providers have begun to express strong interest in its longer-term economic benefits. Some believe that future radio services will provide seamless access across cordless telephones, wireless local loop, cellular/PSC, and mobile satellite modes of communication. Anyone who needs to access even half that many radio modes at once clearly will have to move to a software radio-based infrastructure.

The ultimate goal of SDR engineers is to provide a single radio transceiver capable of playing the roles of cordless telephone, cell phone, wireless fax, wireless e-mail system, pager, wireless videoconferencing unit, wireless Web browser, GPS unit, and other functions still in the realm of science fiction, operable from any location on the surface of the earth, and perhaps in space as well.

4.11 Cognitive Radio

Cognitive radios are intelligent radios (i.e., handsets) that can detect and adapt to their RF environments. With its ability to adapt the radio to its environment, cognitive radio creates a new playing field because it lets a device pick the channel that will allow for the best communication performance moment to moment. Functionally, a cognitive radio is smart enough to find communications opportunities between existing spectra in the environment in which a device is operating. This provides flexibility by allowing the use of licensed spectrum by other users, under acceptable conditions. The CR has the ability to adapt to its environment on the fly, frame by frame. Cognitive radio can build multiple radio technologies into one device and incorporate the endless combinations characteristic of any computer system.

The technical evolutionary predecessor to cognitive radio was SDR. SDR represents an evolution of networking technology achieved by defining a single hardware architecture that is capable of supporting multiple physical layers under software control. Cognitive radio takes SDR even further. In a product where multiple functions are possible— such as UWB wireless, cable, and electrical communications—a combination of discernable traits can be defined and continuously evaluated to automatically determine which networking medium is appropriate or optimal in a given situation. The most appropriate transmission medium is selected automatically and dynamically switched as network conditions and requirements warrant. To understand the cognitive radio concept, it is useful to look at where cognitive radios fit in the continuum of radio technologies, as follows:

- For traditional, fixed radios, the technical characteristics are determined at the time of manufacture and cannot subsequently be easily modified without a trip to RadioShack and a knowledge of radio systems.

- Software-definable radios are both flexible and agile. Software performs all of the signal processing and the device can be reprogrammed to be an analog cellular phone, a digital PCS phone, a cordless home phone, or even a garage door opener. Software-definable radios can transmit and receive on many frequencies and can use any programmed, desired transmission format within the limits of their physical design.

- CRs take this one step further: They incorporate software-definable radio technology with its attendant functionality *and* they can learn. CRS can respond to their RF and geographic environment and do so, according to their experience. On a real-time basis, CRs sense their environment, decide what to do, adopt appropriate operational protocols, and then adjust their operations accordingly. They accomplish this through the use of reasoning algorithms, the application of RF environment modeling, and simulation.

CRs learn through experience over time. Depending on the RF environment, the cognitive radio determines the best frequencies to use, determines how to avoid interference with existing users, and can effectively use available spectrum channels that are not in use.

Cognitive radio technologies offer the potential to become an important tool for enhanced spectrum usage. Driving the development of cognitive radio is the fact that spectrum demand is outstripping spectrum supply particularly because most choice spectrum has already been assigned to one or more parties.

An FCC task force determined that limited access is having more of an impact on spectrum efficiency than limited throughput. They identified several possible ways to improve access to the radio spectrum—through better use of time, frequency, power, bandwidth, and geographic space, as follows:

- First, the task force found that increased opportunities for access can mitigate the apparent scarcity of spectrum. Improved access can permit more intensive use of the radio spectrum if licensees had greater flexibility and if our regulations took into account technological developments and their resulting improved spectrum efficiency.

- Second, the task force also found that more clearly defined interference metrics could improve the utilization of the spectrum. This is because interference management is a cornerstone of any spectrum policy. With more and more intensive use of the radio spectrum in the way of greater density, mobility, variability of RF emitters, an already challenging task has become even more difficult. Users need ever more certainty in determining the applicable RF environment. At the same time, it is becoming increasingly difficult to use predictive modeling, due to variability in waveforms and the more intensive use of the radio spectrum.

- Third, the task force noted that spectrum usage could be parceled according to frequency, geographic space, and the time (or temporal) dimension. But existing spectrum policy does not fully take into account all of these dimensions of spectrum usage.

The implementation of cognitive radio technologies provides the potential to improve access to the spectrum; when access is improved, spectrum efficiency is also increased. Another way in which access to the spectrum can be improved is through providing additional flexibility. For example, the task force also recommended that the Commission consider making the technical rules flexible so that in less congested areas, like rural areas, the rules would not prevent licensees from operating at higher powers on a noninterference basis. The potential exists for spectrum-based devices to operate in both the temporal white spaces (those resulting from variability in the operations of existing spectrum users over time) and the geographic white spaces (those resulting from the geographic separation of existing spectrum users). These task force's observations confirmed some longstanding views regarding actual spectrum usage.

Again, if there is one message to take away relating to the task force report, it is that improved access can mitigate the physical scarcity of the finite spectrum resource. Cognitive radios could operate on a not-to-interfere basis as secondary users of a particular band, perhaps using the geographic and temporal white spaces just described. Or they could operate below some predetermined acceptable interference temperature.

The task force noted the variability in the use of the spectrum by certain users. This example illustrates dynamic spectrum use. Cognitive radios could ultimately be used in coordination with licensed users of the spectrum. One licensee could operate when a particular spectrum is not in use by another licensed user.

Key: Cognitive radios could sense the RF energy in a particular band and, if they sensed that the total RF energy in a particular band was less than the interference "temperature," they could then operate.

Hybridization can also exist within the wireless context, where spectrum users can combine *licensed* links with the use of *unlicensed* hot spots.

Wireless carriers raised interference concerns from the standpoint of licensed users of the spectrum. "Requiring a licensee to endure a nonlicensee's use of the licensee's spectrum on an opportunistic basis with cognitive radios would disrupt the licensee's internal management of its radio network," BellSouth and Cingular said in a joint filing to the FCC. "Such a requirement would also adversely impact the leasing of spectrum in the secondary market. A licensee needs to control how its spectrum is used. Otherwise, 'rogue' devices have the potential to cause interference that will reduce efficiency by degrading quality, capacity, and coverage. Eventually, licensed networks would need to employ higher power for both mobiles and base stations and add cell sites to avoid the loss of coverage." This is a valid argument but an argument that may be mitigated by strict intercarrier standards that govern the use of cognitive radio technology.

▧ Test Questions

True or False?

1._____ The radio transmission from the base station to the mobile phone is known as the uplink, or reverse channel.

2._____ The terms "frequency reuse" and "channel reuse" are synonymous.

3._____ Software-defined radios (SDRs) adapt intelligently to their RF environment and select the least used, most powerful RF signals for processing calls.

4._____ Downtilting base station antennas may compensate for the ducting phenomenon.

5._____ There are four main types of signal fading: absorption, free-space loss, Rayleigh fading, and cross-channel equalization.

6._____ Inorganic materials tend to absorb more RF signals than organic materials.

7._____ Ducting a cellular RF signal is caused by an atmospheric anomaly known as temperature infraction.

Multiple Choice

1. The three key parameters that determine actual RF coverage from any given cell base station are:

   a. RF power levels at the cell base station

   b. The height of the cellular base station antenna

   c. The type of tower use

   d. The type of antenna used

   e. How many buildings are in the area

   f. a, c, and e

   g. a, b, and d

2. The higher the frequency of a radio signal, the greater the:

    a. Absorption rate

    b. Free-space loss

    c. Antenna reverberation

    d. Cochannel interference

    e. None of the above

3. Cochannel interference describes which of the following?

    a. Interference between mobile phones

    b. The inability of the mobile phone to filter out signals of contiguous cellular channels (e.g., channel 331 and channel 332)

    c. Interference between two base stations transmitting on the same frequency

    d. All of the above

    e. a and b only

4. The multipath effect of cellular RF signals bouncing off of many objects resulting in signals arriving at a mobile antenna at different times, in or out of phase with the direct signal, is known as:

    a. Fessenden fading

    b. Uplink fading

    c. Rayleigh fading

    d. Free-space loss

    e. Ducting

    f. None of the above

5. Radio signals in free space travel at:

    a. The speed of sound

    b. Twice the speed of sound

    c. Half the speed of light

    d. The speed of light

    e. None of the above

Antennas, Power, and Sectorization

5.1   Overview

An antenna's function is twofold. To transmit, an antenna must take the radio signal that is applied to it and broadcast that signal as efficiently as possible into the intended coverage area. An antenna transforms alternating electrical current into a radio signal that is propagated through the atmosphere. Conversely, to receive, an antenna transforms a radio signal into an alternating electrical current (AC) that is decipherable to the receiving equipment. Antennas are the ports through which RF energy is coupled from the transmitter to the outside world and, in reverse, to the receiver from the outside world. Radio antennas couple electromagnetic energy from one medium to another (i.e., coaxial cable, waveguide, or wire).

There are two main types of antennas used in the wireless industry. All antennas fall under one of these two categories: *omnidirectional* or *directional* antennas. There are a multitude of omnidirectional and directional antennas available for deployment in a wireless system. One of the largest manufacturers of antennas and base station equipment for the wireless industry is Andrew Corporation, based in Orland Park, Illinois.

The word "decibel" is used to compare one power level to another and is denoted by the symbol dB. Antenna propagation characteristics are measured in decibels. For example, a 9 dB antenna is commonly used at omni base stations in rural areas. Base station antennas range from 2 to 14 feet in length and weigh from 10 to 50 pounds.

5.2   Omnidirectional Antennas and Gain

An omnidirectional antenna is an antenna that radiates an RF signal equally well in all directions. Cells using omnidirectional antennas are known as *omni,* or *single-sectored,* cells. Omnidirectional antennas are sometimes referred to as *sticks* because of their appearance. As discussed in Chapter 4, the single-faceted design of an omni antenna broadcasts omnidirectionally in a patter resembling ripples radiating outward in a pool of water. See Figure 5-1, which shows a picture of an omni antenna.

The term "gain" refers to how the radiation pattern emitted from an antenna is reshaped. Gain is a logarithmic progression. It is represented by the width of a horizontal radiation pattern, or how much the pattern has been electrically compressed vertically and expanded horizontally.

Figure 5-1
An omnidirec-
tional antenna
radiates equally
360 degrees.
(Photo
courtesy
Andrew
Corporation)

Key: Gain is a measure of how much an antenna's vertical radiation pattern is electrically compressed, which
results in propagation to the horizontal plane. This vertical
compression results in pushing the radio frequency (RF)
propagation outward *horizontally* from a radiating antenna
(at a given radio power setting). How far the RF propagates
outward horizontally is approximately equivalent to how
far the coverage extends from any given base station.

Think of the RF coverage area of a cell as being the surface of a balloon. Gain is achieved (and increased) if the balloon is compressed on the top and the bottom simultaneously. As the balloon is squeezed, it flattens out and widens on the sides. This flattening represents gain on the horizontal plane. With omnidirectional antennas that radiate in a 360-degree pattern, the simplest way to achieve gain is by stacking multiple wave antenna elements one on top of the other. Then inside the body of the antenna (the *radome*), the energy emitted by these antennas is fed *together* in such a way as to add each antenna's vertical radiation pattern together in phase. This means there are electrical connections among all the elements. When the radiation patterns of all the antenna elements are added together in phase, their signals *complement* each other, thus contributing to the total composite compression of the signal. This vertical, electrical stacking represents an additive process. The more that radiating quarter-wave antennas are stacked on top of each other (while their patterns are kept in phase), the more gain is produced. In this balloon analogy, as the amount of downward force is increased, the balloon becomes flatter, but its circumference is increased. The amount of bulge represents gain. As the gain is increased, the vertical pattern becomes progressively more compressed. The higher the stated decibel level is for a given antenna, the higher the gain capability of the antenna.

When we speak of stacking multiple-wavelength antennas one on top of the other, this is a reference to installing antennas that were designed and built to a specific gain level by the antenna manufacturer. This type of antenna configuration is known as a *collinear array*, described in the following section. See Figure 5-2 for an illustration of the concept of gain.

▬▬ 5.3 Collinear Array Antennas

The antenna design described in the previous section is known as a *collinear array antenna*, which is a predominant type of omnidirectional and directional antenna used in the wireless industry. As mentioned previously, to produce a collinear array antenna, one or more wave antennas are stacked inside the antenna radome in order to produce gain. The radome is the fiberglass shell that encases the actual antenna elements. It is what we all see and think of as the actual antenna. Each antenna element that is stacked inside the radome represents an increase in gain in increments of 3 dB.

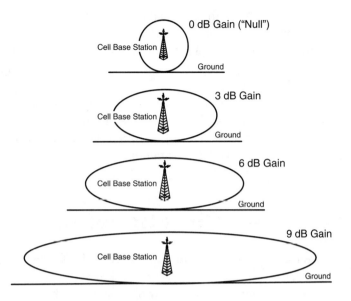

Figure 5-2
The concept of gain

Key: Each time 3 dB of gain is produced, the overall length of the antenna is doubled electrically, or virtually. *This concept correlates directly to the fact that gain is a logarithmic progression.* If an antenna with 3 dB gain is 3 feet in length, an antenna with 6 dB gain will be 6 feet in length. A 9-dB-gain antenna would then be 12 feet long and a 12-dB-gain antenna would be 24 feet long. In the past in the wireless industry (1980s and early 1990s), the actual physical length of the antennas did double. But with the progress in antenna design and manufacturing over the years, this length progression became an electrical (virtual) result of incremental gain increases (3dB) versus a physical length progression.

The antenna elements that comprise the collinear array antenna will be *longer* for antennas that are designed and used for *lower* frequencies and *shorter* for antennas that are designed and used for *higher* frequencies. This length in the antenna elements correlates directly to the fact that lower frequencies emit longer radio wavelengths, and higher frequencies emit shorter wavelengths. These differences are directly attributed to the properties of radio physics. Thus, antennas used for lower frequencies will be physically longer than antennas used for high frequencies, if the gain assigned to both of those antennas is the same. If a wireless carrier deploys a 9 dB antenna that operates at 850 MHz and a 9 dB antenna that operates at 1,900 MHz, the 800 MHz antenna will be physically longer due to the longer wavelength inherent with RF emitted at 850 MHz.

Today, 0 dB (*unity antennas*) or 3-dB-gain antennas are frequently used in urban areas. These antennas are mounted on the sides of buildings or at street level and are used to cover very small areas to support enhancer or microcell deployments.

5.4   Directional Antennas

A directional antenna is an antenna that shapes and projects a beam of radio energy in a specific direction and receives radio energy only from a specific direction, employing various horizontal beamwidths. Directional antennas are effectively omni antennas that use a reflecting element, which directs or focuses the RF signal (energy) over a specified beamwidth. They produce more gain than a typical omni base station antenna produces. The most popular beamwidth used in the United States is a 120-degree beamwidth, which supports three-sectored base stations (see Section 5.8). Other beamwidths are used though, such as 90-degree antennas and even 60-degree antennas.

   Key: Using the analogy of light, as is done many times
   when discussing RF propagation, an omni antenna can be
   compared to a lamp with no shade on it. The light from that
   lamp is emanating equally and freely in all directions; thus,
   360 degrees of light is shone. Directional antennas are
   viewed as being similar to a flashlight: There is an omni
   lightbulb in the flashlight housing, but the mirror-like
   reflecting element in every flashlight where the bulb is
   housed reflects the light into a beam. The bigger, or wider,
   the reflecting element is, the wider the light's beamwidth.
   The same principle applies to directional antennas. The
   size and shape of the reflecting element in a directional
   antenna dictates its beamwidth.

There are many types of directional antennas used by wireless carriers: log periodic, Yagi, phased-array, and panel antennas, which are the most popular. Panel antennas resemble white shoeboxes. Along with being used primarily for sectorization purposes, directional antennas may also be used to keep RF signals out of specific undesired areas, such as the border of a neighboring wireless market. Directional antennas are mostly used in sectorized cells, as described in Section 5.8. See Figures 5-3 and 5-4 for photos of panel directional antennas, the most common and popular form of directional antennas used in the wireless industry today.

Using directional antennas, gain is developed by inserting a reflecting element behind a modified omni antenna. The reflector distorts and

Figure 5-3
Panel
directional
antennas on
monopole
tower (Photo
courtesy Mike
P's UK GSM
and UMTS
Pages)

Figure 5-4
Panel
directional
antennas
(Photo
courtesy
Andrew
Corporation)

compresses the horizontal radiation pattern at the sides, causing it to bulge forward and produce directional gain. The amount of gain developed by a directional antenna depends on the size and shape of the reflecting element and its distance away from the actual antenna.

For instance, a 10-dB-gain omni antenna could be modified with a V-shaped reflector to produce a 60-degree horizontal radiation pattern and 17 dB of gain. This example shows that, all things being equal, a directional antenna will produce more gain than an omni antenna at a given location. In this example, the directional antenna produced 7 dB more gain.

5.5   Downtilt Antennas

At cell sites with a very high tower and a high-gain antenna, coverage shadows may be created near the tower. To compensate for coverage shadows, electrical downtilt antennas and mechanical downtilt kits were developed specifically for the wireless industry by antenna manufacturers.

> *Key:* A downtilted antenna is an antenna whose radiation pattern is *electrically* or *mechanically* tilted a specified number of degrees downward. Downtilting of antennas decreases distance coverage horizontally but increases signal coverage closer to the cell site.

Omni antennas can only be downtilted *electrically*, which is accomplished in the manufacturing process of the antenna by adjusting the phasing of the RF signal that is fed to the collinear antenna elements. Electrical downtilting is the way in which a specific antenna is manufactured, similar to how given antennas are manufactured with specific gains assigned to them. Electrical downtilt antennas are manufactured to downtilt to a preset amount of degrees.

Directional antennas can be downtilted either electrically or mechanically. Mechanical downtilting is accomplished by actually manipulating the antennas so that they tilt toward the ground. Although mechanical downtilting is less expensive then electrical downtilting, it distorts (expands) the side lobes of the radiation pattern and could lead to interference with adjacent sectors. See Figure 5-5 for a photo of mechanically downtilted directional antennas.

A common place to install a downtilt antenna is at a cell site that is on a very tall tower or a hill, or near a large body of water. Downtilt antennas are also used to reduce the impact of what is known as the *far-field effect* in wireless networks. The far-field effect occurs when the radio coverage projected from site A may completely and unintentionally overwhelm the intended coverage area of site B or other nearby sites. Site A may transmit and receive into site B or other sites, theoretically leaving these sites unused. This would not only be terribly inefficient but would be a terrible waste of equipment and frequency resources at cell site B and other nearby base stations. The deployment of a downtilt antenna at cell site A would ensure that the intended radio coverage from site A stays within its designated coverage boundary.

The far-field effect can occur for any of the following reasons:

- RF power level is too high at the base station.
- Downtilt antennas are *not* being used at the base station.
- The tower is too high at the base station or the base station transmit antenna is too high on the tower.
- The antenna gain is too high at the base station, exceeding its intended coverage area.

Downtilt antennas can shrink base station coverage areas although RF power levels are held constant. Downtilt antennas improve portable coverage for subscribers. The antennas can also be used to pull coverage away from a market border to avoid projecting coverage into neighboring

Figure 5-5
Mechanical downtilt of directional base station antennas

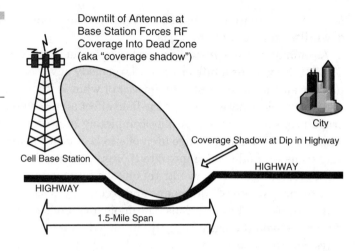

Figure 5-6
Sample
downtilt
antenna
operation

systems or to decrease the potential for ducting to occur near large bodies of water. See Figure 5-6 for an illustration of how a downtilt antenna operates.

5.6   Criteria Used to Select Base Station Antennas

Base station antennas are much more sophisticated and utilize a much wider variety of designs than mobile phone antennas. One reason for this difference is that base station antennas are required to have a higher degree of gain, ordinarily between 6 and 12 dB for omnidirectional antennas and between 4 and 18 dB for directional antennas. In some cases, 0-dB-gain omni antennas are used to support microcell deployments.

The type of base station antenna that is chosen in any situation depends on many factors:

- The size of the area to be covered
- Neighboring cell sites' configurations
- Whether the antenna is omnidirectional or directional
- If it is a directional antenna, the antenna's beamwidth
- How much of the allotted RF spectrum the antenna can utilize

In the wireless industry, carriers can get antennas for the transmit band only, the receive band only, or both the transmit and receive bands. Most antennas used today are wide bandwidth, meaning they can be used as either transmit or receive. This is done for ease of installation and as an option in case a carrier elects to duplex. Duplexing, which allows the antenna to both transmit and receive simultaneously, will be discussed in Chapter 10.

Most cell phone antennas are 0-dB-gain collinear array antennas. These antennas do not provide much signal improvement and do virtually nothing when not extended (if they are the extendable type). Improvements have been made in recent years to the antennas on portable phones, but they are still not very effective. This is done to give the consumer the sense that they have more control over their mobile environment and their overall wireless experience. In cases where this placebo effect exists or where no external antenna comes with a mobile phone, the actual antenna element is thin copper or steel wiring that is wrapped around the inside of the phone on the perimeter of the casing.

Quality varies with base station antennas used in the wireless industry, just like any other service or product. Some antennas can stand up to the elements although some do not. Some will leak and some will literally fall apart. Ultraviolet rays are especially harsh on antennas.

Antenna quality is measured in several ways. One way is to determine how many dissimilar metals are used in the antenna. This is important because the wider the variety of metals used in the manufacturing of an antenna, the more likely the antenna will develop corroded junctions, which can cause intermodulation (IM) interference, or signal mixing. For instance, two signals may go in and four signals may come out. Ultimately, this results in a degradation of signal quality. The metal connectors that connect antennas to coax cables at base stations are the weakest link on a base station antenna and can cause water ingression into the antenna if not properly weatherproofed.

5.7 Radio Frequency (RF) Power

RF power is defined as the amount of RF energy, in watts, delivered by base station radios (and amplifiers) to the base station's transmit antenna. RF amplifiers used at the cell site determine power levels and the amount of RF energy that is delivered to the base station antenna by the base station radios (transceivers). There will also be loss of the signal as it propagates through the coaxial cable from the transceiver up to the transmit antenna due to impedance. This signal loss is factored into RF design by wireless system engineers.

5.7.1 Effective Radiated Power (ERP)

Effective radiated power (ERP) is determined by the gain of the antenna (mobile or base station) times the power delivered to the base of the antenna. ERP is measured in watts and also defines the sum total of power emitted by a base station.

 Key: It is important to remember this rule of thumb: For every 3 dB of gain, the output power is doubled in watts.

The following explanation describes how ERP is measured from a base station. Loss incurred due to cable impedance is factored into this equation to measure true *net* signal strength emitted by the base station. This sample following is a measurement taken from an actual Chicago region base station on a rooftop site.

Consider the following scenario: A carrier is using an 8-watt power amplifier (PA) with a 15-dB-gain antenna and 3 dB of line loss. This equates to a net gain of 12 dB when the line loss is factored in. The number 8 is the constant value in this equation because it is the power delivered by the amplifier to the antenna network (the coaxial cable *and* the antenna). The use of the number 8 will illustrate how gain is a logarithmic function. In this scenario, with the 8-watt amplifier, presume we start with 0 dB of gain. Three decibels of gain would deliver 16 watts ERP (8-Watt amplifier with 3 dB of gain equals a doubling of ERP *if RF power levels remain static*). Six decibels of net gain would equate to 32 watts ERP. Again, every additional 3 dB of gain equates to a doubling of ERP *if RF power levels remain static*. Nine decibels of net gain would equate to 64 watts ERP. Twelve decibels of net gain would equate to 128 watts ERP. Because we know we have a net gain of 12 dB from the antenna (including line loss), the ERP for a base station using the equipment listed previously would be 128 watts. Notice that each time we increase net gain by 3 dB, *the ERP doubles*. The actual formulas used to determine ERP can be very granular, as they represent logarithmic relationships. That information is beyond the scope of this text and could be obtained by reading one of the Clint Smith books on wireless (*Wireless Telecom FAQs* or *Wireless Network Performance Handbook*).

5.7.2 Carrier-to-Interference (CI) Ratio

The carrier-to-interference (C/I) ratio is a measure of the desired signal a cell or mobile phone recognizes relative to interfering signals. The interfering signals could be either other mobile phones or other base stations. The C/I ratio is a key design parameter in wireless systems and is measured in decibels. Each wireless technology has an objective of a given C/I level that should exist throughout a wireless system. A well-maintained C/I level helps to reduce or eliminate cochannel interference. The target levels for the various wireless technologies are as follows:

- AMPS: 17 dB
- IS-136: 19 dB
- GSM: 13 dB
- iDen: 20 dB
- CDMA: N/A (due to N=1 basis of system and the fact that it is a noise-based system; this will be explained in Chapter 6)

There should be an *X* dB difference (see preceding list) between any given cell and all other cells (and mobile phones) throughout a wireless system. The C/I ratio is somewhat analogous to the signal-to-interference ratio. The frequency-reuse plan is the tool that is used to keep the C/I ratio at or above the ideal levels indicated in the preceding list.

Key: As more base stations are added to a wireless system, a migration occurs from a noise-limited system to an interference-limited system. The system has more potential to produce cochannel interference and other types of interference, as it grows in size. In other words, the overall quality of calls is easier to maintain in networks that are smaller than in those that are larger.

5.7.3 Allowable Power Levels

In the wireless industry, carriers are allowed to use up to a maximum of 500 watts ERP at cell base stations. The range of 100 to 500 watts ERP is possible at rural base stations. The average ERP for MSA urban base stations (macrocells) is 20 to 100 watts to cover an area from 2 to 30 miles. The base tranceiver station (BTS) power levels used depend on the frequency-reuse plan in place for any given market, and these power levels can differ throughout the areas of any given market. Like RF coverage itself, power levels that are actually used are a reflection of the real or potential subscriber densities in any given area. Ideally, the maximum power level needed to provide coverage is used, but no more than that, and this level will depend on terrain in the coverage area and antenna gain. The overall range of ERP in a wireless system can be anywhere from ¹/₁₀ of a watt to 500 watts ERP.

Key: The power level of a mobile phone can and will be throttled down by base stations in a wireless network if the mobile phone is emitting too much power. This will avoid overloading equipment at the base station or causing interference with other subscribers in the network. One example of when this could occur is when a mobile phone is located too close to the base station. The base station controller, or the BTS itself, will send a message over the control channel assigned to the base station where the subscriber is located, directing the mobile phone to throttle down its power to a specified level.

Mobile phones have allowable power classes assigned to them by the FCC. These power levels have a range from 0.006 of a watt to 4 watts. Mobile phones installed in autos have higher allowable power levels because they emit more power, as they are powered by car batteries. Table 5-1 depicts allowable power levels in watts for different types of mobile phones.

5.8   Sectorization

As more cochannel cells were added to a wireless system over time (1983 to 1993), it became necessary to develop a means to increase systems' capacity without constantly having to split cells. Cell splits could be very

Table 5-1

Mobile Phone
Power Classes

Power Level	Class 1	Class 2	Class 3
0	4.000	1.600	0.600
1	1.600	1.600	0.600
2	0.600	0.600	0.600
3	0.250	0.250	0.250
4	0.100	0.100	0.100
5	0.040	0.040	0.040
6	0.016	0.016	0.016
7	0.006	0.006	0.006

costly undertakings. The industry developed a way to migrate from an omni antenna configuration at cell base stations and begin to sectorize base stations in order to obtain more capacity from each base station deployment.

Key: Sectorized base stations are created by subdividing an omni cell into sectors that are covered using directional antennas that are mounted in the same base station location (i.e., the same tower or rooftop). Operationally, each sector is treated as a different cell, the range of which is greater than an omni cell. Directional antennas always produce more gain than omni antennas. All subcell directional antennas supporting each sector are collocated at the same base station. All base station/radio equipment for each subcell, or sector, is housed in the same base station. To sectorize a cell, a horizontal, equilateral platform that resembles a triangle is deployed on a tower. Each side of the platform is called a *face* and has three or four directional antennas installed. The directional antennas propagate the different frequencies/channels assigned within each respective face.

Sectorization facilitates wireless engineering and operations in the following ways:

▪ It minimizes or eliminates cochannel interference.

▪ It optimizes the frequency-reuse plan. This is facilitated through another concept known as the front-to-back ratio, which will be reviewed in the next section.

▪ At a minimum, it triples the capacity of any given coverage area when compared to the capacity that would be offered by deploying omni antennas.
Most wireless carriers in the United States usually deploy three sectors per cell site and, in some cases, four. In some parts of the world, six-sectored base stations are used. The type of directional antennas deployed at any given sectorized base stations are dependent on the number of sectors engineered at the base station.

▪ In cases where three-sectored base stations are deployed, the directional antennas mounted in each sector will have 120-degree beamwidths. From a graphical viewpoint in this scenario, sectorization takes a circle (representing an omni base station) and converts it into a three-section pie chart.

▪ In cases where four-sectored sites are deployed, 90-degree beamwidth directional antennas are used.

▪ In cases where six-sectored base stations are deployed, 60-degree beamwidth directional antennas will be used.

Obviously, the objective when implementing sectorized base stations is to support 360-degree coverage from a single location. The amount of sectors will dictate the beamwidth of the directional antennas used within each sector.

Each sector has its own assignment of radio channels and its own channel set or sets. Each sector also has its own control channels and will hand calls off to its adjacent sectors that are housed on the same tower, rooftop, water tank, and so forth. Each sector will also hand calls off to adjacent sectors when required. See Figure 5-7 for an illustration of the sectorization concept. Figure 5-8 illustrates this same coverage.

5.8.1 The Grid Angle

When sectorizing base stations, wireless carriers develop and adhere to what is known as the *grid angle*, also known as the *azimuth* of the system.

Figure 5-7
The sectorization concept (using 120-degree beamwidth directional antennas). Most sectorized base stations in the U.S. use 3-sided, 120-degree sectors.

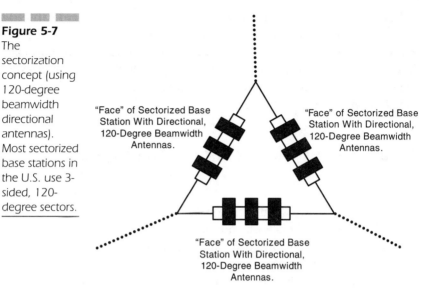

"Face" of Sectorized Base Station With Directional, 120-Degree Beamwidth Antennas.

"Face" of Sectorized Base Station With Directional, 120-Degree Beamwidth Antennas.

"Face" of Sectorized Base Station With Directional, 120-Degree Beamwidth Antennas.

Figure 5-8
Sectorized antenna pattern and resulting coverage (Figure courtesy www.iec.org)

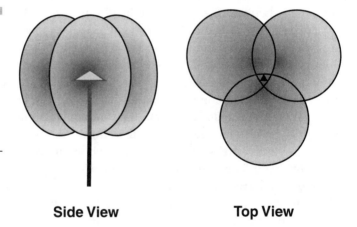

Side View   **Top View**

Key: The grid angle is the orientation, in degrees, of the first face of all sectorized cells from true north. The grid angle allows for all base station faces throughout a sectorized system to be laid out symmetrically in order to avoid interference between cochannel sectors.

In other words, a wireless carrier in city X would have the antenna platform faces at all of their base stations oriented in the exact same way, at the exact same angle, throughout the area. Some exceptions may exist for cases where there is reason to have unique azimuths due to interference with other systems or to concentrate a signal in a defined direction, such as a highway.

5.8.2 Tower Mounting of Directional Antennas

The positioning of directional base station antennas is quite different in sectorized cells than it is with omnidirectional cells.

Directional antennas are used at sectorized base stations, which are much more prevalent than omnidirectional sites and much easier to identify. Sectorized cells have a triangular platform mounted on a tower, rooftop, water tank, and so forth. Directional antennas are most often arranged in three sectors, or directions, with three to four antennas per sector. A typical sectored site will utilize 9 to 12 panel antennas, which are usually between 3 and 8 feet in height and about 12 inches wide.

Like omnidirectional cells, directional cells use space diversity. There are two receiving antennas mounted at each end of the platform, where this spacing serves as the diversity receive path. Of the three directional antennas mounted per face, the two outer antennas are the base station diversity receive antennas (RX 0 and RX 1), and the middle antenna is the base station transmit antenna. The transmit antenna transmits the downlink, base-to-mobile signal. Like omni cells, the diversity receive antennas receive RF signals from the mobile phone, the uplink signal. (Sites utilizing four antennas per sector are using the fourth antenna for additional transmit frequencies outside of the normal spectrum. These are usually sites operated by traditional 850 MHz cellular carriers.) Diversity is reviewed in Chapter 10. Refer back to Figure 5-3 for an illustration of tower mounting of directional antennas at a sectorized base station.

 Key: The length of the separation that exists between the two diversity receive antennas is roughly equal to 20 times the wavelength of the frequency being used.

For instance, the wavelength of 1,900 MHz (the U.S. Personal Communication Services [PCS] frequency band) is approximately 6 inches. Therefore, the diversity receive antennas in a sectorized PCS cell site will be approximately 120 inches apart, or about 10 feet.

Platform faces on towers also vary in length depending on the height of the tower. However, each platform on any one given tower will be the same length on all sides. For example, the three platform faces on a sectorized cell site will all be equilateral, but the platform faces on all the towers throughout a wireless system may be different.

In many cases, when looking at a duplexed base station where there are only two antennas per face, one of those antennas is used for both transmit (base-to-mobile signals) as well as receive (mobile-to-base signals, RX 0/RX 1). The second antenna is used for diversity receive only. In these scenarios, it might appear as if there actually is a third antenna in the middle of the platform faces. What looks like a third antenna is actually only a *mounting post* for a third antenna.

As with omni cell base stations, each sector in a sectorized cell can be analyzed separately for design and planning purposes. This relates to the radio channels assigned to it, the percentage of traffic it carries, century call seconds (CCS) traffic engineering data, and so on.

Sectorization allows for the development of traffic studies to determine where customers are actually located.

In some instances, for one or more sectors in a sectorized cell site, wireless carriers will install directional antennas of different gains in a given sector or sectors to achieve a higher level of gain into a specific geographic area. This may be done because of zoning restrictions in certain localities that may prevent the wireless carrier from installing a full-fledged base station in a certain area. To compensate for this situation, the carrier may pump up coverage in that area from a specific sector in the nearest cell site by using higher-gain antennas in that sector only.

5.8.3 Front-to-Back Ratio

As directional antennas are used in sectorized cells, RF engineers must take the front-to-back ratio into account to reduce the potential of picking up signals from cochannel sectors that are 180 degrees to the rear. This reduces the possibility of cochannel interference caused by the *back lobe* of the antenna propagation. The back lobe is RF that's projected 180 degrees to the rear of the antenna's front.

The front-to-back (F-B) ratio is defined as the ratio of the forward gain of a given cell's sector (based on the placement of the directional antenna) to the gain 180 degrees to the rear of cochannel sectors. Antennas with high F-B ratio ratings are sought by engineers and purchasing managers when trying to reduce cochannel interference in a sectorized environment.

Directional antennas are assigned an F-B ratio by manufacturers, based on the electrical, frequency, and gain characteristics assigned to any particular antenna.

5.9 Smart Antenna Systems

In the late 1990s, the limitations of broadcast antenna technology on the quality, capacity, and coverage of wireless systems prompted an evolution in the fundamental design and role of the antenna in wireless systems. Sectorization was the first step toward increased spectral efficiency in wireless networks. The next step in this evolution has been the development of the *smart antenna*, also known as the *intelligent antenna*.

Although directional antennas and sectors multiply the use of radio channels, they do not overcome the major disadvantage of standard antenna broadcast cochannel interference. Standard antennas also compensate for a lack of knowledge of end user whereabouts by simply boosting the RF power levels of signals they broadcast. This approach could generate interference with signals in the same or adjoining cells. Sector (directional) antennas provide increased gain over a restricted range of azimuths when compared to an omni antenna. This is commonly referred to as *antenna element gain* but should not be confused with antenna processing gains (i.e., software management of signals) with smart antenna systems.

The management of cochannel interference is the number one limiting factor in maximizing the capacity of a wireless system. To combat the effects of cochannel interference, smart antenna systems focus directionally on intended users and in many cases even direct intentional nulls toward known, undesired users. Think of a null as an empty signal. The goal of a smart antenna system is to increase the signal quality of the radio-based system through more focused transmission of radio signals while enhancing capacity through increased frequency reuse.

Key: When spectrally efficient solutions are a business imperative, as they are today, smart antenna systems can provide greater coverage area for each cell, higher rejection of interference, and substantial capacity improvements. Generally speaking, each type of smart antenna system forms a main lobe toward individual users and attempts to reject interference or noise from outside of the main lobe.

Smart antennas are designed to help wireless operators cope with variable traffic levels and the network inefficiencies they cause. These systems also allow carriers to change gain settings to expand or contract coverage in highly localized areas all without climbing a tower or mounting another custom antenna. Wireless carriers can then tailor a cell's coverage to fit its unique traffic distribution. If necessary, carriers can modify a cell's operation using smart antennas based on the time of day or the day of the week or to accommodate an anticipated surge in call volume from a sporting or community event.

Usually collocated with the base station, a smart antenna system combines an antenna's array with a digital signal-processing capability to transmit and receive in an adaptive, spatially sensitive manner. In other words, this type of system can automatically change the directionality of its radiation patterns in response to its signal environment. They can increase the performance of a wireless system dramatically.

An analogy can be used to explain how smart antenna systems operate. Imagine you are in a room sitting in a chair. Someone in the room is talking to you and, as they speak, they begin moving around the room. Your ears and brain have the ability to track where the user's speech is originating from as they move throughout the room. This is very similar to how smart antenna systems operate: They locate users, track them, and provide optimal RF signals to them as they move throughout a base station's coverage area.

Terms that are commonly associated with various aspects of smart antenna system technology include phased *array*, spatial division multiple access (SDMA), *spatial processing, digital beamforming, adaptive antenna systems*, and others.

Smart antenna systems fall into two main categories: switched-beam systems and adaptive-array systems. Generally speaking, each approach directs a main lobe (or radio beam) toward individual users and attempts to reject interference or noise from outside of that main lobe.

5.9.1 Switched-Beam Smart Antenna Systems

Switched-beam systems use a finite number of fixed, predefined patterns, or combining strategies (sectors). These antenna systems detect signal strength, choose from one of several predetermined, fixed beams, and switch from one beam to another as the mobile phone moves throughout a base station's sector. Switched-beam systems communicate with users by changing between preset directional patterns, largely on the basis of the signal strength received from the mobile phone. In terms of radiation patterns, switched beam is an extension of the current microcellular or cellular sectorization method of splitting a typical cell: The switched-beam approach further subdivides macrosectors into several microsectors as a means of improving range and capacity. Each microsector contains a predetermined fixed-beam pattern with the greatest sensitivity located in the center of the beam and less sensitivity elsewhere. The design of such systems involves high-gain, narrow-azimuth antenna elements. The switched-beam system selects one of several predetermined fixed-beam patterns with the greatest output power. The system switches its beam in different directions throughout space by changing the phase differences of the signals received from or used to feed the antenna elements. When the mobile user enters a particular macrosector, the switched beam system selects the microsector containing the strongest signal. Throughout the call, the system monitors signal strength and switches to other fixed microsectors as required.

The switched-beam (lobe) design is the simplest smart antenna design. The main idea of this design is to measure the received signal on a finite number of predefined azimuths and choose the setting that gives the best performance. See Figure 5-9 for an illustration of switched-beam microsector coverage, as explained in the preceding passage.

5.9.2 Adaptive-Array Smart Antenna Systems

Adaptive-array technology represents the most advanced smart antenna approach to date. Using a variety of signal-processing algorithms, the adaptive system takes advantage of its ability to effectively locate and track various types of signals to dynamically minimize interference and maximize intended signal reception. Adaptive-antenna systems

Figure 5-9
Switched-beam
antenna
coverage
patterns
(Figure
courtesy
www.iec.org)

approach communications between a user and base station in a different way, in effect adding a dimension of space.

Key: By adjusting to an RF environment as it changes via the origin of signals, adaptive-antenna technology can dynamically alter the signal patterns to near infinity to optimize the performance of the wireless system.

Adaptive arrays utilize sophisticated signal-processing algorithms to continuously distinguish among desired signals, multipath signals, and interfering signals, as well as calculate their directions of arrival. Adaptive-array systems attempt to understand the RF environment more comprehensively and transmit more selectively. This approach continuously updates its transmission strategy based on changes in both the desired and interfering signal locations.

Among the most sophisticated utilizations of smart antenna technology is an adaptive-array technology known as SDMA. SDMA employs advanced processing techniques to basically locate and track fixed or mobile terminals, adaptively steering transmission signals toward users and away from interferers. This adaptive-array technology achieves

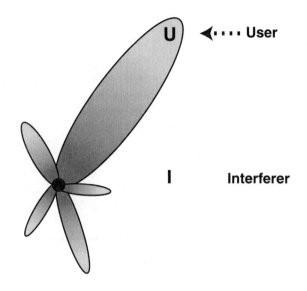

Figure 5-10
Adaptive-array coverage; note main lobe (signal) directed toward user. (Figure courtesy www.iec.org)

U ◀····· User

I Interferer

superior levels of interference suppression, making possible more efficient reuse of frequencies than standard fixed hexagonal reuse patterns. In essence, SDMA can adapt the frequency allocations to where the most users are located. Utilizing highly sophisticated algorithms and rapid-processing hardware, spatial processing takes the reuse advantages that result from interference suppression to a new level. Spatial processing dynamically creates a different sector for each user and conducts frequency/channel allocation in an ongoing manner in real time.

To process information that is directionally sensitive requires an array of antenna elements (usually 4 to 12 where the inputs from each element are combined to control signal transmission adaptively). Antenna elements can be arranged in linear, circular, or planar configuration and are most often installed at base stations (though they could be used in mobile phones or laptop computers as well).

See Figure 5-10 for an illustration of the operation of an adaptive-array smart antenna system and how it minimizes interfering signals while steering coverage lobes intelligently toward the end user.

5.9.3   Summary of Smart Antenna Systems

The distinctions between the two major categories of smart antennas relate to the choices in transmit strategy: Switched-beam antennas use a finite number of fixed, predefined patterns or combining strategies (sectors). As a mobile moves throughout a macrosector, wireless calls are switched from microsector to microsector. Conversely, adaptive array antennas use an infinite number of scenario-based patterns that are adjusted in real time, steering the lobe with the user as they move through the sector.

Both systems attempt to increase gain according to the location of the user. Only the adaptive system provides optimal gain while simultaneously identifying, tracking, and minimizing interfering signals. Switched beam and adaptive-array systems share many hardware characteristics and are distinguished primarily by their adaptive intelligence. With some modifications, smart antenna systems are applicable to all major wireless protocols and standards.

Switched-beam solutions work best in environments with minimal to moderate cochannel interference and have difficulty distinguishing between a desired signal and an interferer. Adaptive-array technology offers more comprehensive interference rejection. Because it transmits an infinite number of combinations, its narrower focus creates less interference to neighboring users than a switched-beam approach.

Simple antennas work for simple RF environments. However, smart antenna solutions are needed as the number of users, instances of interference, applications, and propagation complexity grow. Their smarts are contained in their digital signal-processing facilities.

Although smart antenna technologies are promising many benefits, they do have some disadvantages. The main drawbacks of smart antenna systems are increased base station complexity, increased need for computational power, and more complex resource management schemes. But the single biggest disadvantage to these systems today is their cost. The average cost of standard (or dumb) base station antennas is around $500. In contrast, smart antennas can cost up to around $4,500 each. This high cost has precluded their widespread deployment in wireless systems today. Smart antennas have not been extensively field tested either. But, as the cost comes down, carriers will begin to deploy smart antenna systems with regularity.

5.10   Gas Plasma Antennas

In May 2004, a company known as Markland Technologies announced that it had developed a gas plasma technology that can be utilized to create secure Wi-Fi data transmission for use in business and military applications. The technology allows for adjustments to bandwidth, beamwidth, and packet direction in wireless transmissions.

The company believes that to create better security in Wi-Fi networks, carriers could incorporate gas plasma transmission antennas within a wireless network. Gas plasma antenna technology would allow for highly directional and electronically steerable digital data transmissions. Because the gas plasma can be rapidly enabled and disabled in less than one microsecond, it can be repositioned to point in any required direction or can scan at a very high speed. Not only can a plasma antenna reposition itself rapidly, but it can also change its beamwidth and bandwidth as well, thereby creating spatial and spectral security features that are not available with conventional Wi-Fi antenna technology. Keep an eye on the news for developments in this area.

Test Questions

True or False?

1. _____ The radome is the white or gray fiberglass tube that encases antennas.

2. _____ Gain is a logarithmic progression.

3. _____ When the wattage output of the amplifiers/radios is doubled, a 10 dB increase in gain is achieved.

4. _____ Switched-beam smart antenna systems use an infinite amount of combining strategies to direct RF lobes toward intended users.

5. _____ Directional antennas produce more gain than omnidirectional antennas produce.

Multiple Choice

1. Gain is defined as the compression of RF at which plane?
   a. Horizontal
   b. Bidirectional
   c. Directional
   d. Vertical

2. What term is used to compare one radio power level to another?
   a. Hertz
   b. EMI
   c. Decibel
   d. Channelization
   e. None of the above

Digital Wireless Technologies
(TDMA/IS-136, CDMA, GSM)

Until 1992, 850 MHz cellular carriers had always used analog technology in their networks. Transceivers within base stations were analog. Mobile phones were analog, and the radio communication between base stations and mobile phones transmitted an analog FM signal.

Key: As the demand for commercial mobile telephone service has increased dramatically since its inception in 1983, service providers found that basic engineering assumptions borrowed from wireline, or landline, networks did not hold true in mobile systems. This became evident as tens of thousands of people per day signed up for wireless service in the 1990s. By the mid-1990s, the critical problem for 850 MHz cellular carriers became system capacity.

Cellular providers had to find a way to derive more capacity from the existing multiple-access method currently in use, namely *frequency division multiple access* (FDMA). A multiple-access method defines how the radio spectrum is divided into discrete, usable channels and how channels are allocated to the many users of the system. Multiple-access technology allows a large number of users to share a common pool of radio channels, and any user can gain access to any channel.

The spectacular growth in the number of wireless customers had to be accommodated through a continual increase in system capacity. The most costly method is to reduce cell sizes in given areas and introduce additional base stations; this is what's known as implementing *cell splits*. However, in most large cities (MSAs), it has become increasingly difficult to obtain the necessary permits to erect base stations and towers. Therefore, 850 MHz cellular carriers wanted a solution that made it possible to increase system capacity significantly without requiring more base stations. The solution was to introduce digital-radio technology, which allows for increased spectral efficiency without having to construct new base stations. In March 1988 the Telecom Industry Association (TIA) set up a subcommittee to produce a digital cellular standard. In 1992, cellular carriers began deployment of IS-54 systems using *time division multiple access* (TDMA) in order to increase the capacity of their systems.

There are three major digital-radio technologies developed and in use today: IS-136 (TDMA), Global System for Mobile Comunications (GSM), and Code Division Multiple Access (CDMA). IS-136 is a rapidly dying technology. Other digital-wireless technologies are also being deployed, but they fall under the *Third-Generation Wireless* category (see Chapter 7). Figure 6-1 illustrates how these digital-wireless systems compare functionally and how each technology utilizes radio spectrum. Note that FDMA analog division of spectrum is also illustrated.

6.1   Digital-Wireless Systems Versus Analog Wireless Systems

Digital-wireless technologies offer the following advantages, which analog cellular technology simply could not provide due to its very nature:

- Given equal amounts of spectrum, all of the digital-wireless technologies allow for increased use of radio spectrum compared to analog systems. A key objective of deploying digital-wireless technologies is to maximize the use of allotted radio spectrum. To

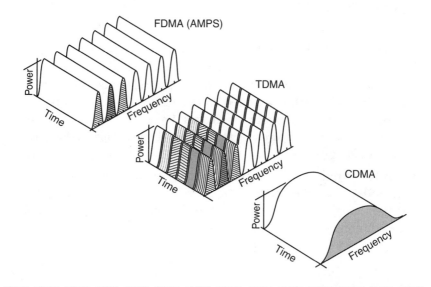

Figure 6-1

Multiple-access methods for wireless technologies. FDMA: Users are assigned different frequencies and different time slots. TDMA: Users are assigned different frequencies and different time slots. CDMA: Users are channelized by specified codes within a frequency band that is 1.25 MHz wide.

that end, all of the digital-radio technologies allow for the use of a substantially larger amount of radio channels than do legacy analog (AMPS) systems.

- Digital-wireless base stations cost far less than analog base stations. This is because digital base stations employ advanced, smaller-scale equipment that has a smaller physical footprint as well as a smaller price tag.

- Digital systems produce cleaner, quieter signals than legacy analog (AMPS) systems. As a matter of fact, some wireless carriers have been known to introduce *white noise* into their digital systems because customers could not tell if their call was still active due to the lack of noise on the system. Because users of telecom services are accustomed to some degree of noise or slight hiss on their phone calls, wireless carriers found benefit in injecting a little white noise into their radios to ease customer's minds.

- Digital systems provide greater security; they are harder or close to impossible to hack.

- Digital-wireless systems have all but eliminated the potential for cloning fraud.

- Digital systems deliver the ability to produce and market smaller, more lightweight handsets.

- Some digital-wireless systems (GSM, CDMA) employ handoff techniques where the mobile handset has a role in determining if and when a handoff is required and to what cell or cells. This improved handoff capability increases the efficiency of the wireless network because it decreases the potential for dropped calls during the handoff process.

- Due to the nature of transmissions in digital systems, instead of using dB measurements to gauge wireless interference within the system, carriers use *bit error rate* (BER) and *frame error rate* (FER) measurements to assess interference.

6.2   Vocoders

All speech begins and ends as analog information; it is the only way the human ears and brain can understand information.

Key: The deployment of digital-wireless systems has been enabled through the development and deployment of equipment known as *vocoders*. Vocoder stands for *voice coder/decoder*.

Human speech is easy to reproduce. Vocoders are firmware and chip sets embedded in mobile handsets and in base station transceivers that digitize human speech. Vocoders are what actually enable digital-wireless systems to exist; they allow for increased use of radio spectrum. Vocoder chipsets are where the specific digital-wireless technology parameters exist. The manner in which voice is sampled and channelized in digital-wireless systems is housed inside the vocoder.

Vocoders sample transmissions of human speech, packetize the samples, and transmit the digitized voice from the handset to the base station over the assigned carrier frequency. Packetize means assembling information into digital packets. A packet is a unit of data that is routed between an origin and a destination on the Internet or any other network. Distant-end vocoders at the base station decode the pulses and route the calls over the backhaul network to the MSC.

In GSM systems, a full-rate vocoder allows for eight users (conversations) over a 250 KHz channel (carrier), where each channel occupies 25 KHz. Therefore, each GSM channel transports eight calls simultaneously.

TDMA/IS-136 systems operate in a similar manner. A full-rate vocoder allows for three users (channels) per carrier.

CDMA systems are very unique in terms of allocating calls per carrier, and this applies as well to vocoder operations.

Half-rate vocoders can sample human speech at half the rate of standard, full-rate vocoders. With half-rate vocoders, there is a 2:1 ratio in increased utilization of available spectrum compared to full-rate vocoders. Half-rate vocoders sample human speech with four bits per sample versus eight with full-rate vocoders, thus theoretically doubling system capacity. This assumes, of course, that half-rate vocoders are used system-wide. In GSM systems, a half-rate vocoder would therefore allow for 16 conversations (users) per GSM carrier. In TDMA systems, a half-rate vocoder would allow for six conversations (users) per carrier. Many wireless carriers today are using half-rate vocoders in their wireless networks.

Key: Half-rate vocoders may produce higher BERs, but modern half-rate vocoder quality has reached great heights since the late 1990s, and many carriers are using these vocoders to obtain better use of spectrum. When half-rate vocoders are used, they are required in both the BTS radios as well as the cell phone (handset), too.

6.3   Time Division Multiple Access (TDMA) Technology

6.3.1.   History

TDMA is the generic name for an air interface technology that is actually used by several digital-wireless radio technologies, such as IS-136 and GSM. TDMA was adopted by the TIA in 1992, and it was first used commercially in 1993. It has been deployed in both 850 MHz and 1,900 MHz systems.

TDMA was the first version of digital-wireless technology deployed. IS-54 was the earliest standard to use TDMA technology, defining the migration path from analog to digital-radio systems. The IS-54 standard, also known as the D-AMPS (digital AMPS) standard, referred to TDMA and digital radio in very generic terms and contained few references to digital-radio feature sets. It focused mainly on the migration from AMPS analog systems to digital-wireless systems.

When TDMA/IS-54 was first deployed as a digital-wireless standard, it had to coexist in an environment where AMPS was still the predominant technology. There was a major, persistent problem with TDMA interfering with analog radio channels, or *analog interferers*. Combined with the poor quality of the earliest vocoders that were offered at the time, this made for a high level of customer dissatisfaction as calls were dropped or broken up easily by static and choppy speech. However, major improvements in vocoders since the late 1990s allowed for vast improvements in dual-mode TDMA/AMPS transmissions. The overall quality of today's vocoders—regardless of the technology in use—is much, much improved.

The updated standard for TDMA systems, in particular IS-136, contains information on full-featured TDMA digital systems with references to feature sets such as caller ID and SMS, which provides text messaging. IS-136 supports three voice calls per carrier.

6.3.2   TDMA Operations and Deployment

TDMA systems use the concept of *time division multiplexing*, which enables multiple users to share the same radio channel. TDMA systems assign both different frequencies and different time slots to each conversation on a wircless system. In TDMA IS-136 systems, spectrum is divided into 30 KHz increments. Separate channels exist for uplink and downlink transmissions.

Each 30 KHz channel is divided into six timeslots, where two timeslots are used on the uplink channel and two timeslots are used on the downlink channel. This means that TDMA (IS-136) is a frequency division duplexed (FDD) system, because different frequencies are used for uplink and downlink transmissions. So, three subscribers can use one channel simultaneously.

The only major wireless carrier who was deploying TDMA/IS-136 on a widespread basis was AT&T Wireless, which is now a part of Cingular Wireless. AT&T Wireless had a long history with TDMA technologies, due to their purchase of McCaw Cellular Communications in the early 1990s. McCaw's network had extensive deployments of TDMA technology (both IS-54 and IS-136). So for economic reasons, AT&T chose to stay with IS-136-based TDMA technology. It was a decision they would ultimately regret, due to the technical limitations of IS-136 technology compared to other technologies such as CDMA and GSM.

TDMA as a technology is being supplanted today by CDMA and GSM because these other digital-wireless technologies offer many benefits over TDMA, as follows:

- CDMA and GSM are more cost-effective technologies because more capacity can be gleaned from an equivalent amount of radio spectrum, compared to TDMA.

- All other major wireless carriers other than AT&T are using either GSM technology or CDMA technology. Surely, the remainder of the industry cannot be wrong about their technical direction.

- Many articles have been written over the years pitting GSM against CDMA from a technical "religious" perspective. None of these articles even make mention of or include TDMA/IS-136. The reason is because it has many technical limitations, which stem mainly from its position as an upshot of the first digital-wireless technology developed.

6.4 Code Division Multiple Access (CDMA) Technology

6.4.1 Spread-Spectrum Technology

Digital spread-spectrum (DSS) technology has its roots back in World War II. Spread-spectrum technology can be implemented in several different ways, two of which are *frequency hopping* and *direct sequencing*. Frequency hopping was the earliest version of spread spectrum, and this technique was coinvented by actress Hedy Lamarr during World War II to help direct American torpedoes to their targets and prevent them from getting jammed and sent off course. Lamarr and her musical arranger invented the frequency-hopping concept. Incidentally, Lamarr did not use her stage name on the patent, and it never made her any money. World War II electronics were primitive. Lamarr's arrangement used a mechanical switching system, similar to a player piano roll, to shift control frequencies faster than the enemy could keep up. The concept was applied by engineers in 1957 and became the basic tool for secure military communications. DSS radios were installed on the ships sent to blockade Cuba in 1962, about three years after the patent expired.

In 1985, the FCC allocated three frequency bands for the radio transmission technique known as spread-spectrum communications, which has much greater immunity to interference and noise compared to conventional radio transmission techniques. The only requirement for business entities that use spread-spectrum technology is that the manufacturers of spread-spectrum products must meet FCC spread-spectrum regulations. The FCC rule alterations in 1985, combined with the continuing evolution of digital technology, catalyzed the development of spread-spectrum communication radios. These radios offer significant performance and operational benefits to end users and are used not only for cellular communications, but also many unlicensed microwave systems (see Chapter 15).

6.4.1.1 Spread-Spectrum Radios Versus Traditional Radios

Conventional radio signals are referred to as *narrowband*, which means that they contain all of their power in a very narrow, specific portion of RF bandwidth. Due to the relatively small portion of the radio band that an individual radio transmission occupies, the FCC has traditionally favored conventional narrowband radio. But, as a result of the very narrow frequency, these radios are prone to interference. A single interfering signal operating at or near its frequency can easily render a radio inoperable.

There are at least two problems with conventional wireless communications that can occur under certain circumstances. First, a narrowband signal whose frequency is constant is subject to major interference. This occurs when another signal is transmitted on, or very near, the frequency of the desired signal, resulting in cochannel interference. Major interference can be accidental as in amateur-radio communications, or it can be deliberate as in wartime. Second, a constant-frequency signal is easy to intercept and is therefore not well suited to applications in which information must be kept confidential between the source and destination.

To minimize problems that can arise from the vulnerabilities of conventional communications circuits, the frequency of the transmitted signal in a spread-spectrum system can be deliberately varied over a comparatively large segment of the electromagnetic spectrum. This variation is made according to a specific but complicated mathematical function. In order to intercept the signal, a receiver must be tuned to frequencies that vary precisely according to this function. The receiver must recognize the frequency-versus-time function employed by the transmitter and must also know the starting-time point at which that function begins. This is the basis of today's modern spread-spectrum systems.

Spread-spectrum signals, which are distributed over a wide range of frequencies and then collected onto their original frequency at the receiver, are so inconspicuous they are transparent. To make them more noise-like, they are intentionally made to be much wider than the information they are carrying. Spread-spectrum transmitters use transmit power levels that are similar to narrowband transmitters. But because spread-spectrum signals are so wide, they transmit at a much lower power level than traditional narrowband transmitters. This characteristic gives spread-spectrum signals a big advantage because spread and narrowband signals can then occupy the same frequency band, with little or no interference with each other. Just as they are unlikely to be intercepted by a military opponent, they are unlikely to interfere with other signals intended for business and consumer users—even ones transmitted on the same frequencies. Such an advantage opens and expands crowded frequency spectra. Besides being hard to intercept and jam, spread-spectrum signals are also hard to exploit or *spoof*. *Signal exploitation* is the ability of an enemy (or a nonnetwork member) to listen to a network and use information from the network without being a valid customer. *Spoofing* is the act of falsely or maliciously introducing misleading or false traffic or messages into a network. Spread-spectrum signals are naturally more secure than narrowband radio communications.

The spread of energy over a wide band, or lower spectral power density, makes spread-spectrum signals less likely to interfere with narrowband communications. All spread-spectrum systems have a threshold or tolerance level of interference beyond which useful communication ceases. This tolerance or threshold is related to the spread-spectrum processing gain. *Processing gain* is essentially the ratio of the RF bandwidth to the information bandwidth.

Key: In a spread-spectrum system, a radio transmitter takes the input data and spreads it in a predefined method over a carrier signal. The input data in the wireless world can be voice, data, image, video, or a combination of one or more of these media types. Each receiver must understand this predefined method and de-spread the signal before the data can be interpreted.

6.4.1.2 Two Types of Spread-Spectrum Technology There are two basic methods to perform the signal spreading: frequency hopping and direct sequencing. Direct sequencing spreads its signal by expanding the signal over a broad portion of a radio band. This is how CDMA wireless technology works—by using direct-sequence spread-spectrum.

Frequency hopping is the easiest spread-spectrum modulation to use. Any radio with a digitally controlled frequency synthesizer can, theoretically, be converted to a frequency hopping radio. This conversion requires the addition of a *pseudo-noise* (PN) code generator to select the frequencies for transmission or reception. Most hopping systems use uniform frequency hopping over a band of frequencies. This is not absolutely necessary, if both the transmitter and receiver of the system know in advance what frequencies are to be skipped. A frequency-hopped system can use analog or digital carrier modulation and can be designed using conventional narrow-band radio techniques. Dehopping in the receiver is done by a synchronized PN code generator that drives the receiver's local frequency synthesizer.

The basic idea behind frequency hopping is simple: Instead of transmitting a signal on one frequency, a spread-spectrum system switches rapidly from one frequency to another. The choice of the next frequency is random, so it is nearly impossible for someone to eavesdrop or jam the signal. The challenge is to keep both the transmitter and receiver synchronized. Using an accurate clocking system and pseudo-random number generators, it is relatively easy to keep the transmitter and receiver synchronized. Today, spread-spectrum communication has been combined with digital technology for spy-proof and noise-resistant battlefield communications. In the nonmilitary world, it is most often used in cordless phones, wireless large area networks (LANs), and wireless networks that employ CDMA technology.

Key: Other devices do not see a phone transmitting over spread-spectrum systems because the spectrum is transmitting for only a fraction of a second on any part of its allocated frequency. Therefore, the average power on any given channel is extremely low, and other devices using that channel do not even notice its transmissions. The phone creates the equivalent of a low-power *noise pattern* across all of the frequency bandwidth it uses. This is why spread-spectrum technology is known as a *noise-based system*.

The other, more practical, all-digital version of spread spectrum is direct-sequence spread spectrum. A direct-sequence system uses a locally generated PN code to encode digital data to be transmitted. The local code runs at a much higher rate than the data rate. Data for transmission is simply logically added with the faster PN code. Carrier modulation other than *binary phase shift keying* (BPSK) is possible with direct sequence. However, BPSK is the simplest and most often used spread-spectrum modulation technique.

A spread-spectrum receiver uses a locally generated replica PN code and a receiver correlator, to separate only the desired coded information from all other possible signals. The desired coded information is either voice, data, video, and so forth. A spread-spectrum correlator can be thought of as a very special matched filter; it responds only to signals that are encoded with a PN code that matches its own code. So, a spread-spectrum correlator can be tuned to different codes simply by changing its local code. This correlator does not respond to man-made, natural, or artificial noise or interference. It responds only to spread-spectrum signals with identical, matched-signal characteristics and encoded with an identical PN code.

6.4.1.3 Spread-Spectrum Operations Spread-spectrum signals use fast codes that run at a rate many times the information bandwidth or data rate. These special spreading codes are called Pseudo Random Noise (Walsh) codes, called PN codes for short. They are called pseudo because they are not real gaussian (white) noise. The use of these special PN codes in spread-spectrum communications makes signals appear wideband and noise-like. It is this very characteristic that makes spread-spectrum signals possess the quality of having a low probability of interception. Spread-spectrum signals are hard to detect on narrowband equipment because the signal's energy is spread over a carrier bandwidth that is 10 to 100 times the information bandwidth itself. The information bandwidth is what is actually being transmitted, such as voice, data, video, or image.

To qualify as a spread-spectrum signal, two criteria need to be met:

- The signal bandwidth must be much wider than the information bandwidth.
- Some code or pattern, other than the data to be transmitted, determines the actual on-air transmit bandwidth.

See Figure 6-2 for an illustration of a real spread-spectrum signal. This is how spread-spectrum carriers are actually drawn in educational settings to depict the nature of the technology itself. This is also reflected by the nature of the CDMA illustration in Figure 6-1.

In summary, the advantages of spread-spectrum radio technology are

- Interference immunity: Spread-spectrum radios are inherently more noise immune than conventional radios. They operate with higher efficiency than conventional radio technology.
- Multichannel capability: Conventional radios operate on a specific frequency controlled by a matched crystal oscillator. The specific frequency is allocated as a part of the FCC site license, and the equipment must remain on that frequency (except for very low-power devices such as cordless phones). This is not a requirement with spread-spectrum systems, which gives them additional flexibility and effectiveness.
- Spread-spectrum data radios offer the ability to have multiple channels, which can be dynamically changed through software. This

Figure 6-2
A spectrum analyzer photo of a direct sequence (DS) spread-spectrum signal

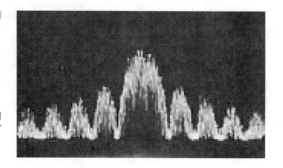

allows for the use of many applications such as repeaters, redundant base stations, and overlapping cells.

6.4.2 CDMA Basics

CDMA is an American digital standard that was developed for commercial use by a company named Qualcomm, based in San Diego, California. CDMA technology has since been licensed by Qualcomm to Ericsson, and possibly others. CDMA was originally deployed as a battlefield-communications system because it is very hard, if not impossible, to intercept CDMA transmissions due to their spread-spectrum nature. CDMA has experienced many advancements since its introduction into the commercial wireless world. It is now defined by multiple standards: IS-95, J-STD-008 (PCS band), and CDMA 2000. CDMA has mainly been deployed in the United States and in several Asian countries, including Hong Kong.

CDMA is a wideband, spread-spectrum technology. A unique code is assigned to all speech bits (conversations). Signals for all calls are spread across a broad frequency spectrum, hence the term "spread spectrum." The dispersed signals are pulled out of what appears as background noise by a receiver that knows the code for the call it must handle. This technique allows numerous phone calls to be simultaneously transmitted over one RF carrier or channel.

Key: In *narrowband* CDMA systems, the standard is to use 1.25 MHz blocks of radio spectrum to carry many conversations, using pseudorandom noise codes known as *Walsh codes*. Because of the nature of spread-spectrum technology, CDMA systems employ an N=1 frequency-reuse format. Wideband CDMA will be reviewed in Chapter 7.

Each CDMA base station can use the same 1.25 MHz carrier at the same time. The only change between each block of 1.25 MHz spectrum at each base station is the pseudorandom Walsh noise codes that are in use. There are a maximum of 64 allowable pseudorandom Walsh noise codes per 1.25 MHz carrier in a CDMAOne system (IS-95A/B).

When a mobile phone call is made using CDMA technology, the sound of the user's voice is converted into a digital code. This digital signal is first correlated with a noise-like code known as a pseudorandom noise (PN) code. As discussed in section 5.4.1, the correlator yields an encrypted digital representation of the original signal. This encrypted signal is then spread over a very wide frequency spectrum (1.25 MHz). At the receiving terminal, the signal is demodulated back to a narrow bandwidth and then fed into a decorrelator. This decorrelator uses its unique PN code to extract only the information intended for it. A signal correlated with a given PN code and decorrelated with the same PN code returns the original signal. Decorrelating the signal with the wrong PN code would result in pure noise, containing no discernible information or sound.

There is an analogy that is frequently used to explain how CDMA works. Imagine you are at a reception being held in a room at the United Nations. There are four people around you, each speaking a different native language: Spanish, Korean, Chinese, and English. Your native language is English. You understand only the words of the English speaker and tune out the Spanish, Korean, and Chinese speakers. You hear only what you know and recognize. The same is true for CDMA. Multiple users share a frequency band at the same time, yet users only hear their own conversations because each is assigned a unique Walsh noise code.

CDMA networks have *pilot channels*, which carry no data but are used by the subscriber unit (mobile phone) to acquire the system and assist in the process of soft handoffs (see Section 6.4.4.3) and synchronization. A separate pilot channel is transmitted for each sector of a cell site and is uniquely identified by its own PN code (like other users).

6.4.3 CDMAOne and CDMA 2000

CDMAOne refers collectively to three CDMA standards, which were the first formal specifications assigned to this technology: IS-95A, IS-95B, and J-STD-008. The J-STD-008 specification is compatible with both IS-95A and IS-95B, with the exception of the frequency band of operation. The primary difference between IS-95A and IS-95B is that IS-95B enables ISDN-like (or Integrated Services Digital Network) data rates to exist.

When deploying IS-95 in either the 850 MHz or 1,900 MHz bands (legacy cellular and PCS, respectively), the same functionalities apply, with the exception of frequency band issues such as increased free-space loss in the PCS spectrum at 1,900 MHz. The primary difference between these two bands is that CDMA has some idiosyncrasies that are directly applicable to the channel assignments in an existing 850 MHz cellular band.

CDMA2000 falls under the specification known as IS-2000, which is backward compatible with IS-95A/B as well as J-STD-008. CDMA2000 is a 3G specification (the North American version of wideband CDMA), and since it is backward compatible with IS-95 systems, wireless carriers can make strategic deployment decisions in a graceful manner. In other words, if they choose to upgrade to a 3G CDMA platform (i.e., CDMA2000), they can do so in a phased manner. CDMA20000 also enables the introduction of high-speed data for mobility, where *high speed* is defined by the IMT-2000 specification. CDMA 2000 is covered extensively in Chapter 7.

6.4.4   CDMA Architecture

Prior to the widespread deployment of CDMA systems, mainly spurred by the broadband PCS carriers, there were concerns that CDMA systems could not handle heavy traffic loads. Therefore, there were concerns in the industry that CDMA systems could not handle a huge acquisition of customers in a short period of time. These concerns were laid to rest as PrimeCo PCS (now defunct), Verizon Wireless, and Sprint PCS proved that CDMA could operate with many thousands (and millions) of subscribers per market.

 Key: Theoretically, there can be nine 1.25 MHz CDMA carriers assigned per BTS. Theoretically, each of these 1.25 MHz carriers (channels) can handle 22 to 40 voice calls.

Today's reality is as follows: Wireless carriers are deploying three to four 1.25 MHz CDMA carriers per base station. This capacity is allocated amongst all the sectors of the base station. Today, each CDMA carrier can support around 22 voice calls, as noted previously. 3G CDMA2000 systems may deploy more carriers per base station, possibly six to eight carriers to accommodate the additional bandwidth requirements.

6.4.4.1 Power Control CDMA base stations control the power of all mobiles within their coverage area for interference reduction purposes. All mobile signals must arrive at the base station at exactly the same power level so that the signals can be properly coded. Power control is a required operational parameter of CDMA digital systems. For example, if a mobile station that is right next to the base station is transmitting at very high power, and a mobile station 10 miles away from the base station is transmitting at very low power, the power of the mobile next to the base station is throttled down to a given level while the power of the mobile 10 miles away from the base station is raised to a given level. This strategy maintains equivalent power levels for all mobiles within a BTS coverage area.

Power control is an absolute necessity to maintain system capacity in CDMA networks. A beneficial by-product of power control is reduced power costs at the base station as well as increased battery life in the mobile phone. Power control as described here does exist in AMPS, TDMA, and GSM systems, but it is simply a benefit that can be utilized to make the systems perform better and avoid possible damage to BTS equipment.

6.4.4.2 Rake Receivers Multipath fading can actually be a benefit in CDMA systems. Another unique attribute of CDMA systems is the existence of *rake receivers* in the mobile phones and the BTS transceivers. Rake receivers are radio receivers that have multiple receiver elements. The name originates from the fact that rake receivers resemble a lawn rake. The signals from each receiving element are combined together to form a composite signal. This is the opposite of space diversity, where the strongest signal is selected for signal processing.

There are four rake receivers within each base station transceiver and three rake receivers within each mobile phone. The function of the rake receiver within both the mobile phone and base station transceivers is to aggregate the multiple received signals within each receiving element in the rake receiver. The strongest signal is combined with the multipath, weaker, reflected signals from the other two or three rake receivers to form the composite signal that is used to process the mobile call. The multipath signals are additive to the direct signal to obtain the cleanest, strongest signal possible. This is known as *diversity combining*, as opposed to *switched diversity* (space diversity), which is provided in non-CDMA systems.

6.4.4.3 Soft Handoff *Hard handoffs* are defined as call handoffs that occur where there is a split second of time (likely milliseconds) where a mobile call technically is not being handled, or managed, by *any* BTS. There is a moment, usually not discernable, where the call is literally in the air as it makes the transition from one base station to another base station during the call-handoff process. Analog AMPS systems and IS-136 TDMA systems used hard handoffs, which are also known as a *break-before-make* switching function in relation to call-handoff. In this context, switching simply refers to the call-handoff process itself.

Key: Soft call-handoffs are different from hard call-hand-offs in that a soft handoff allows both the original cell and *one to two new cells* to service a call during the handoff transition. The handoff transition is from the original cell carrying the call to one or more new cells and then to the final new cell. With soft handoff, the wireless call is actually carried by two or more cells simultaneously. *This is achieved because all of the cell sites are transmitting the same frequency, unlike an analog system where each cell site has a unique set of frequencies.* In contrast to AMPS and IS-136 TDMA systems, CDMA-based soft handoffs provide a *make-before-break* switching function with relation to call-handoff. In CDMA wireless systems, 35 to 45 percent of the capacity of each base station can be reserved for soft hand-off processing.

CDMA systems require a GPS receiver and antenna at every cell base station. The GPS antennas synchronize all the cell sites to one timing source: the GPS. This is an absolute necessity for soft handoffs because timing is critical among the multiple sites that may simultaneously handle a call during the soft handoff process. See Figure 6-3, which shows a photo of a GPS receiver.

The soft-handoff process is analogous to a trapeze artist flying in the air between two trapeze platforms. In the analog AMPS world, the artist in the middle is transferred from the trapeze at one end to the trapeze at the other end by being flung through the air. This equates to a break-before-make switching function. Because of an RF anomaly or an error in signaling, calls could get dropped during the handoff process in analog

Figure 6-3
GPS antenna

AMPS and TDMA systems. In a CDMA world, instead of flinging the middle trapeze artist through the air, the first trapeze artist does not let go of the middle artist until he is sure that the opposite-end trapeze artist has a firm grip on the middle artist.

Not only does soft handoff greatly minimize the probability of a dropped call, but it also makes the handoff virtually undetectable to the user. Soft handoffs are directed by the mobile telephone. As such, soft handoff is also known as *mobile-directed handoff.*

The sequence of events in a soft handoff is as follows:

1. After a mobile call is initiated, the mobile station continues to scan the neighboring cells to determine if the signal from another cell becomes stronger than that of the original cell.

2. When this happens, the mobile station knows that the call has entered a new cell's coverage area and that a handoff can be initiated.

3. The mobile station transmits a control message to the MSC, which states that the mobile is receiving a stronger signal from the new cell site, and the mobile identifies that new cell site.

4. The MSC initiates the handoff by establishing a link to the mobile station through the new cell while maintaining the link to the old cell that was managing the call.

5. Although the mobile station is located in the transition region between the two cell sites, the call is supported by communication through both cells. This eliminates the ping-pong effect of repeated requests to hand the call back and forth between two cell sites.

6. The original cell site will discontinue handling the call only when the mobile station is firmly established in the new cell.

In CDMA systems that coexist with analog/AMPS technologies (i.e., 850 MHz cellular carriers), hard handoffs do exist. A hard handoff in a CDMA system describes a call handoff from a CDMA carrier to an analog/AMPS RF channel. It should be noted that in this scenario, a CDMA wireless call can step down from a CDMA carrier to an analog channel. Yet the reverse can never happen: A wireless call can never be handed off from an analog RF channel to a CDMA-based channel. This situation rarely, if ever, occurs today as the vast majority—or all—of analog infrastructure is now removed from all wireless networks. The only possible exception might be certain rural areas.

6.4.4.4 CDMA Cell Breathing   Because CDMA is a noise-generated technology, it is important to monitor the noise floor, or ambient noise, in the coverage area of a cell site. Too much intrusive noise from a source other than the intended mobile or base station will cause interference and lead to a degraded or dropped call. With that in mind, CDMA base stations are designed to constantly adjust their output levels and the output levels of the mobile phones operating within that cell. When the noise floor level gets too high, the base station will begin to throttle down its power in an attempt to eliminate the interference. The coverage area of the site will then actually shrink, causing mobile phones on the perimeter of the base station coverage area to be forced to another site or to get dropped in order to prevent more interference from intruding on the coverage area of the site. When the noise floor begins to drop, the site will allow itself and mobiles on it to transmit at higher levels. Thus, the net effect of this phenomenon is known as *CDMA cell breathing:* Based on interference levels and the overall number of mobile users in any given CDMA cell site's coverage area, the actual coverage area itself can expand or contract based on system commands.

6.4.5   The Benefits of CDMA Technology

CDMA radio technology offers the following benefits to wireless carriers that implement this technology:

- Increased capacity over other technologies (i.e., FDMA, TDMA) by allowing for reuse of the same carrier frequency in all sectors and cells. Studies show that CDMA technology offers an estimated increase of about 6 to 18 times the capacity of legacy analog AMPS systems. This number may rise significantly over time with the honing of current technology and the development of newer technologies such as CDMA 2000.

- Simplified RF engineering, due to the N=1 reuse pattern in CDMA systems. This reduces the time and effort required to expand or modify CDMA systems.

- Increased performance over the weakest link in the wireless system, the air interface. This is mainly due to the use of rake receivers to resolve multipath fading (Rayleigh fading).

- Lower-transmitted power levels. This equates to lower power bills at the base station level and longer battery life for CDMA handsets. Power adjustments are constantly being made in the handset to reduce the amount of interference introduced to other conversations. Because CDMA systems are noise limited, the less interference introduced by one conversation, the greater the system capacity that remains.

- Greater security due to the encoding of CDMA signals.

- Enhanced performance and voice quality due to soft-handoff capability.

6.5 Global System for Mobile Communication (GSM) Technology

During the early 1980s, the Conference of European Posts and Telecommunications (CEPT) formed a group entitled Groupe Special Mobile (GSM) to develop a standard for the European mobile wireless market. The goals of the GSM were to develop one consistent wireless network throughout Europe to provide roaming capabilities across the continent, security against wireless fraud, good speech quality, and ISDN compatibility. At the time GSM was developed, there were six incompatible cellular systems in operation throughout Europe. A mobile phone designed for one system could not be used with another system. This situation served as the catalyst for the development of an all-European system.

Phase one of the GSM specifications were completed in 1989 and the first systems were activated in 1991. Commercial service began in 1992 and, by 1996, there were over 35 million GSM customers being served by over 200 GSM networks.

GSM is a digital-wireless standard in its own right, though it uses TDMA-like technology as its air interface foundation. This is because GSM has many special features and system attributes that make it a distinct radio standard unto itself.

GSM has been deployed in the 900 MHz, 1,800 MHz and 1,900 MHz bands. GSM's original design specified operation at 900 MHz (in Europe). Yet the United Kingdom licensed the 1,850 MHz frequencies for second-generation (2G) cellular. Instead of developing their own standards, they developed *DCS 1800* based on the GSM standard. DCS 1800 is typically used in the United Kingdom, Russia, and Germany. When the FCC issued the 1,900 MHz frequency range for PCS in the United States, it was again agreed that this frequency would be based on GSM and deployment targeted at the United States and Canada. DCS 1900 was then considered the GSM standard for North America. Today, this standard is now called *North American GSM*.

Along with becoming the standard pan-European digital-wireless system, over time GSM has become a worldwide digital-wireless standard. This is evidenced by the fact that many American PCS carriers chose GSM as their digital-radio technology standard in late 1995. Since then, many countries the world over have embraced the GSM standard.

Because the performance of wireless radio systems is restricted primarily by cochannel interference, a digital standard was sought to obtain improvements in spectral efficiency and increase capacity. GSM systems use TDMA technology as their air interface standard. However, the standardized approach, distinctive features (e.g., *subscriber identity module* [SIM] cards), and subsystem architecture of GSM make it a completely distinct digital-wireless technology. GSM operates under a strictly controlled series of standards known as the *Memorandum of Understanding* (MoU). These standards dictate how the system should be designed and operated. By developing the system in accordance with these standards, the original developers of the GSM standard could be assured that GSM systems around the world would be completely compatible.

6.5.1   GSM Architecture and Subsystems

The GSM standard offers an open architecture according to the *Open Systems Interconnect* (OSI) model, for layers 1, 2, and 3. This approach represents a significant departure from legacy-analog cellular systems. Because the architecture is a fully open system, a key benefit is afforded to GSM carriers as they can go to any supplier of GSM equipment to build out their systems; they are not beholden to any one equipment supplier due to proprietary schemes.

Key: GSM systems where full-rate vocoders are used employ a 200 KHz channel where each channel is subdivided into eight time slots, or subchannels. Therefore, it is said that GSM systems have a 25 KHz channel spacing and offer eight channels per radio (transceiver). Multiple 200 KHz channels will be assigned to each base station. One timeslot must be allocated for control channel purposes; therefore, up to seven subscribers can use a channel simultaneously.

According the GSM standard, GSM networks are divided into four main subsystems:

1. The base station subsystem
2. The network subsystem
3. The operations and support subsystem
4. The mobile station subsystem (the mobile phone)

The *base station subsystem* is comprised of the base station controller (BSC), base transceiver station (BTS), and the air interface. The BTS consists of the antenna and the radio transceiver at the base station. The radio transceiver defines a cell coverage area and controls the radio link protocol(s) with the mobile station. The BSC is the control computer that manages many BTSs and is a required component of GSM systems. The controller is usually housed at the MSC location and manages which radio channels are being used by which BTSs; it also manages the call-handoff process between BTSs. The BSC also regulates the transmit power levels of the towers and handsets. As a handset gets closer to the tower, the BSC lowers the transmitter power levels. If a handset travels away from the tower, the BSC raises the transmitter power levels. In summary, the BSC serves as a type of front-end processor for the MSC. It handles much of the overhead and administrative burdens associated with frequency management, call setup, and call handoff. This allows the MSC to concentrate its processing power for what it does best: processing and switching mobile calls.

In the *GSM network subsystem*, the switch is the central component of the network. The MSC provides connection to the PSTN or the ISDN.

Today, this is really SS7-based interconnection. The MSC also provides subscriber management functions such as mobile registration, location updating, authentication, and call routing to a roaming subscriber. The MSC in a GSM network also houses the *home location register* (HLR). There is only one HLR in GSM networks, unless the HLR reaches capacity and a second HLR is installed. For every geographical area controlled by an MSC, there is a corresponding *visitor location register* (VLR). The VLR is cleared every night and starts each day with an empty database. The security infrastructure of GSM systems is also part of the network subsystem infrastructure. This includes the authentication center, which provides the parameters needed for authentication and encryption functions. The *equipment identity register* (EIR) is a database used for security as well. Every GSM handset manufactured has a unique identification number known as the *international mobile equipment identity* (IMEI). The IMEI number of each handset is stored in the EIR. If the handset has been lost or stolen, the IMEI is placed on the blacklist of the EIR and will not work on the network. The EIR is one more way to prevent fraud on the GSM network.

The *operations and support subsystem* (OSS) is the command center that is used to monitor and control the GSM network. If an emergency occurs at a base station, the OSS can determine where that BTS is located, what type of failure occurred, and what equipment the site engineer will need to repair the failure.

The *mobile station subsystem* consists of two main components: the handset and the SIM. The handset in a GSM system is different from analog phones in that the identification information of the subscriber is programmed into an SIM module and not the handset itself. The handset's main functions are receive/transmit and encoding and decoding the voice transmission according to the GSM standard. This means that the chipset in the GSM handset reflects the GSM standard itself. The SIM is a microcontroller embedded into a small piece of plastic, which holds the GSM operating program and customer and carrier-specific data storage. The SIM is a tiny piece of plastic with a small memory chip embedded into it; it is barely 1 square inch. When subscribers purchase service from a GSM carrier, the sales office will program the SIM module with the user's identification information and plug the SIM module into the phone. The SIM card provides authentication, information storage, subscriber account information, and data encryption. SIM chips and handsets are swappable in the sense that any SIM card works with any GSM handset. When the customer uses their GSM phone, the SIM is activated and will recall everything the subscriber has stored: speed dial numbers, messages that were received via short message service, handset preferences, custom downloaded ring tones, and so forth.

Key: The key concept of the smart card is to associate a wireless subscription with a memory module, versus a mobile phone. Figure 6-4 shows a picture of a modern SIM.

Figure 6-5 shows a SIM alongside a GSM cell phone for a perspective on the size of today's SIMs.

In Europe, there are known instances where subscribers have had their health histories embedded into their SIM modules. When emergencies occurred, medical authorities in emergency centers simply plugged the SIM card into a GSM reader to obtain critical medical information on the person, such as whether they were diabetic or had heart disease, asthma, allergies, or high blood pressure. SIM cards may even be used as charge cards in the near future.

From its inception, the GSM system used authentication technology to thwart wireless fraud. Authentication was included into the development of the GSM standard, and subsequently into all GSM systems that have been implemented since 1992, all around the world. North Ameri-

Figure 6-4
A SIM within a GSM phone (Photo courtesy Howstuffworks .com)

Figure 6-5
SIM module within GSM phone (Photo courtesy Howstuffworks .com)

can GSM security systems use a complex series of algorithms, secret keys, and random numbers. To date, there has been no cloning fraud on GSM systems, mainly due to the fact that authentication technology was deployed from the start.

GSM also uses a distinctly digital approach to call-handoff, which is known as *call-handover* in Europe. GSM employs what is known in the industry as mobile-assisted handoff. In GSM systems, the mobile phone plays an active part in the handoff process: The mobile phone, not the MSC, continuously monitors other base stations in its vicinity, measuring signal strength through the MSC. The identities of the six base stations that are the best prospects for a clean call handoff are then transmitted back to the MSC and the network then decides when to initiate handover.

GSM is described as the first true instance of a wireless intelligent network (IN), as it was designed with close reference to the IN model. This is because it exhibits the following network traits:

- An open, distributed architecture
- Separation of switching and service control functions
- Full use of SS7 as the signaling infrastructure
- Clearly defined and specified interfaces
- IN structure

GPRS and EDGE are both GSM-based wireless data standards, which will be discussed in Chapter 18.

6.5.2   GSM Adjunct Systems

The GSM standard defines the use of multiple ancillary systems in conjunction with the use of GSM technology. These functions are listed here:

- The *gateway MSC (GMSC)*, whose purpose is to query the HLR to determine the location of subscribers. Calls from another network (i.e., the PSTN) will first terminate into the GMSC.
- *SMSC* is the short message service center. Technically, the proper term is SMS-SC. This is the node that stores and forwards short messages to and from mobile stations. These are typically messages up to 160 characters in length (maximum), per the SMS standards.
- The *equipment identity register (EIR)* identifies what equipment (handsets) are acceptable in a GSM network.

■ The *interworking function (IWF)* is used for circuit-switched data and fax services and is basically a modem bank.

6.6   GSM Versus CDMA

For many years, because it was inevitable that IS-136 was dying a slow death, there has been a steady debate about the merits of GSM technology versus CDMA technology. Each technology has its own benefits and disadvantages. The debates center around one being better than the other. The truth of the matter is that 3G-based technologies are actually making use of the best aspects of both technologies. The objective of the EDGE wireless data standard is to serve as a bridge between legacy GSM systems and wideband CDMA systems.

Carriers using both technologies have already begun launching interim versions of 3G. CDMA carriers are currently launching networks utilizing EV-DO (or evolution of voice, data only) services. This allows the carrier to utilize certain channels and bandwidth for data only, increasing data throughput speeds to 256 KHz and better. Eventually, these carriers will move to EV-DO, allowing both voice and high speed data to be run on the same channel.

GSM carriers have countered by launching GPRS networks allowing high-speed data of up to 256 KHz to broadcast over GSM spectrum using data-only channels.

In the end, both of these excellent wireless technologies will likely coexist to a large degree. They both have many positive things to offer.

Test Questions

True or False?

1. ____ The TDMA digital technology offers a 10:1 ratio in increased system capacity.

2. ____ Power control is not necessarily a key parameter of CDMA digital operations.

3. _____ Rake receivers choose the best signal from one of the three mobile phone or four base station receivers and use that signal to process a CDMA digital cellular call.

Multiple Choice

1. How many 1.25 MHz carriers are actually being used in a CDMA cell today by wireless carriers?
   a. 18
   b. 22
   c. 9
   d. 3 to 4
   e. 64
   f. None of the above

2. The key concept of the smart card in GSM systems is to associate:
   a. Multiple mobile phones for one MIN
   b. Multiple MINs for one mobile phone
   c. A cellular subscription with a card instead of a specific mobile phone
   d. A credit card for all mobile phone users
   e. None of the above

3. The CDMA soft-handoff capability reflects:
   a. A break-before-make switching function
   b. An ache-before-break switching function
   c. A make-before-break switching function
   d. A rake-receiving switching function

4. What digital wireless technology uses a pilot channel to access the system?
   a. IS-54
   b. IS-136
   c. CDMA
   d. GSM

3G: Third-Generation Wireless

Technically, the generations of wireless technology are defined as follows:

- 1G, or first-generation, networks were the first analog cellular systems, transmitting only voice, and were launched in the late 1970s and the early 1980s. These systems represented a huge leap in mobile communications technology, mainly in the areas of capacity and mobility. Semiconductor technology and microprocessors made smaller, lighter-weight mobile systems a practical reality for many more users. The most prominent 1G cellular networks were American AMPS (Advanced Mobile Phone System), Nordic NMT (Nordic Mobile Telephone), and European TACS (Total Access Communication System). 1G systems were differentiated from previous wireless services by their employment of the frequency reuse concept.

- 2G, or second-generation, networks were the first all-digital cellular systems and were launched in the early-to-mid 1990s, depending on the area of the world in question. The development of 2G cellular systems was driven by the need to improve transmission quality, system capacity, and coverage. Further advances in semiconductor technology brought digital transmission to mobile communications. 2G systems also ushered in the era of mobile-assisted handoff capabilities, where the mobile phones themselves took an active part in the call-handoff process. 2G networks included GSM, CDMAOne (IS-95A/B), and D-AMPS, which eventually became IS-136. GSM is the most successful 2G technology, serving roughly half of the world's mobile subscribers.

- 2.5G networks are enhanced versions of 2G networks, offering improved wireless data capabilities. Data rates up to about 144 Kbps are available in many systems, when the full potential of these networks is reached (but most carriers advertise average rates around 60 Kbps). Examples of 2.5G technology are systems such as GSM-based GPRS, and CDMA2000 1x.

- 3G, or third-generation, networks are the latest wireless networks that offer data rates of 384 Kbps and higher (depending on the technology in question). Examples of 3G network technologies include Universal Mobile Telecommunications Service (UMTS), Frequency Division Duplex (FDD), Time Division Duplex (TDD), CDMA2000 1x, EV-DO, CDMA2000 3x, TD-SCDMA, Arib WCDMA, Enhanced Datarate for Global Evolution (EDGE; for more on EDGE, see Chapter 18), and IMT-2000 DECT. FDD means that paired channels of different frequencies are used in the uplink and downlink, respectively. TDD means that one frequency is used in both the uplink and the downlink.

- 4G, or fourth-generation, is mainly a marketing buzzword at this time. Some basic 4G research is being done, but no frequencies have been allocated. The 4G networks could be ready for implementation around 2012. Some industry buzz is occurring around making orthogonal frequency division multiplexing (OFDM) technology the centerpiece of a 4G wireless technology. This topic will be further explored in Section 7.9.

7.1   3G Systems and IMT 2000

3G wireless systems are intended to provide a global mobility with a wide range of services including telephony, paging, messaging, Internet access, and broadband data transport. The International Telecommunications Union (ITU) began the process of defining the standard for 3G systems, referred to as International Mobile Telecommunications 2000 (IMT-2000). In 1998, the Third Generation Partnership Project (3GPP) was formed to continue the technical specification work. In Europe, the European Telecom Standards Institute (ETSI) was responsible for the UMTS standardization process. The UMTS standard is the centerpiece of the 3G evolution, using GSM standards for its network architecture and wideband CDMA (WCDMA) as its air interface technology.

The 3GPP has five main UMTS standardization areas:

- Radio-access network
- Core network
- Terminals
- Services and system aspects
- GERAN (or the GSM/EDGE radio access network)

The 3GPP Radio Access group is responsible for defining the following:

- Radio layer 1, 2, and 3 specifications
- UMTS Terrestrial Radio Access Network (UTRAN) operation and maintenance requirements
- BTS radio-performance specification

- Conformance test specification for testing radio aspects of base stations
- Specifications for radio performance from the system perspective

The 3GPP Core Network group is responsible for defining the following:

- Mobility management and call, connection-control signaling between end user equipment and the core network
- Core-network signaling between core-network nodes
- Definition of interworking functions between the core network and external networks (i.e., the Internet and/or the PSTN)
- Packet-related issues
- Core-network aspects and operation and maintenance requirements

The 3GPP Terminal group is responsible for defining the following:

- Service-capability protocols
- Messaging
- Services end-to-end interworking
- The framework for terminal interfaces and services execution
- Conformance test specifications of terminals, including radio aspects

The 3GPP Services and System Aspects group is responsible for the following:

- Definition of services and feature requirements
- Development of service capabilities and service architecture for cellular, fixed wireless, and cordless applications
- Billing and accounting
- Network management and security
- Definition, evolution, and maintenance of overall architecture

There are five types of 3G radio interfaces:

- IMT-2000 CDMA direct sequence spread spectrum, also known as UTRA FDD, which includes WCDMA in Japan.
- IMT-2000 CDMA multicarrier, also known as CDMA2000 3X being developed by 3GPP2 (IMT-2000 CDMA2000 includes CDMA 1X components, such as CDMA2000 1X EV-DO).
- IMT-2000 CDMA TDD, also known as UTRA TDD and TD-SCDMA. TD-SCDMA is being developed in China and supported by the TD-SCDMA Forum.

- IMT-2000 TDMA single carrier, also known as UWC-136 (EDGE).
- IMT-2000 DECT supported by DECT Forum.

Data rates for 3G services are as follows:

- 2.048 Mb/s for picocell (and microcell) applications
- 384 Kb/s for medium-size cells (micro and small macro cells)
- 144 Kb/s and 64 Kb/s for large cell applications (large macro cells)
- 14.4 Kb/s for continuous low-speed data applications in very large cells
- 12.2 Kb/s for speech (4.75 Kb/s–12.2 Kb/s)
- 9.6 Kb/s globally (satellite)

Key: In February 1992, the World Radio Conference allocated frequencies for UMTS use. The spectrum for UMTS has been identified as frequency bands 1885 to 2025 MHz for future IMT-2000 systems. Frequency bands 1980 to 2010 MHz and 2170 to 2200 MHz have been identified for the satellite portion of UMTS systems.

All 3G standards are still under constant development. Most European countries and some other countries around the world have already issued UMTS licenses either by *beauty contest* or auctions. Beauty contest means that the government asks all license applicants to develop a plan explaining how they would build a network and manage their future 3G business. A plan typically includes the following: How many new jobs are created, what kind of services will be available and when, how many domestic products are used, how will rural areas benefit from the service, and what kind of financial plan is in place to guarantee the success and avoid bankruptcy. The objective is to enable the government to decide what is best for the country, not who is willing to pay the most (in the United States and Australia, some of the highest-bidding carriers have gone bankrupt, and the government has had to pay unemployment insurance and other payments). By charging high fees for 3G licenses, governments impose an indirect tax that mobile users have to bear by paying more for making calls once the networks are built and turned up. By giving away spectrum, governments can create a good environment for technology start-up companies because the company has more money to invest and less up-front fees.

7.2 Universal Mobile Telecommunications System (UMTS)

UMTS is one of the 3G mobile systems being developed within the ITU's IMT-2000 framework. UMTS is a 3G broadband, packet-based standard that defines the transmission of text, digitized voice, video, and multimedia at data rates up to 2 Mbps. UMTS offers a consistent set of services to mobile computer and phone users, no matter where they are located in the world. UMTS is based on the GSM standard and is endorsed by major standards bodies and manufacturers. It is the planned standard for mobile users around the world and is, at present, still being rolled out worldwide. Once UMTS is fully available geographically, mobile computer and mobile phone users can have continuous access to the Internet as they travel. When roaming, they will have the same set of capabilities no matter where they travel. *Users will have access to UMTS systems through a combination of terrestrial wireless and satellite transmissions* when the system's design is fully realized. Until UMTS is fully implemented, users can have multimode devices that switch to the currently available technology (such as GSM 900 and 1800) where UMTS is not yet available. The coverage area of UMTS service provision is to be worldwide in the form of future land mobile telecommunications Services (FLMTS), now called IMT2000.

Today's cellular systems are mainly circuit-switched systems whose connections are always dependent on circuit availability. A packet-switched connection, using the Internet Protocol (IP), means that a virtual connection is always available to any end point in the network. This capability will also make it possible to provide new services, such as alternative billing methods (pay-per-bit, pay-per-session, flat rate, asymmetric bandwidth, and others). The higher bandwidth of UMTS also promises new services.

Key: The difference between CDMA2000 and UMTS is that CDMA2000 and UMTS were developed separately and are two separate ITU-approved 3G standards.

UMTS is not necessarily intended to be a replacement for 2G technologies (e.g., GSM, CDMA, DCS1800, etc.), which will continue to evolve to their full potential. Some 2.5G systems (GSM GPRS, IS-95B, and CDMA2000 1X) will be able to deliver 3G services, so it will be difficult for users to see the difference.

The coverage area for UMTS systems will be provided by a combination of cell sizes ranging from in-building picocells to global cells provided by satellite, giving service to remote regions of the world.

UMTS was developed mainly for countries with operational GSM networks because these countries have agreed to open up new frequency ranges for UMTS networks. Because it is a new technology using a new frequency band, a whole new radio access network must be built. The advantage is that a new frequency range offers plenty of new capacity for wireless carriers. The 3GPP is overseeing the development of the UMTS standard and has wisely kept the core network as close to a GSM core network as possible. UMTS phones are not meant to be backward compatible with GSM systems. However, hopefully subscriptions in the form of the SIM card can be backward compatible with GSM systems, and dual-mode phones could solve the compatibility problems. UMTS also has two types: FDD, which will be implemented first, and TDD.

UMTS is different from current 2G networks because of the following characteristics:

- Higher speech quality than current digital wireless networks. Speech traffic over a UMTS network, together with advanced data and information services, will be a multimedia network.

- UMTS is an improvement over 2G mobile systems because it has the potential to support 2 Mbps data rates.

- UMTS is a truly global system, consisting of both terrestrial and satellite components.

- UMTS offers a consistent service environment even when roaming via its Virtual Home Environment (VHE) feature. When roaming from home network to other UMTS networks, end users will experience a consistent set of services, therefore feeling like they are on their home network, independent of the location or access mode (satellite or terrestrial).

UMTS networks can operate with GSM/GPRS networks. The systems use different frequency bands, so base stations and mobiles should not interfere with each other. Some vendors claim their core network (MSC/HLR/SGSN, etc.) and BSC/RNC are UMTS compatible, but most wireless carriers will prefer to build a totally separate and independent UMTS network. Some of the latest GSM base stations can also have

UMTS radio parts and share the same rack. Figure 7-1 shows how GSM technology fosters a migratory evolution to UMTS, via GPRS, then EDGE, finally into UMTS.

As of early 2003, the UMTS specification is designed so that there is maximum compatibility between GSM and UMTS systems. Dual or multiband phones that can be used in both GSM and UMTS networks exist. Eventually, phones that will be able to make handovers between these types of networks will be available, likely by 2006.

> *Key:* To ultimately reach the full potential of UMTS per its design intent, dual mode/dual band wireless phones that operate via terrestrial frequencies, as well as satellite frequencies, will be required in the marketplace.

There will probably not be a transition period to UMTS in a literal sense because GSM systems will continue operating at least into 2012 (some old 1G networks are still running around the world). The only limitations for carriers are the GSM license terms and customer preferences. UMTS networks will simply be overlaid onto the mobile landscape.

7.2.1 UMTS Network Architecture

A UMTS network consists of three interacting domains: the Core Network (CN), the UTRAN, and User Equipment (UE).

The primary function of the core network is to provide switching, routing, and transit for user traffic. The core network also contains the data-

Figure 7-1
GSM migration
path to UMTS

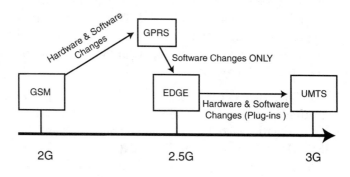

bases and network management functions (i.e., home location register [HLR], visitor location register [VLR], and element management systems[EMS]).

The IMT-2000 family of 3G systems includes three types of CN technology:

- GSM-based technology using Mobile Application Part (MAP) protocols, encapsulated into SS7 protocols for signaling
- ANSI-41–based technology (IS-634 protocols for signaling)
- Internet Protocol–based technology (to be specified in the future)

Key: The basic core network architecture for UMTS is based on GSM network technology, with GPRS as the data network foundation. All GSM-based equipment has to be modified to support UMTS operation and services. The UTRAN provides the air interface access method for end-user equipment (i.e., mobile phones and terminals).

The base station in UMTS standards is referred to as *Node-B*, and the RNC is the control equipment for Node-B. The RNC is analogous to the base station controller (BSC), as described in Chapter 12.

It is necessary for the network to know the user's approximate location so the user can be located by the system in order to terminate calls. UMTS divides geographic boundaries into "system areas," which from largest to smallest are as follows:

- UMTS systems (including satellite)
- Public land mobile network (PLMN)
- MSC/VLR or serving GPRS support node (SGSN)
- Location area
- Routing area
- UTRAN registration area (packet-switched domain)
- Cell
- Subcell

7.2.2 The UMTS Core Network

The core network is divided into circuit-switched (CS) and packet-switched (PS) domains. Some of the CS elements are the MSC, the VLR, and the Gateway MSC (the Gateway MSC is used in GSM networks to interconnect the GSM network to other outside networks). The PS elements are the Serving GPRS Support Node (SGSN) and the Gateway GPRS Support Node (GGSN). The SGSN and GGSN are the enabling network elements supporting GPRS technology. Because UMTS network architecture is based on GSM technology, this approach is sensible (see Chapter 16). Some network elements, like equipment identity register (EIR), HLR, VLR, and Authentication Center (AuC), are shared by both the CS *and* PS domains. All of these elements are standard GSM (and GPRS) network elements.

Key: Asynchronous Transfer Mode (ATM) is defined in the standard for UMTS core network transmission. ATM Adaptation Layer Type 2 (AAL2) handles CS connections, and ATM Adaptation Layer 5 (AAL5, the ATM packet connection protocol) is designated for data transport.

The architecture of the core network may change when new services and features are introduced. A number portability database (NPDB) will be used to allow users to change network providers while keeping their phone number. A device called the Gateway Location Register (GLR) may be used to optimize subscriber management between network boundaries. The MSC, VLR, and SGSN network elements can merge to become a UMTS MSC.

UMTS specifications do not have any special interface planned for interconnection to other telecom and mobile networks, but all telephone networks can be connected to a UMTS core network with standard SS7 (or other) signaling system using E1 or T1 circuits. This enables voice calls to be made to all other telephone networks. If other networks support additional services like call forwarding, caller ID, fax, and slow-speed data, among other things, technically it is possible to implement this between networks.

All telephone networks are designed to work with each other and UMTS networks will use standard interfaces toward all other networks. SS7 and IP will be the most commonly used standard interfaces, but all UMTS vendors can offer many different country-specific interface protocols if required.

7.2.3   UMTS Radio Access

In addition to channelization, UMTS codes are used for synchronization and scrambling (i.e., encryption). WCDMA has two basic modes of operation: FDDTDD.

> *Key:* WCDMA technology has been selected for the UTRAN air interface. UMTS WCDMA is a direct sequence spread-spectrum CDMA system where user data is transmitted using pseudo-random bits derived from WCDMA spreading codes.

The functions of the UMTS base station, also known as Node-B, are as follows:

- Air interface transmission and reception
- Modulation and demodulation
- CDMA physical channel coding
- Error management
- Power control

The functions of the RNC are as follows:

- Radio resource control
- Admission control
- Channel allocation
- Power control settings
- Handover control
- Ciphering (encryption)
- Segmentation and reassembly (of packet-based transmissions)
- Broadcast signaling

See Figure 7-2 for a high-level illustration of UMTS UTRAN Phase 1 network architecture.

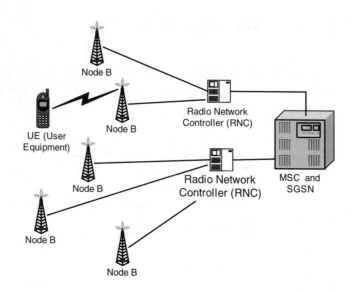

Figure 7-2
UMTS Phase 1:
UMTS
Terrestrial Radio
Access
Network
(UTRAN)

7.2.4 UMTS Mobile Terminals

The UMTS standard does not restrict the functionality of end-user equipment in any way. Mobile terminals can have many different types of identities, and most of these UMTS identity types are obtained from GSM specifications:

- The International Mobile Subscriber Identity (IMSI)
- The Temporary Mobile Subscriber Identity (TMSI), used in roaming scenarios
- Packet Temporary Mobile Subscriber Identity (P-TMSI)
- The Temporary Logical Link Identity (TLLI)
- Mobile station ISDN (MSISDN)
- The International Mobile Station Equipment Identity (IMEI), analogous to the ESN
- The International Mobile Station Equipment Identity and Software Number (IMEISV)

UMTS mobile stations can operate in one of three modes of operation:

- PS/CS mode: The mobile is attached to both the PS domain and the CS domain, and the mobile unit is capable of simultaneously using both types of services.

■ PS Mode (only): The mobile is attached to the PS domain only and may operate only the services of the PS domain. This does not prevent CS-type services to be offered over the PS domain, such as voice over IP (VoIP).

■ CS Mode (only): The mobile is attached to the CS domain only and may only operate services of the CS domain.

The UMTS IC card has the same purpose and physical characteristics as a GSM SIM card. Its functions are as follows:

■ Support of one User Service Identity Module (USIM) application (optionally more than one)

■ Support of one or more user profiles on the USIM

■ Update USIM-specific information over the air

■ Provide security functions

■ Provide user authentication

■ Offer optional inclusion of payment methods

■ Offer optional secure downloading of new applications

Some of the 3G mobiles are dual-band UMTS-GSM handsets and will be able to perform UMTS-GSM handovers. Current GSM phones will not work in 3G networks. Several SIM card manufacturers now offer cards compatible with 2G and 3G systems.

7.2.5   The History of UMTS/3G Deployments

A condensed history of some of the major milestones in the development and deployment of 3G technology is listed here. The milestones will continue, until worldwide saturation is achieved, likely by 2007.

■ In February 1999, Nokia said that it had completed what it claimed to be the first WCDMA call through the PSTN in the world. The calls were made from Nokia's test network in Finland using a WCDMA terminal, WCDMA base station subsystem, and Nokia GSM mobile switching centers connected to the PSTN.

■ In March 1999, the ITU approved radio interfaces for 3G mobile systems. Also, Ericsson and Qualcomm agreed to share access to each other's technology, ending a 2-year patent dispute.

- In April, 1999 Lucent Technologies, Ericsson, and Nippon Electric Company (NEC) announced that they have been chosen by Nippon Telephone and Telegraph (NTT) DoCoMo to supply WCDMA equipment for NTT DoCoMo's next-generation wireless network in Japan. This was the first announced WCDMA 3G infrastructure deal.

- In October 2000, SK Telecom of Korea launches the first commercial CDMA2000 network.

- In March 2001, the 3GPP approves UMTS Release 4 specification.

- In September 2001, NTT DoCoMo announced that three 3G phone models are commercially available.

- In October 2001, NTT DoCoMo launched the first commercial WCDMA 3G mobile network.

- In December 2001, Telenor in Norway launched the first commercial UMTS network.

- In December 2001, Nortel Networks and Vodafone in Spain (formerly Airtel Movil) completed first live international UMTS 3GPP standard roaming calls between Madrid (Vodafone network) and Tokyo (J-Phone network).

- In January 2002, SK Telecom in Korea launched the world's first commercial CDMA2000 1xEV-DO.

- In February 2002, Motorola unveils the company's first GSM/GPRS and 3G/UMTS product, the A820. Motorola is one of the first vendors to introduce a dual-mode enabled UMTS mobile phone.

- In September 2002, Ericsson announces the first live, dual-mode WCDMA/GSM calls with seamless handover between the two modes and high data rate in live networks.

- In November 2002, Nokia introduces the world's first GSM/EDGE 3G mobile phone: Nokia 6200.

- In January 2003, Ericsson conducts the world's first IPv6 over 3G UMTS/WCDMA network demonstration.

- On July 1, 2003, Cingular Wireless announced the world's first commercial deployment of wireless services using EDGE technology (in Indianapolis, Indiana).

- On August 27, 2003, Nokia announced that the world's first CDMA2000 1x EV-DV high-speed packet data phone call, achieving a peak data rate of 3.09 Mbps, was completed in San Diego, California.

■ On October 13, 2004, Nokia and TeliaSonera Finland successfully conduct the world's first EDGE-WCDMA 3G packet data handover in a commercial network.

7.3   CDMA2000

The Third Generation Partnership Project 2 (3GPP2) was formed to support the technical development of CDMA2000 technology, which is a member of the IMT-2000 family.

Most, if not all, CDMA2000 networks would be upgrades from existing CDMAOne networks running IS-95A/B. Network upgrades from CDMAOne to CDMA2000 would involve BTS changes via installation of multimode channel element cards, the BSC with IP routing capability, and the introduction of the packet data serving network (PDSN). The radio channel bandwidth is the same for CDMA2000-1X as for existing CDMAOne channels, again leading to a graceful upgrade when required. And of course, dual-mode handsets that can operate in both modes— CDMAOne and CDMA2000—are required in order to support such an upgrade.

Key: CDMA2000 1xRTT, CDMA2000 1x EV-DO, and future CDMA2000 3x were developed to be backward compatible with CDMAOne (IS-95A/B). Both 1x types have the same bandwidth and chip rate, which can be used in any existing CDMAOne frequency band and network. Backward compatibility was a requirement for successful deployment for the U.S. market. It is easy to implement because operators do not need new frequencies.

CDMA2000 enables a doubling of the voice-carrying capacity of a CDMA carrier through the introduction of more Walsh codes—from 64 to 128, and in the future, another doubling to 256 codes.

There are several terms used to describe CDMA2000 for different radio platforms, or versions of the specification. Some of these substandards exist today, and some of them are still under development:

CDMA2000-1X (1XRTT)

1. 1XEV (enhanced voice only and the first version of 1X)

2. 1XDO (one carrier supporting data services only)

3. 1XDV (one carrier supporting both data and voice services)

The 1XRTT platform still uses a 1.25 MHz carrier, but it uses a different vocoder than CDMAOne platforms and introduces more Walsh codes. The additional Walsh codes allow for higher data rates and more voice conversions than what is possible with CDMAOne platforms. Under 1XRTT, the three primary methods (1XEV, 1XDO, and 1XDV) are *not* mutually exclusive—they can be used in combination with each other if necessary or desired.

The future version of CDMA2000 is known as CDMA2000-3X (3XRTT). 3XRTT is intended to use 3.75 MHz of spectrum, or three times the 1.25 MHz carrier that CDMAOne uses (narrowband CDMA).

An other unique attribute of CDMA2000 is that it supports not only ANSI-41 system connectivity like CDMAOne (IS-95), but it also supports GSM mobile application part (GSM-MAP) connectivity requirements. What this offers is the ability to deploy dual-system implementation in the same market: WCDMA with CDMA2000 at the same time. WCDMA uses 5 MHz-wide CDMA carriers, hence the term "wideband."

The introduction of CDMA2000 into a network requires new or upgraded platforms, including the BTS and the BSC, which involves module additions or swap-outs, depending on the BTS infrastructure vendor being used.

CDMA2000 offers other enhancements over IS-95 and J-STD-008 wireless systems:

- Improved power control
- Diversity-transmit capability
- Modulation-scheme changes
- New vocoders
- An uplink pilot channel and a downlink pilot channel (where the pilot channel is similar to the control channel)
- Expansion in the amount of Walsh codes available
- Channel-bandwidth changes (from 1.25 MHz to 3.75 MHz)

The vision of a typical CDMA2000 network architecture is shown in Figure 7-3.

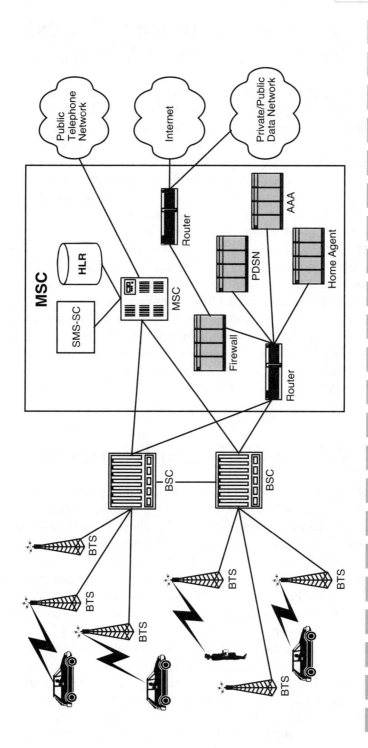

Figure 7-3
Typical CDMA2000 system (Figure courtesy Wireless Network Performance Handbook, Clint Smith, McGraw-Hill, 2004)

7.4 Wideband Code-Division Multiple Access (WCDMA)

WCDMA is an ITU standard derived from CDMA and is officially known as IMT-2000 direct-spread spectrum. WCDMA is a 3G mobile wireless technology that delivers much higher data speeds to mobile and portable wireless devices than previous 2G technologies. WCDMA is the air interface standard for UMTS 3G systems.

WCDMA can support mobile/portable voice, images, data, and video communications at up to 2 Mbps (local area access) or 384 Kbps (wide area access). The input signals are digitized and transmitted in coded, spread-spectrum mode over a broad range of frequencies.

3G WCDMA systems use 5 MHz bandwidth carriers. This 5 MHz is neither wide nor narrow; it is just the bandwidth. The reason that this adjective is used may be because new 3G WCDMA systems have *wider* bandwidths than existing 2G CDMAOne systems that operate using 1.25 MHz of bandwidth. That is why it is called wide. There are even some commercial CDMA systems operating with 20 MHz bandwidth.

Other than its 5 MHz-wide carrier, WCDMA systems are designed and function just like other CDMA systems using spread-spectrum wireless technology.

7.5 U.S. 3G Deployments

New 3G frequency allocations have been postponed in the United States, so wireless carriers are using their existing frequencies, which naturally limits the available capacity for 3G deployments. Nevertheless, most of the major wireless carriers in the United States have launched or are launching 3G networks, as indicated in this section.

Verizon Wireless, a joint venture of Verizon Communications and Vodafone Group Plc., is offering 3G services based on CDMA2000 1x technology with initial data rates up to 144 Kbps. Verizon has said that users should see speeds of 40 to 60 Kbps on average.

Verizon made big headlines in October 2003 when they rolled out CDMA 1xEVDO in San Diego and Washington, DC. Later it launched the service in Las Vegas, and it is receiving rave reviews in all three cities. The broadband access service offered with this launch was seen by some as the catalyst that would propel the U.S. wireless data industry several years into the future. Verizon's new networks boasted average download speeds of 400 Kbps and was billed as the first true 3G service deployed in a major U.S. market. Verizon viewed these deployments as test beds that would determine 3G's market viability. If deemed successful, it could spur Verizon to launch EVDO services nationwide, which would likely prompt competitors to follow suit.

Verizon's initial market segment targeted with the 1xEVDO rollout was the enterprise market, where they hyped the service's ability to link with corporate VPNs and work with any security protocol. A key attraction for larger enterprises would be nationwide coverage. With just three markets launched, it's hard for Verizon to make that claim. But the 1xEVDO networks do handoff to the carrier's nationwide 1xRTT network. By this time, most major national carriers had completed CDMA1000 1X and GPRS rollouts, but the data speeds produced by those technologies usually clocked in between 32 Kbps and 64 Kbps.

In September 2004, Verizon said it would expand its high-speed wireless service (CDMA 1xEVDO) to 14 cities, including New York, Miami, and Los Angeles. Verizon said it plans to spend over $1 billion in 2005 to extend its high-speed wireless service to most corners of the United States.

Cingular Wireless, a joint venture between BellSouth Corporation and SBC Communications, committed to rolling out EDGE technology in late 2003.

Key: EDGE, the fastest nationwide wireless data network, is available through Cingular Wireless with average data connection speeds up to 135 Kbps, depending on the device used. EDGE is available along 30,000 miles of interstate highways in the United States, and in the near future, UMTS devices will be backward compatible with EDGE.

In December 2004, Cingular announced its 3G strategy, which involves a UMTS and high-speed downlink packet access (HSPDA) contract signed by Ericsson, Lucent, and Siemens that is worth an estimated $3 to $5 billion over four years. The buildout will include a number of major urban and suburban markets, beginning in 2005.

Cingular's acquisition of AT&T Wireless provides them with the spectrum necessary to build the 3G networks. The new network will make many new applications available for both businesses and consumers, including multiple types of multimedia streaming and high-speed Internet access.

AT&T Wireless, now part of *Cingular*, deployed GPRS networks in about 40 percent of their markets in 2001.

In July 2004, AT&T Wireless launched 3G in Detroit, Phoenix, San Francisco, and Seattle. The WCDMA technology (UMTS) was also rolled out in San Diego and Dallas in late 2004. Bear in mind that by virtue of its acquisition by Cingular Wireless, any 3G developments planned by AT&T Wireless were effectively defunct and subject to the assessment and business decisions of Cingular Wireless.

Sprint PCS began its CDMA2000 1X network rollouts nationwide in 2002, offering data rates of up to a maximum of 144 Kbps with average speeds of 40 to 60 Kbps. In December 2004, Sprint announced its plans for a massive rollout of CDMA 1xEVDO, investing $3 billion in the network upgrade. It signed contracts with Lucent Technologies, Nortel Networks, and Motorola to support the project. Another $2 billion will be allocated toward upgrading the network core and expanding the existing 1X network, bringing new base station transport, PDSNs, BSCs, softswitches, gateways, and servers forming the foundation of the IP multimedia subsystems platform. The remaining $1 billion will cover the actual EVDO access network upgrades, which will involve installing new cards in existing CDMA 1x base stations and new CDMA RF carriers at each cell site.

T-Mobile, (previously VoiceStream), a unit of Germany's Deutsche Telekom, launched its GPRS service in November 2001 under the brand *iStream*. VoiceStream's data network can run at speeds up to 56 Kbps and averages up to 40 Kbps.

In December 2004, T-Mobile's CEO Robert Dotson told analysts that his firm does not expect to offer UMTS for at least the next two years, possibly not until 2007. Dotson said T-Mobile would delay 3G deployment until it has enough spectrum to support it. Until such time, T-Mobile will continue its EDGE rollout, launching service in 2005, and continue buildout of its Wi-Fi hot spot footprint.

Nextel Communications, which has been bid upon by Sprint PCS, has not yet disclosed plans for a data-capable next-generation network. Analysts have speculated that Nextel will switch to CDMA and upgrade its networks with CDMA2000 technologies.

The stakes here are huge from a revenue and positioning standpoint: billions of dollars in potential 3G sales, especially among corporate customers. But, as with any new technology, competition is likely to grow fierce and profit margins could fall rather quickly. Verizon is charging $80 a month for their 3G service, and it requires customers to buy an expensive (around $200) adapter card for their laptops. So, manufacturers will profit from these 3G endeavors, too.

Key: Of all the major wireless carriers, only T-Mobile and Nextel have not committed to a 3G migration pathas of early 2005. By the end of 2006, the other major wireless carriers could feasibly have fully operational nationwide 3G networks, as well as the first of a series of augmentations in place to carry them beyond 3G. This will place T-Mobile and Nextel at a considerable competitive disadvantage, which may be a moot point for Nextel if they are indeed acquired by Sprint PCS. That would leave T-Mobile hobbled and alone in the 2G wilderness.

7.6 Vendors

No network vendor can supply all the equipment and components necessary to fully outfit a 3G network, but quite a few can become main contractors to build turnkey 3G networks. Normally, network vendors can subcontract with partners such as service and applications providers, handset manufacturers, civil work, and acquisition companies to build and turn up networks.

Here is a current short list of main vendors for turnkey UMTS networks:

- Alcatel
- Ericsson
- Lucent
- Motorola
- Nokia
- Nortel
- Siemens/NEC

The following is a current short list of main vendors for turnkey CDMA2000 networks:

- Ericsson
- LG Electronics
- Lucent
- Motorola
- Nortel
- Samsung

Since 2001, over 30 UMTS networks have been deployed worldwide, mostly in Europe.

7.7   3G Applications

Some say that there will not be a single application but a palette of services that will be the 3G *killer app*. As mentioned earlier in this book, voice transmission is still the killer app for now, at least in the United States. The United States has been very, very slow to adopt nonvoice applications using wireless service. Most likely there will not be only a single application that becomes very popular and at the same time makes a lot of money for the wireless carrier. E-mail, voice, messaging, and music/video streaming are likely 3G, broadband-wireless money-making applications. Mobile commerce, or m-commerce, and location-based services are also predicted to become very popular. Pricing will have a lot to do with driving what becomes popular as well.

3G *bearer services* have different quality of service (QoS) parameters for maximum transfer delay, delay variation, and bit error rate. The targeted data rates are as follows:

- 44 Kbps for satellite and rural outdoor environments
- 384 Kbps for urban outdoor environments
- 2.048 Kbps (2 Mbps) for indoor and low-range outdoor environments

UMTS network services have different QoS classes for four types of traffic:

- Conversation class (voice, video telephony, video gaming)
- Streaming media class (multimedia, video-on-demand, webcast)
- Interactive class (web browsing, network gaming, database access)
- Background class (e-mail, SMS, downloading)

UMTS will also have a VHE capability. It is a concept for personal service-environment portability across network boundaries and between terminals. Personal service environment means that users are consistently presented with the same personalized features, user interface customization, and services for the network or terminal in which the user may be located. UMTS also offers improved network security and location-based services.

UMTS Forum's Market Aspects Group has identified seven common lifestyle attributes for mobile multimedia applications. Here is a list of possible services that will be available in 3G networks:

Work: rich call with image and data stream, IP telephony, business-to-business ordering and logistics, information exchange, personal information manager, scheduler, note pad, two-way video conferencing, directory services, travel assistance, work group, telepresence, FTP, instant voicemail, and color fax.

Enjoyment: Internet access, video transmissions, postcards, snapshots, text, picture and multimedia messaging, personalization applications (ring tone, screen saver, desktop), and jukebox.

Education: online libraries, search engines, distance learning, and field research.

Security: remote surveillance, location tracking, and emergency use.

Health: telemedicine, remote diagnostics, and heath monitoring.

Mass media: push newspaper and magazines, advertising, and classified ads.

Shopping: e-commerce, e-cash, e-wallet, credit card, telebanking, automatic transactions (i.e., point of sale [POS]), and auctions.

Entertainment: news, stock market, sports, games, lottery, music, video, concerts, and adult content.

Travel: location-sensitive information and guidance, e-tour, location awareness (when vacationing), time tables, and e-ticketing.

Adjuncts: TV, radio, PC, access to remote computer, MP3 player, camera, video camera, watch, pager, GPS, and remote control unit.

7.8   High-Speed Downlink Packet Access (HSPDA)

High-Speed Downlink Packet Access (HSDPA) is a WCDMA, packet-based data service in the downlink with data rates of 8 to 10 Mbps (and

20 Mbps for multiple-input, multiple-output, or MIMO, systems) over a 5 MHz carrier in a WCDMA downlink.

In 3GPP standards, Release 4 specifications provide efficient IP support, which enables provision of services through an all-IP core network. 3GPP Release 5 specifications focus on HSDPA to provide data rates up to approximately 10 Mbps in support of packet-based multimedia services. MIMO systems are the work item in Release 6 specifications, which will support even higher data-transmission rates up to 20 Mbps. HSDPA is evolved from and backward compatible with WCDMA systems. 3GPP began a feasibility study on high-speed downlink packet access in 2002.

7.8.1   Benefits of HSDPA

As 3G networks roll out across the world, mobile subscribers are just now beginning to experience the true capabilities of 3G wireless services. Some in the industry already have concerns that the demand for high-speed data access will exceed the capability of today's UMTS networks. Fortunately, the wireless industry already has plans in place for enhanced data networks.

Among the options for next-generation high-speed wireless data access are Flarion's Flash OFDM (F-OFDM), WiMAX (IEEE802.16e), and CDMA2000 1XEV-DO. But HSDPA is what enables a smooth, cost-efficient upgrade to existing WCDMA networks at minimal cost.

HSDPA's incremental UMTS network upgrade seeks to improve peak data rates, QoS, and spectral efficiency—similar to how EDGE and 1XRTT improved the effectiveness of 2G. Even though UMTS enables streaming video, broadband Internet access, and video conferencing, HSDPA offers peak downlink data rates of up to 14 Mbps—dramatically more than the 384 Kbps that is typical of today's UMTS systems and the highest data rate of any available mobile wide area network (WAN) technology.

HSPDA works by moving important processing functions closer to the air interface itself. Although current UMTS networks perform network scheduling and retransmission in the radio network controller, HSDPA moves this function to the base station (called Node B in UMTS systems), which allows scheduling priority to take account of channel quality and mobile terminal capabilities. Retransmission also benefits from hybrid automatic retransmission request (ARQ) in which retransmissions are combined with prior signal transmissions to improve overall reception. HSDPA uses a channel-sharing mechanism that allows several users to share the high-speed air interface channel and other technological advances such as adaptive modulation and coding, quadrature amplitude modulation (QAM), and feedback related to channel quality. These enhancements allow HSDPA to approximately double the total throughput capacity of a UMTS network.

Key: For consumers, this means shorter service response times, less waiting, and faster connections. Mobile subscribers will experience true wireless convergence as they talk on the phone while simultaneously downloading packet data. Most importantly, they can use their wireless handsets to download Web pages, audio, or video at speeds well above the performance they are accustomed to with landline-based DSL or even cable Internet connections.

What's good for the consumer ultimately is good for the wireless carrier in most cases, provided the costs and obstructions to deployment are not too great. HSDPA significantly enhances WCDMA with little hardware investment. It operates in 5 MHz channels and is backward compatible with current WCDMA networks. This allows network operators to introduce greater capacity and higher data speeds on the same CDMA carriers as with existing Release 1999 WCDMA services. A wireless network may be upgraded incrementally to enhance performance for users of the latest handsets, without losing network capacity or interrupting service to subscribers who rely on older handset technology.

Because it's an extension of GSM, HSDPA can be deployed readily in the United States, Europe, and Japan—where GSM networks have wide footprints. In fact, some service providers may introduce pilot projects in 2005. By 2007, HSDPA likely will be a leading technology worldwide. This means that operators will be able to offer global roaming capabilities based on infrastructure already in place today and serving a sizable handset base, including older WCDMA units.

Among future developments is *high-speed uplink packet access*, a companion standard of HSDPA, known as *HSUPA*. HSUPA will supplement HSDPA to create a more symmetrical high performance system. Ongoing improvements in high-speed technology should boost network efficiency, reduce latency, and increase overall network throughput. Figure 7-4 illustrates the evolution of GSM-based wireless data technologies and

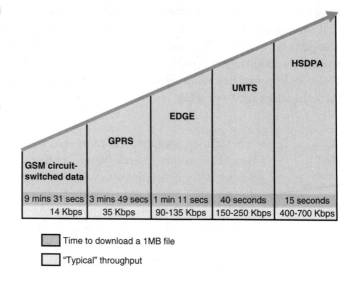

Figure 7-4
Wireless data
network
evolution and
speeds
comparison

GSM circuit-switched data	GPRS	EDGE	UMTS	HSDPA
9 mins 31 secs	3 mins 49 secs	1 min 11 secs	40 seconds	15 seconds
14 Kbps	35 Kbps	90-135 Kbps	150-250 Kbps	400-700 Kbps

Time to download a 1MB file

"Typical" throughput

clearly shows how HSDPA tops out its predecessors in data rate and throughput.

7.9   4G Wireless

Up until 2004, the term "4G" held little or no meaning in the wireless world. It was simply a buzzword—a catchphrase to demonstrate the wireless industry's desire to show that it was always trying to stay ahead of the technology curve. Even though 3G is still being developed and rolled out across the world, the next big thing was something that we all needed to start thinking about.

The time has come when many in the telecom industry see what 4G can really be all about: It is a technology called OFDM, or orthogonal frequency division multiplexing. OFDM is a method of digital modulation in which a signal is split into several narrowband channels at different frequencies. The technology was first conceived in the 1960s and 1970s during research into minimizing interference among channels near each other in frequency.

In some respects, OFDM is similar to conventional frequency division multiplexing (FDM). The difference lies in the way in which the signals are modulated and demodulated. Priority is given to minimizing the interference, or crosstalk, among the channels and symbols comprising the data stream in OFDM technology. Less importance is placed on perfecting individual channels.

OFDM is used in European digital audio broadcast services. The technology lends itself to digital television and is being considered as a method of obtaining high-speed digital data transmission over conventional telephone lines. It is also used in wireless local area networks (LANs).

Vendors and standards bodies are starting to warm up to the idea of OFDM as a mobile cellular technology. More and more vendors are developing prototype gear based on OFDM and submitting proposals to the GSM-based 3GPP and its CDMA counterpart, 3GPP2.

OFDM's supporters believe the technology will be the key to cutting down on the multipath distortion that's inherent in signal-carrier wireless networks today, which will lead to greater spectral efficiency and ultimately to broadband speeds unthinkable on today's networks. OFDM splits a single carrier signal into multiple signals, dividing the transmitted data among them. In a single-carrier systems, a data call is sent on a single signal stream from transmitter to receiver. But the individual signals in that transmission do not go directly to the receiver. Because RF bounces off all inorganic objects in its path, the result is that two bits of data sent simultaneously from a transmitter do not arrive at the receiver at the same time. This mixing of bits results in distortion and degradation of the original transmission.

Today's wireless technologies easily sort out this distortion with equalization technologies, but as capacity increases over any given network, more data is being packed into the same amount of spectral bandwidth, compounding the effects of multipath distortion (i.e., Rayleigh fading). Using OFDM technology, the objective is to divide one extremely fast signal into numerous slow signals, each spaced apart at precise frequencies. Even though each of these individual subcarriers is subject to the same multipath interference faced by a single-carrier transmission, the data is traveling slowly enough that the effects of the distortion become negligible. The numerous slow transmissions are then all collected at the receiver and recombined to form one high-speed transmission.

So, will OFDM be the answer to future wireless networks? Will it be the new 4G? Hard to tell, as WiMAX, a wireless metro area network (MAN) technology, will also offer high-speed transmissions and data rates (about 100 Mbps), which may be sufficient for most users. Only time, technology evolution, marketing, and competition will tell.

7.10   Conclusion

The telecommunications world is changing as the trends of media convergence, industry consolidation, Internet and IP technologies, and mobile communications merge into one conceptual—and in many ways tangible—entity. Significant change will be brought about by this rapid evolution in technology, with 3G mobile Internet technology a radical departure from that which came before in the first, and even the second, generations of mobile technology. Some of the changes we will witness are as follows:

- People will look at their mobile phone as much as they hold it to their ear and talk into it.
- Data (nonvoice) uses of 3G will be as important as, and very different from, traditional voice communications.
- Mobile communication will be similar in capability to fixed communications, to the point where many people will only have a mobile phone and nothing else (i.e., no laptop computers, no PDAs).
- The mobile phone will be used as an integral part of the majority of people's lives—it will not be an accessory, but a core part of how they conduct their daily lives.

As with all new technology standards, there is the FUD factor: fear, uncertainty, doubt. 3G mobile is newsworthy and controversial for several reasons:

- Because the nature and form of mobile communications is so radically changed, many businesspeople do not understand how to make money in the nonvoice world and do not understand their role in it.
- 3G licenses have started to be awarded around the world, necessitating that existing mobile communications companies in the 2G world think about and justify their continued existence.
- Many industry analysts and other pundits have questioned the return on an investment in 3G technology, asking whether wireless carriers will be able to earn an adequate return on the capital deployed in acquiring and rolling out a 3G network.
- Many media and Internet companies have expressed an interest in bidding for and using 3G technology as a new channel to distribute their content, opening the opportunity for new entrants, new partnerships, and value chains. Content truly will be king in the 3G wireless world of the future.

▮▮▮ Test Questions

True or False?

1. _____ As of late 2004, Nextel Communications and T-Mobile were the only two major wireless carriers who did not have a published 3G migration plan.

2. _____ The UMTS version of the base station controller is known as the RNC.

3. _____ Wideband CDMA (WCDMA) uses CDMA carriers that are 3.75 MHz wide.

4. _____ The UMTS standard calls for the eventual use of both terrestrial wireless and satellite wireless technologies.

Multiple Choice

1. What technology is being touted as a potential 4G technology?
   a. TDM
   b. SDMA
   c. OFDM
   d. FDM
   e. None of the above

2. What is the base station called in a UMTS network?
   a. BTS
   b. BST
   c. 3GPP2
   d. The Base
   e. Node-B
   f. Node-Z

3. Which 3G technology will use 3.75 MHz of spectrum?
   a. UMTS
   b. CDMAOne
   c. OFDM
   d. CDMA2000
   e. None of the above

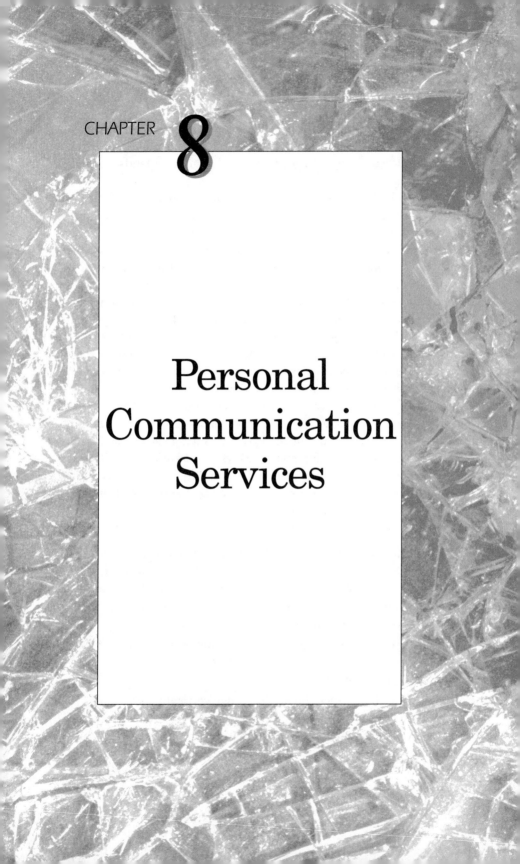

CHAPTER **8**

Personal Communication Services

8.1   Overview

The information contained in this chapter is largely a historical primer. But it is important to know and understand this information to gain a complete understanding of the current wireless market landscape in the United States.

In the United States, personal communication services (PCS) is referred to as the second-generation, or 2G, wireless service. Cellular systems were the first-generation commercial mobile telephone service, their unique attribute being the first deployment of widespread frequency reuse in an analog environment. Remember that for cultural and network-design purposes, as mentioned earlier, all *broadband* wireless service today can still be known as *cellular* service. This terminology is what people are accustomed to, and the design of all wireless systems— even those where the spectrum was originally obtained under the PCS banner—still uses hexagonal cells for design and management purposes.

The FCC has defined PCS as radio communication that encompasses mobile and fixed communications to individuals and businesses that can be integrated with a variety of competing networks. The term "competing networks" means transport to other mobiles using a nationwide, landline-based SS7 network, use of the PSTN to support seamless roaming (in conjunction with the aforementioned SS7 network), and ultimately integration with wireless fidelity (Wi-Fi) or WiMAX systems. Over time, the definitions of PCS service will be formed by the economic and technological interests of the license holders.

There are other ways to define the mobility PCS networks:

- Personal mobility defines the ability of a user to access telecommunications services at any terminal (wireless subscriber unit) on the basis of a personal identifier, and the ability of the network to provide services according to the user's service profile.

- Terminal mobility defines the ability of a terminal (wireless subscriber unit) to access telecommunications services from different locations while in motion, and the ability of the network to identify and locate the terminal.

- Service mobility defines the ability to use vertical features (e.g., custom local area signaling services, or CLASS, features) provided by today's landline network, by users at remote locations, or while in motion. Some examples of CLASS features include call waiting, caller ID (ANI delivery), call forwarding, and automatic callback.

In contrast to telecommunication services that have evolved over the years, PCS refers to services and technologies that are user-specific instead of location-specific. This architecture reflects the desire of today's wireless carriers to implement as much intelligence into the network as possible. It also explains why it is now possible to reach individuals at any time in just about any place by using a single phone number. This is referred to as *find me/follow-me* service and is an example of a wireless intelligent network (WIN) capability such as automatic call delivery (ACD).

Unique attributes of PCS in the United States, making it different than traditional 850 MHz cellular, are as follows:

- PCS is the first all-digital wireless network from its inception. In the United States, every company that won PCS licenses in March 1995 had to select which type of digital wireless technology it wanted to deploy to build out its system: IS-136 TDMA, GSM, or CDMA.

- It uses different frequencies than cellular. Cellular service uses the 850 MHz range. PSC services operate in the 1,900 MHz, or 1.9 GHz, range.

- PCS service has the objective of enabling true multimedia transmissions to and from wireless handsets. This is rapidly becoming a reality with the advent of Internet access from today's mobile handsets, the huge popularity of picture phones, the integration of PDA functionality into handsets, the use and integration of the Blackberry, and ultimately the advent of full-motion video. Convergence will apply to wireless industry applications just as it will apply to landline-based applications.

Today's wireless carriers now offer service that reflects their vision and the FCC's vision of the service: allowing voice, data, image, and soon full-motion video transmissions to occur through a small, lightweight phone.

To foster the rapid development and evolution of PCS, the FCC wisely placed few restrictions on the type of services that can be offered under the PCS banner. The FCC allocated four times as much spectrum for PCS compared to the original amount of spectrum allocated for cellular communications in the early 1980s. In doing so, the intent was to provide sufficient capacity for the estimated 60 to 90 million subscribers projected to use PCS by the year 2005. The industry is well on its way to achieving this growth forecast. As of September 2004, there are about 190,000,000 wireless subscribers in the United States.

Key: Although PCS carriers may have had different marketing strategies than their forerunners, the cellular carriers, most of the fundamental concepts and attributes that were/are inherent to cellular system design also apply to the PCS industry: frequency reuse, hexagon grid design, a required backhaul network, RF propagation theory and modeling, interconnection to the PSTN, and roaming among other things. All the same types of equipment are used at PCS base stations as they are for cellular base stations: radios, amplifiers, duplexers, combiners, and multi-couplers. The core design difference between cellular and PCS networks is that, given the same geographic coverage area, ultimately more PCS base stations will need to be deployed to cover a market area, compared to the amount of cellular base stations required.

8.2 PCS Types

The PCS family of licenses is divided into two main groups: narrowband licenses and broadband licenses.

8.2.1 Narrowband PCS

By design, 3 MHz of radio spectrum that was set aside for narrowband PCS was to be used primarily for data transmissions. New services expected to be provided on the narrowband PSC airwaves include advanced voice paging, two-way acknowledgment paging, data messaging, and both one-way and two-way messaging and facsimile. In July 1994, the FCC auctioned licenses for the rights to provide narrowband PCS service on a nationwide basis. Six companies paid a total of $617 million for these rights. In November 1994, the FCC auctioned off 30 regional narrowband PCS service area licenses as well.

As of 2004, no company had made inroads into the wireless marketplace by deploying narrowband PCS services of any sort because, by definition, any service marketed under this banner would need to be paging or SMS-type service. As standalone offerings, these services are effectively dead, having been supplanted by traditional broadband wireless service.

Even if a spunky company were to attempt to market narrowband PCS services, the wireless world is going in a different direction. Broadband wireless, ultra-wideband (UWB) wireless, Wi-Fi, and WiMAX can or ultimately will support the same applications that narrowband PCS could ever attempt to launch.

8.2.2 Broadband PCS

In December 1994, the FCC began the auctions for broadband PCS licenses. In March 1995, the FCC concluded the auctions for 99 broadband metropolitan trading area (MTA) PCS licenses. Broadband PCS is intended to be used for multimedia transmissions: voice, data (Internet, SMS), image, and, soon, full-motion video.

Voice transmission traditionally requires greater channel capacity than data transmission, so the FCC set aside a total of 140 MHz of radio spectrum for broadband PCS usage. This will change as compression and codec technology improves. One of the reasons that more spectrum was allocated for PCS than cellular (140 MHz versus 50 MHz) is because of the high hopes that the FCC and the industry have for PCS services.

Key: A majority of the 140 MHz spectrum (120 MHz) was allocated for use by licensed wireless service providers. The remaining 20 MHz of spectrum is reserved for unlicensed broadband PCS applications, such as advanced cordless telephones, wireless PBXs, and low-speed computer data links.

8.3 PCS Markets

To encourage the participation of both large telecommunications companies and smaller businesses in the development of PCS, the FCC issued two types of market licenses: metropolitan trading areas (MTAs), which cover huge geographical areas spanning many states, and smaller basic trading areas (BTAs) where many BTAs exist within each state. These market boundaries are based on the natural flow of commerce in much the same way that cellular MSAs and RSAs were created. Nationwide, there are 51 MTAs and 492 BTAs.

8.3.1 Metropolitan Trading Areas (MTAs)

The MTAs are the largest type of PCS market, covering the largest expanses of geographic territory. Similar to the initial regulatory framework for the cellular industry where an A and B carrier were the designated market operators, two broadband PCS carriers are operating in each MTA: an A band carrier and a B band carrier.

Like 850 MHz cellular markets, the MTA markets are labeled according to the largest city within their boundaries. However, in contrast to cellular MSAs, MTA boundaries encompass huge swaths of territory that extend far beyond the boundaries of the largest city within the MTA border. For example, the Chicago MTA encompasses almost the entire state of Illinois, a portion of northwest Indiana, small portions of Ohio, and portions of southern Wisconsin. The Minneapolis MTA encompasses the entire state of Minnesota, western Wisconsin, eastern Montana, and part of North Dakota. This contrasts sharply with the division of cellular markets, as Illinois has many RSAs and MSAs within its state boundary.

The MTA auctions began in November 1994 and concluded in March 1995. There are 51 MTAs in the United States. A total of 99 MTA licenses were issued to serve the 51 MTAs across the country. Four special, preauction *pioneer preference* licenses were already awarded for experimental systems in Washington, DC. The MTA licenses were assigned a total of 60 MHz of the PCS spectrum allotted by the FCC. The A and B bands each occupy 30 MHz of that spectrum, as follows:

A band (30 MHz total): 1850–1865 MHz, 1930–1945 MHz

B band (30 MHz total): 1870–1885 MHz, 1950–1965 MHz

See Figure 8-1 for a map showing the broadband MTA PCS markets in the United States.

8.3.2 Basic Trading Areas (BTAs)

There are 492 BTAs in the United States. In contrast to the initial regulatory framework for the cellular industry, there are multiple BTA licenses assigned within each MTA, instead of adjacent to each MTA.

The BTA auctions began in 1996 and concluded in spring 1997. Although smaller than the MTAs, BTAs are still larger than most cellular MSAs and much more expansive than cellular RSAs. There are four

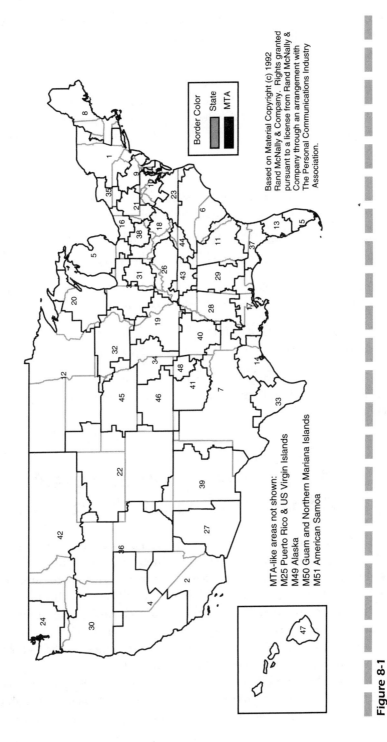

Figure 8-1
PCS MTA Market Map: The 51 MTAs (Source www.fcc.gov)

Border Color
State
MTA

Based on Material Copyright (c) 1992
Rand McNally & Company. Rights granted
pursuant to a license from Rand McNally &
Company through an arrangement with
The Personal Communications Industry
Association.

MTA-like areas not shown:
M25 Puerto Rico & US Virgin Islands
M49 Alaska
M50 Guam and Northern Mariana Islands
M51 American Samoa

distinct blocks of PCS spectrum allocated to BTAs, totaling 60 MHz of the FCC-assigned broadband PCS spectrum. They are the C, D, E, and F blocks.

Key: The total amount of licensed, broadband PCS spectrum is evenly divided between MTAs and BTAs: 60 MHz to the MTA license blocks, 60 MHz to the BTA license blocks.

There were a total of 986 licenses issued under the D and E BTA spectrum blocks. Their spectrum is allocated as shown:

D block (10 MHz): 1865–1870 MHz, 1945–1950 MHz

E block (10 MHz): 1885–1890 MHz, 1965–1970 MHz

There were 493 licenses issued for both the C and F BTA spectrum blocks. The spectrum reserved for these blocks is shown:

C block (30 MHz): 1895–1910 MHz, 1975–1990 MHz

F block (10 MHz): 1890–1895 MHz, 1970–1975 MHz

See Figure 8-2 for a map showing the broadband BTA PCS markets in the United States. See Figure 8-3 for an illustration of the entire PCS spectrum breakdown.

8.4 Licensing Mechanisms and the Spectrum Auctions

In the past, the FCC granted licenses for radio spectrum to commercial wireless carriers (i.e., cellular carriers) without requiring payment to the government for that spectrum. In some instances, cellular carriers made windfall profits in the early to mid-1980s by selling their licenses soon after they were awarded. To put an end to such profiteering and raise revenues at the same time, Congress mandated in 1993 that the FCC release licenses for PCS through an auction procedure and issued rules governing postauction sale of the licenses as well.

In March 1995, the U.S. government obtained over $7 billion by auctioning off what amounts to the right to project radio waves through the air at a certain frequency. This amount of money was paid just for PCS

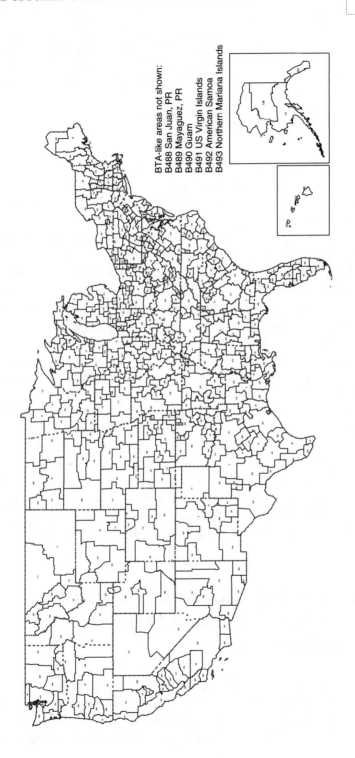

BTA-like areas not shown:
B488 San Juan, PR
B489 Mayaguez, PR
B490 Guam
B491 US Virgin Islands
B492 American Samoa
B493 Northern Mariana Islands

Figure 8-2
PCS BTA market map (Figure courtesy www.fcc.gov)

Figure 8-3
Broadband PCS
frequency
spectrum plan

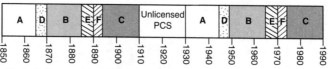

- Blocks A, B are major trading areas (MTAs)
- Blocks C, D, E, F are basic trading areas (BTAs)
- Unlicensed PCS is nationwide

licenses, at the conclusion of the auctions in March 1995. This did not include the huge multibillion dollar cost required to build out the PCS networks themselves.

Never before in U.S. history had radio spectrum been sold; it has always been granted by the FCC. In some circles, the auctioning of national radio spectrum was viewed as a dangerous precedent because it amounted to the selling of a national resource. In some quarters it was decreed as tantamount to selling off a national park, wildlife refuge, or national museum. By another perspective, the thinking of the government was that because companies would have to pay enormous sums of money for the PCS spectrum, only well-heeled and competent carriers would enter the PCS marketplace. The government's thinking in this regard appears to have been on target.

The U.S. government also auctioned off Multichannel Multipoint Distribution Service (MMDS) and Local Multipoint Distribution Service (LMDS) licenses in the late 1990s; the trend became normal. The Europeans took a lesson from the United States in this regard and began auctioning off European spectrum for certain radio services in a similar fashion, even into the early twenty-first century. The value of these licenses and cost per POP today may be different as all markets have evolved differently since PCS systems launched in late 1996 and 1997.

8.5 PCS Market Buildouts and Evolution of the Marketplace

Many industry sources speculated that the cost to fully build out the PCS systems would be double what the licenses themselves cost. Because of the huge cost of obtaining the PCS licenses and the subsequent massive investment required to build out these systems, many large telecommunications companies formed partnerships before going into the auctions. These partnerships were formed so that the burden of capitalizing the development of the system buildouts was shared. In an effort to establish brand identities, many of the PCS bidders underwent name changes from 1995 to 1999. A few of the major PCS partnerships that formed were (many of these partnerships have since dissolved) as follows:

- *Sprint Spectrum.* A company known as WirelessCo, L.P., at the time the auctions began in late 1994 changed its name to Sprint Telecom Venture (STV) in 1995. This company is now finally known as Sprint PCS. It was originally a partnership of the nation's third-largest long-distance carrier and three major cable television companies: Cox Communications, Telecommunications Incorporated (TCI), and Comcast Communications. Sprint now wholly owns Sprint PCS.

- *Primeco Personal Communications.* This partnership, initially known as PCS Primeco, consisted of several regional Bell operating companies' cellular companies and a cellular carrier spun off from Pacific Telesis. They were Nynex Mobile, U.S. West's New Vector cellular company, Bell Atlantic Mobile Systems (BAMS), and Airtouch Cellular. PrimeCo selected CDMA as their digital wireless technology of choice. At that time, only one other carrier had selected CDMA: Sprint. PrimeCo was broken up in 2001 as they were purchased by a venture capital firm that later broke up PrimeCo market by market and sold off the licenses (and the customers). One of PrimeCo's most expensive and valuable markets was the Chicago MTA license. The Chicago MTA was the most expensive broadband PCS market to obtain, and it cost PrimeCo and AT&T each about $31 per POP to acquire the A and B MTA licenses for the Chicago MTA market. The PrimeCo license was later purchased by U.S. Cellular Corp, whose headquarters has always been in Chicago. U.S. Cellular's parent company, TDS, had always sought a real market presence in their headquarters market, and, with the breakup of PrimeCo, they got their wish.

The following former GSM carriers were little known at the time of the auctions, but they also built out and launched markets. The regions where they obtained licenses and built out their systems were as follows:

- *PowerTel*: Southeastern United States.

- *Omnipoint*: Eastern seaboard and portions of the midwestern United States (Indianapolis, Indiana).

- *Western Wireless*: Western and northwestern United States. Western Wireless spun off its PCS subsidiary and renamed that new company Voicestream Wireless.

The only other major player in the PCS auction and buildout picture was AT&T. AT&T and Sprint each spent around $2 billion on PCS licenses.

The names, ownership, and makeup of PCS and traditional 850 MHz cellular carriers have changed dramatically since the late 1990s. Most of these changes were the result of mergers and buyouts, as all wireless carriers—both 850 MHz traditional cellular and 1,900 MHz PCS—sought to develop nationwide U.S. footprints. Following is a description of the industry's evolution.

- *Aerial Communications*: A GSM operator and license holder of eight MTA markets, it was purchased by Voicestream Wireless, based in Bellevue, Washington. Aerial was originally a subsidiary of TDS, based in Chicago, Illinois. Going into the PCS auctions in 1994, Aerial used the name American Portable Telecom, in keeping the patriotic-themed naming convention preferred by the parent company, TDS (TDS also owns U.S. Cellular). Aerial was eventually purchased by Voicestream Wireless (which was eventually purchased by Deutsche Telekom).

- *Voicestream Wireless*, a division of Western Wireless, a legacy 850 MHz cellular carrier operating in the northwest United States, proceeded to develop a nationwide GSM footprint by purchasing both Omnipoint and Aerial Communications. Voicestream's objective was to attain a nationwide GSM footprint to compete with AT&T Wireless and Sprint PCS, because those carriers had nationwide coverage and simple rate plans in IS-136 (TDMA) and CDMA systems, respectively. Voicestream itself was eventually purchased by German telecom powerhouse Deutsche Telekom in June 2001 and renamed T-Mobile. Deutsche Telekom also purchased Powertel at the same time, patching together a huge U.S. GSM footprint with these two major purchases.

- Southwestern Bell Mobile entered into a merger with Bellsouth Mobility Wireless in October 2000, forming the company known today as *Cingular Wireless*. In early 2004, Cingular stated its intention to purchase AT&T Wireless, subject to FCC approval. In early November 2004, the FCC granted its approval, creating a wireless behemoth with over 46 million subscribers. Executives at AT&T have stated that they may consider relaunching a new wireless carrier under the AT&T brand name because Cingular will only be allowed to use the AT&T brand name in its advertising (as part of the merger conditions) for a period of around 6 months.

- GTE's merger with Bell Atlantic and Nynex (former Bell companies) created the company known as Verizon, with the subsidiary known as *Verizon Wireless*. Verizon had been the largest wireless carrier in the United States for many years until the approved merging of Cingular and AT&T.

- In December 2004, Sprint PCS stated its intent to purchase Nextel Communications. Although this may seem like an odd pairing because Nextel uses proprietary Motorola iDen TDMA technology and Sprint uses CDMA in their network, Sprint's purchase is about getting easy access to cash and getting Nextel's rich customer base. But most important of all, Nextel has been quietly obtaining about 95 percent of the U.S. MMDS licenses over the last few years. So, this purchase is also about gaining access to more spectrum, too. The additional spectrum could be sold, or it could be used by Sprint (with FCC permission).

In summary, the United States now has four major wireless carriers as of January 2005: Cingular Wireless, Verizon Wireless, T-Mobile, and Sprint PCS (presuming the FCC approves Sprint's purchase of Nextel Communications). Other large wireless carriers include U.S. Cellular and Alltel Mobile, with large footprints in second- and third-tier markets. The core objective of all major wireless carriers is to obtain a nationwide footprint. Nationwide footprints offer major economies of scale from marketing, cost, operational, and pricing perspectives.

8.6   FCC License Restrictions for PCS Markets

PCS licensees are not restricted in the number of MTAs or BTAs they can operate. However, they are restricted in the amount of radio spectrum they can own and operate in any one service area.

Key: The FCC ruled that no wireless carrier could control more than 45 MHz of spectrum in any one market (55 MHz in rural markets). This restriction was lifted by the FCC in January 2003. The industry speculated that the cap was lifted to encourage the industry to consolidate.

An 850 MHz (cellular) carrier who was bidding for a broadband PCS service area MTA license that overlapped its existing cellular service area was restricted in the amount of spectrum for which it could bid. The FCC ruled that if 10 percent or more of the population of the PCS service area was within the cellular carrier's existing service area, the cellular carrier could bid for only one of the smaller 10 MHz BTA broadband spectrum licenses. One option for the 850 MHz cellular carrier was to sell its markets in those cases where they overlapped 10 percent or more of a PCS MTA purchased-license area's POPs. This option was proven with the spinoff of Sprint Cellular from Sprint Corporation. *Note*: FCC activity has softened this rule over time.

Sprint Cellular became an independent company in the first quarter of 1996. It also was required to change the name of the company, which became 360 Degrees Communications. Alltel Mobile then purchased 360 Degrees Communications in 1998. This divestiture became necessary because Sprint PCS won the second-largest amount of nationwide broadband PCS licenses, serving the second highest number of POPs in the United States, next to AT&T Wireless. Sprint PCS's MTAs overlapped most of Sprint Cellular's existing markets—way more than 10 percent—forcing Sprint Corporation to divest themselves of Sprint Cellular.

To discourage successful PCS license winners from making a quick profit, the FCC required that PCS licensees adhere to strict construction requirements. The FCC also prohibited certain licensees from selling their operating rights for a given period of time, or the FCC would impose stiff penalties.

After winning a PCS license and making the required down payment, licensees were expected to provide service to a certain percentage of the population of the service area within 5 years. Licensees of the 30 MHz broadband PCS spectrum blocks were required to provide coverage to at least one-third of the population of their markets within 5 years and to two-thirds of the market population within 10 years.

Licensees of the BTA 10 MHz broadband blocks had less stringent requirements. The FCC mandated these license winners to provide coverage to at least one-fourth of the population of their service area within 5 years, or to at least demonstrate to the FCC that they were providing *substantial* service to the populace in their service areas. It was a given that the PCS license winners would build the markets out as fast as possible in order to generate revenue to start recovering the enormous capital expenditures they incurred by purchasing the licenses themselves, having, at that time, spent approximately double that amount to actually construct the PCS networks. The point is that the FCC did not really need to impose these rules; the license winners had every incentive in the world to build out the markets as fast as possible.

8.7 RF Propagation and Cell Density

Because radio propagation at higher frequencies fades more quickly due to free-space loss, many more cells are required overall to build out PCS networks when compared to the deployment of 850 MHz cellular networks. Most wireless engineers state that the required ratio of PCS cells operating at 1,900 MHz to cellular cells operating at 850 MHz will need to be 3:1 (average) to deliver equivalent coverage within the same market in PCS markets.

Key: It was estimated in 1995 that one hundred thousand PCS base stations would need to be installed nationwide with the rollout of PCS services. Many of these sites will place antennas on existing towers or on tall buildings, but nearly 30 percent of these sites were estimated to require new tower structures more than 200 feet tall. In 1996, there were 24,802 cell sites in service. As of 2004, there were over 174,000 cell sites in service. Thus, the original projection was a bit conservative.

See Figure 8-4 for an illustration of 1,900 MHz PCS base station requirements to deliver coverage equivalent to legacy 850 MHz cellular base stations.

Finding suitable tower and antenna sites and obtaining zoning approval for them ranks as a main challenge for PCS providers. The NIMBY perspective applies to 1,900 MHz PCS carriers just as it does for 850 MHz cellular carriers. The zoning boards of many municipalities now require wireless carriers to attempt to collocate onto existing towers as a prerequisite for zoning approval of new PCS base stations.

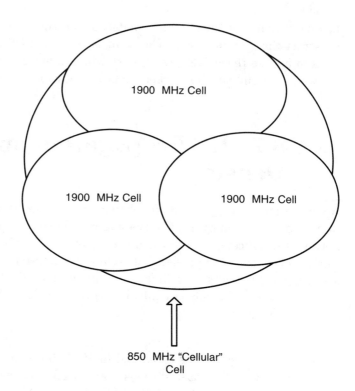

Figure 8-4
Typical cell density requirements: 850 MHz BTS versus 1,900 MHz BTS coverage requirement. Three 1,900 MHz BTSs are required for every 850 MHz BTS.

8.8 Digital Wireless Technology Selection

The winners of the broadband MTA licenses spent the last half of 1995 determining which digital wireless technology they would select so they could begin building out their networks. Their selections varied among the multiple choices available at the time:

- TDMA (IS-54, soon IS-136)
- GSM, a popular international wireless standard
- CDMA, which was a technology that was popular yet still not 100 percent tested for major commercial applications at the time

This was a very serious decision for all the carriers who won licenses because there were far-reaching implications:

- The carriers had to select a technology that they thought would thrive and remain operationally viable for a long time (i.e., CDMA).

■ They had to select a technology that was standardized, which would be technically popular and to enable them to implement widespread roaming capabilities, even internationally (i.e., GSM). This would also enable them to learn from the many existing deployments of the technology to date. By 1995, GSM deployments were occurring at a rapid pace across Europe and some other parts of the world.

■ They could select a technology that was proven and already had a large embedded base (i.e., IS-54/IS-136TDMA). But it is interesting to note that the only major PCS license winner to select TDMA as their digital wireless technology of choice was AT&T Wireless. This was odd because one would think that with the vast technical background of this company, they would select a digital wireless technology with more promise—one that had far more potential than IS-54 or IS-136. This is because IS-136, when compared to GSM and CDMA, has little to offer in terms of being a digital wireless technology of the future.

At launch in the 1996 to 1997 timeframe, PCS providers were required to provide dual-band, dual-mode handsets to their customers in order to compete with and coexist with their 850 MHz cellular counterparts. The digital wireless mode chosen depended on which digital wireless technology any given PCS carrier selected. This meant the handsets they sold to customers had to operate in both the 800 and 1,900 MHz frequency bands *and* in analog or digital mode. Because all PCS carriers built their markets out at different paces, or certain portions of their markets at different paces, PCS phones needed the ability to operate in an analog cellular mode so they could roam in AMPS systems, if necessary. PCS providers also needed to develop roaming agreements with cellular carriers so that PCS customers could roam onto cellular systems.

■ 8.9 The Microwave Relocation Requirement

There was a large obstacle to the process of deploying PCS systems nationwide. The spectrum between 1,850 and 1,990 MHz that was allocated to PCS carriers by the FCC was occupied by more than 474 microwave *incumbents*, who were operating more than 5,000 point-to-point microwave radio systems nationwide. The microwave incumbents include public, private, and semiprivate entities such as utilities, railroads, public safety administrations, governments, and 850 MHz cellular carriers (fixed network). These systems sometimes spanned multiple states and covered rural, urban, and suburban areas.

> *Key:* The FCC established rules in 1991 that defined how microwave incumbents would share this spectrum with incoming PCS carriers. The FCC stated that the newly licensed PCS operators could not interfere with the currently operating microwave systems. The FCC rules established broad guidelines for eventually relocating the microwave users to new spectrum or alternative technologies but, ultimately, to comparable facilities.

The guidelines concerning relocation of microwave facilities were as follows:

- For incumbent microwave facilities that were licensed before January 1992, the incoming PCS operator will have to bear the cost of moving the incumbent to comparable facilities.

- For incumbent microwave facilities that were licensed after January 1992, the incumbent operator of the microwave facility would have to bear the cost of relocation themselves. This is because the incumbent operator was warned ahead of time that they should not implement microwave systems in the 1900 MHz frequency range that was to be occupied by PCS carriers.

The rules governing the burden of the cost of relocation were set up in this manner because, in late 1991, the FCC orders gave fair warning to all parties who were applying for microwave licenses at that time. The orders stated that the 2 GHz spectrum was being set aside for PCS services to be auctioned off in several years. So around that point in time, microwave license applicants were made aware that they would most likely have to relinquish their microwave license by the late 1990s when PCS networks were turned up. They would then install 2 GHz microwave systems at their own risk. Although some microwave incumbents were aware of this proposed resolution for sharing spectrum, many others received their introduction to the process when a PCS licensee initiated a call to them to discuss relocation.

The cost to relocate microwave systems was a major issue relating to the deployment of PCS systems. This is because the cost to relocate a single microwave path could easily run into the hundreds of thousands of dollars, depending on the type of system in place. Some incumbent microwave operators stooped to forms of extortion when incoming PCS operators requested relocation of the incumbent's microwave system. For example, if the estimated cost to relocate the incumbent's system was around $400,000, the incumbent may have demanded a comparable facility that was actually worth $850,000. The FCC and the CTIA received numerous complaints about this situation. In 1996, the FCC filed a Notice of Proposed Rulemaking, stating their intent to implement a cost-sharing plan for the microwave relocation process. Newly proposed rules would create a clearinghouse to administer the costs and obligations of PCS licensees that benefited from the spectrum-clearing efforts of other licensees. However, the CTIA stated that such rules should also permit parties to individually negotiate cost-sharing arrangements. In addition, the CTIA stated that the bargaining power of microwave incumbents and PCS licensees had to be equalized by creating incentives for microwave incumbents to relocate during a voluntary negotiation period.

Every relocation of a microwave path had to be addressed differently because microwave incumbents had tremendously varied needs, interests, resources, and incentives. Because this was a unique process, negotiations between PCS licensees and microwave incumbents were often complex. Not all microwave paths occupying the 2 GHz spectrum were an immediate interference problem. PCS licensees approached this issue in different ways, depending on their market strategy:

- Some PCS licensees, eager to deploy their systems, adopted a first-to-market strategy, and were aggressive in beginning discussions with microwave incumbents.
- Other PCS licensees took a more deliberate approach, concentrating their resources on prioritizing the deployment of multiple markets, or various areas within a market.

As with microwave incumbents, resources, strategies, and goals varied widely among PCS licensees, contributing to the complexity of the relocation process. Because microwave relocation was a new process, every resolution that was reached set an industry precedent.

During the relocation process, some PCS licensees attempted to realize economies of scale where there may have been overlapping relocation discussions. They would then hold simultaneous talks that addressed relocation needs in multiple markets. Some PCS licensees sought to determine whether microwave incumbents were interested in partnership opportunities, such as providing PCS services together, sharing microwave towers, or leasing excess system capacity. Many PCS licensees also anticipated the magnitude of the relocation process, and they created special teams devoted solely to managing the process.

Some PCS systems were online in the third quarter of 1996 (e.g., Primeco PCS in Chicago). Most PCS carriers launched service in Spring 1997. At launch, the PCS carriers developed and distributed detailed maps showing customers where coverage was at launch as well as areas where they planned to have expanded coverage within 1 year. By late 1997, serious competition between 850 MHz cellular and PCS providers had begun.

Test Questions

True or False?

1. _____ Because of the nature of RF propagation in the PCS spectrum, PCS carriers will need to install far fewer cells than cellular carriers have had to install.

2. _____ T-Mobile is a wireless carrier that is owned by parent company Nokia.

3. _____ For incumbent microwave facilities that were licensed after January 1992, the incoming PCS operator will have to bear the cost of moving the incumbent to comparable facilities.

Multiple Choice

1. Why did the FCC issue licenses for both MTA-type markets and BTA-type markets?

   a. To implement a random system of licensing

   b. To foster close cooperation with cellular companies

   c. To foster the participation of both large telecommunications companies and smaller businesses

   d. To ensure there were many PCS markets

   e. All of the above

2. How many major wireless carriers are there in the marketplace as of January 2005?

   a. Six

   b. Four

   c. Two

   d. Ten

   e. Five

3. In contrast to telecommunications services that have evolved over the years, PCS refers to services and technologies that are:

   a. Location specific instead of user specific

   b. MSC specific

   c. LEC central office specific

   d. User specific instead of location specific

   e.   None of the above

4. The FCC rules established broad guidelines for eventually moving microwave users to:

   a. Fiber-optic facilities only

   b. Infrared technologies

   c. Comparable facilities

   d. None of the above

Towers

9.1 Overview

There are three basic types of tower available for deployment at a cell site: *monopole towers*, *free-standing towers* (also known as self-supporting or lattice towers), and *guyed towers*. The type of tower that is actually installed at any given cell base station may be dictated by company policies, operational needs (such as a minimum tower height), or local zoning restrictions. Towers are only deployed at *raw land* cell sites. Each choice in tower design has advantages and disadvantages that must be weighed against its intended use.

Tower design and construction methods are predicated on the tower's purpose, location, average weather conditions at the site, projected equipment load, and future expansion needs. With proper research and commonsense planning, a properly constructed and maintained tower can last for decades. How much space a tower structure will occupy is determined by the type of tower, its maximum height, and its support design.

Key: In the wireless industry, towers may be used not only for mounting base station antennas, but also for mounting microwave antennas. The microwave antennas are used as part of a wireless carrier's fixed network, also known as the backhaul network (see Chapter 12). Whether an extensive amount of microwave radio systems are used for the fixed network usually depends on the policies or strategy of the wireless carrier.

Local municipalities and their zoning boards have become much more stringent in approving tower locations. In some communities (usually urban or suburban areas), wireless carriers sometimes meet with very stiff resistance when attempting to install cell base stations with physically obtrusive towers.

When building base stations, the goal when determining tower heights is to get above the ground clutter (trees and buildings). The tower height for any given base station depends on how many cell sites a carrier installs overall throughout a market, taking into account frequency reuse and RF power-level factors. The minimum height depends on the area in question (i.e., the overall height of the ground clutter) and the extent of the territory where the carrier wants to provide coverage.

■■■■ 9.2   Site Surveys

Site surveys are a requirement for tower building. Many of the factors that are used as input in a site survey are simply common sense. When conducting a site survey for a communications tower, a wireless carrier includes all the applicable criteria that were used regarding cell placement in general (see Chapter 3). For example, the carrier takes into account the geographic area to be covered and whether land for the cell base station can be obtained at a reasonable price (lease or own). The site for any type of communications tower should be level and, ideally, easily accessible by a semitrailer or a heavy-duty crane. This access is needed in order to bring the tower components to the base station construction site.

After a location is selected for a cell base station, a contractor is chosen to erect the tower. The next step in the tower construction process is the evaluation of the proposed location. It is the responsibility of the party erecting the tower to research deed restrictions and zoning regulations in order to avoid conflict with neighboring property owners and any local and state agencies.

The tower site should also be free of overhead obstructions such as power lines and trees, and the foundation area should be void of utility lines or pipes. The preferred site should undergo a comprehensive soil analysis to determine whether the ground will adequately support the structure and what type of system will be required to meet electrical grounding standards. Soil testing should be conducted by a soil engineer who is familiar with tower and foundation construction. Testing procedures should encompass multiple geological core samples from across the site. Detailed soil analysis will weed out undesirable locations such as rocky terrain, swampy or sandy areas, or even old, plowed-under farm fields. Soil testing should also be done to determine if there is any toxic waste on the site. If the site is purchased by a wireless carrier and toxic waste is found afterward, the new owner (the carrier) must pay to have the waste cleaned up.

The site selection project manager for one wireless carrier did not do thorough soil testing at a proposed location for a cell site before he purchased the land. Once excavation for the cement pads for the shelter and tower had begun, it was discovered that the site was a plowed-under dump. During the remainder of the construction phase, the workers had to wear gas masks. A much deeper hole also had to be excavated for the cement pads, resulting in a much higher overall cost to build that particular cell site.

When necessary, helicopters or large tracked vehicles have been used to transport base station shelters and towers to remote site locations.

9.3 Monopole Towers

Monopole towers are constructed of tapered steel tubes that fit together symmetrically. They are simply stacked one section on top of another and then bolted together. Many versions of the monopole tower use 40-foot tube sections. The base of the monopole is bolted onto concrete pads, and no support cables are required. Tower construction companies use giant augers around 8 feet in diameter to drill deep into the ground to create the hole that concrete is poured into, which forms the base of the tower. The bottom section of the monopole tower is then bolted to this concrete base. See Figure 9-1.

Depending on soil conditions at any given location and the intended height of the tower, the hole that is drilled for the base (concrete pad)

Figure 9-1
Monopole
tower

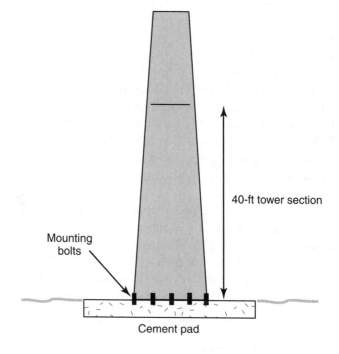

40-ft tower section

Mounting
bolts

Cement pad

could be anywhere from 8 to 40 feet deep. Monopole towers are predominantly located in urban areas and in many cases are installed for aesthetic purposes. These towers are usually not more than 150 feet tall but can reach heights of 250 feet. Of all the tower types, they have the smallest physical footprint from a size perspective.

Monopole towers are the most aesthetic of the available types of tower and require the least amount of land area for installation. Of the three tower types, these towers are the easiest to get through zoning hearings because of their less offensive appearance.

Some municipalities have been known to insist that wireless carriers paint monopole towers a sky-blue color. The thinking is that the tower will then blend in with the sky. In reality, this strategy may make the tower stand out even more starkly against the landscape and the horizon.

In some rare, extreme cases, some carriers have even been known to actually cut off the tops of monopole towers and install new mounting structures for the antennas. However, three reasons make this activity not a good idea: It is expensive, it could void the warranty on the tower, and it is risky because it could affect the structural integrity of the pole. This usually does not occur because it has become much easier to lower antenna platforms on monopole towers. See Figures 9-2 and 9-3 for photos of typical monopole towers.

9.3.1 Stealth Towers

To overcome the opposition of communities to tower erection, tower manufacturers began offering disguised monopole towers around the mid-1990s. The most popular types of disguised tower are the models that

Figure 9-2
Typical
monopole
tower

Figure 9-3
Typical
monopole
tower

look like trees. These camouflaged towers can be made to look like almost any type of tree, complete with fake trunks, fake branches, and fake leaves. Some of them look like redwood trees, palm trees, or even oak trees. Monopole towers can also be disguised as parking lot light poles, monuments, or flagpoles. These towers are more expensive than a standard monopole tower. Early versions of this type of tower were poor quality, but over the years the quality rapidly improved. See the photos in Figures 9-4 through 9-7 for samples of disguised towers.

Key: In many cases, the ability to construct a disguised tower can mean the difference between getting zoning approval for a base station in a particular area or having the request rejected.

Figure 9-4
Monopole
tower
disguised as a
tree (Photo
courtesy Mike
P's UK GSM
and UMTS
Pages)

Figure 9-5
Monopole
tower
disguised as a
tree (Photo
courtesy
kramerfirm.
com)

Figure 9-6
Full shot of
monopole
tower
disguised as
an art structure

Figure 9-7
Monopole tower disguised as a flagpole (Photo courtesy kramerfirm. com)

9.4 Free-Standing Towers

The free-standing tower, also known as a self-supporting or lattice tower, is the median tower type from both a cost and structural perspective. Free-standing towers are three- or four-sided steel structures, constructed of criss-cross sections that increase in diameter as they approach the ground. Like an elongated pyramid, free-standing towers are widest at the base and narrowest at the top. No cabling is used to support these towers; they are supported by having their bases anchored in cement with large bolts, similar to monopole towers. Free-standing towers may be installed almost anywhere. They are frequently seen next to major highways in urban areas (e.g., toll plazas). Free-standing towers require minimal land area for installation (more land than monopoles but less land than guyed towers). See Figures 9-8 and 9-9 for photos of free-standing towers.

If base station or microwave antennas need to be moved higher or lower because of changes in network design, this can be accomplished with minimal effort on a free-standing (or guyed) tower simply by moving the antennas up or down the tower structures.

Free-standing towers are usually not more than 300 feet high. They become very expensive at heights of 150 feet or greater. This is because as free-standing tower heights increase, they require deeper foundations and thicker structural elements.

Figure 9-8
Free-standing
tower (view
from BTS
shelter)

Figure 9-9
Free-standing
tower: Note
three carriers
use this tower
(three antenna
platforms).
(Photo
courtesy
kramerfirm.
com)

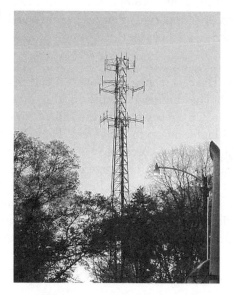

9.4.1 Free-Standing Tower Anchoring Options

There are two methods by which free-standing towers can be anchored: spread footing and pier footing.

The *spread footing* method involves pouring a large single slab of concrete, usually about 5 feet deep across the entire diameter of the base of the tower. The tower structure is then anchored to the slab of concrete with bolts. This method of footing is built to withstand a great amount of twist and sway due to wind conditions. It is also the preferred method for anchoring free-standing towers. See Figure 9-10.

The *pier footing* method involves boring deep holes in the ground, one for each leg of the tower, and filling them with concrete. Bolts fasten each leg of the tower into the concrete piers. This method is used more often with smaller free-standing towers (i.e., less than 100 feet tall).

9.5 Guyed Towers

Guyed towers are constructed of steel cross arms of equal lengths, which are installed at 45-degree angles in a criss-cross fashion (like free-standing towers) from the top to the bottom of the tower. Unlike free-standing towers, guyed towers are supported or *guyed* by tensioned support cables held in place by concrete anchors on the ground.

Key: The cost of guyed towers is linear in relation to their height: The higher the tower, the more expensive it becomes. Guyed towers are the tallest tower type and can reach heights of up to *2,000 feet.* They are more predominant in rural areas because of fewer zoning restrictions and the need to provide coverage across a larger area due to lower subscriber densities. Remember: One of the determinants of coverage is the height of the antenna—that is, the radiation source. All other things being equal, the higher the antenna (the bigger the diameter of the base station), the larger the coverage area.

Guyed towers for wireless carriers operating in RSAs are usually built to heights of around 250 to 300 feet.

Guyed towers are the same width at the base and the top. They are supported by thick guy cables that anchor the tower in place from three sides. The guy cables are placed at designated intervals throughout the height of the tower, these points being dictated by physics of tower design (overall tower height) and tower loading factors (see Section 9.7). Guyed

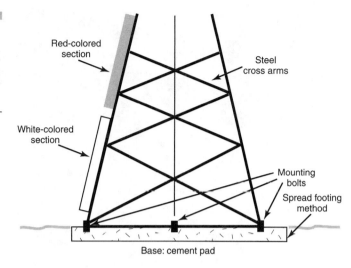

Figure 9-10
Spread footing options for mounting free-standing towers

Red-colored section

Steel cross arms

White-colored section

Mounting bolts

Spread footing method

Base: cement pad

towers can frequently be seen on farm land along the interstate highway system in the United States. See Figure 9-11 for a photo of a guyed tower.

Overall, guyed towers are the cheapest of all tower types to install, especially when a very tall tower is required at a given location. A wireless carrier building a guyed tower can make a deal with the land owner to purchase *only* the small guy anchor plots of land plus the tower's base area. When this is done, fences can be placed around all the guy anchor plots to prevent farm machinery from damaging the guy anchors and the tower itself.

The main drawback of guyed towers is that this tower type requires more land than other tower types because it is necessary to install the guyed support cables to hold the tower up. Cabling for guyed towers can require a radius that is 80 percent of the tower height. The guy cables extend beyond the base of the tower for up to 200 feet in direction. It is possible that guy cables can cause interference to base station antennas that are not *mounted on top* of the tower, but there is no way to model this potential interference.

The installation of a guyed tower can become costly if the wireless carrier cannot strike a deal to buy just the land necessary for the base of the tower plus the guy anchor plots. Because guyed towers require more land, when the wireless carrier must buy the entire plot of land necessary to construct the guyed tower, the land purchase (or lease) alone can become a major cost. Guyed towers also require more maintenance than all other tower types because guy-wire tension needs to be checked, and possibly adjusted, at regular intervals.

Figure 9-11
Guyed tower:
Note BTS
shelter at base
of tower.

Figure 9-11
Guyed tower:
Note BTS
shelter at base
of tower.

9.6 Structural Design Options for Free-Standing and Guyed Towers

When ordering any type of tower, it is important for wireless carriers to specify the wind loading, or the maximum wind speed the carrier estimates the tower will encounter projected for the tower at a given location. The carrier also needs to specify the number and size of base station antennas and the number and size of microwave antennas—if used—to be mounted on the tower. These factors will determine the size and thickness of the steel for towers, and for guyed towers they can determine the total number of guy wires that will be required to support the structure.

Key: Wireless carriers can choose among solid-leg, tubular-leg, and angular-leg tower designs for the infrastructure of the actual cross-bar tower elements. These three design options apply to both free-standing towers as well as guyed towers.

9.6.1 Solid-Leg Design

Solid-leg towers are those towers whose structural crossbar elements are solid, circular steel. Solid-leg towers withstand damaging conditions better than other designs. Towers built solid from mill-certified steel are more resistant to sway, wind, rust, and freezing than hollow, tubular-leg designs. A solid-leg tower will sway less in high winds, reducing the potential for damage to the antennas or the tower itself, and it is also less vulnerable to damage that may result from antenna installation. This tower type can be built in various sizes.

With solid-leg towers, there are no parts that can collect moisture and subject the structure to ice-induced cracking or splitting. Corrosion is less likely to occur because manufacturing standards require 25 percent more galvanizing on solid-bar legs than on tubular legs. Galvanizing means that the steel is coated with rust-resistant zinc.

9.6.2 Tubular-Leg Design

Tubular-leg towers are those towers whose structural crossbar elements are circular and hollow. These towers, like solid-leg towers, can also be manufactured in assorted sizes. However, the tubular-leg tower's hollow structure can collect moisture, risking ice-related damage. Rust and corrosion can also occur internally from condensation or pooling of water. Therefore, effective drainage must be monitored and maintained. Drainage holes, or *weep holes*, at the bottom of towers must be periodically checked to ensure there is no blockage, and that no rust is evident.

Ice can and will break a tubular tower if it is allowed to form extensively within the tower structure.

Construction costs for this type of tower are similar to those for solid-leg towers, but repair bills are more expensive. If the thickness of the tubes is substantial enough, tubular-leg towers work best with light loads in locations where no expansion of tower load is expected.

9.6.3 Angular-Leg Design

Angular-leg towers are those towers whose structural elements are solid L-shaped steel beams. The components of these towers are not tube-based but instead are composed of solid steel sections where the actual cross-arm sections themselves are at 45-degree angles. The size and thickness of the angular-leg tower elements themselves is dictated by the tower-loading factor. Tower loading is explained in Section 9.7.

With angular towers, freezing is not a concern because of their solid nature. Of the three structural models, angular-leg design is the most expensive to build. Depending on the thickness of the steel, it may also be the most susceptible to damage sustained from swaying because the thin flat surfaces may be less sturdy than solid circular crossbars. The thinner the angular legs are, the less robust they are. But thicknesses are calculated into the designs to ensure optimal sturdiness. Most deployed guyed towers use the angular-leg design.

9.7 Tower Loading

Tower loading describes the weight and wind force a tower is expected to withstand. The amount of support a tower will need is dictated by projected tower-loading factors, which are based on the following:

- The tower's height
- The total number of antennas to be mounted on the tower and whether they will be base station antennas or point-to-point antennas, such as microwave, or both
- Existing weather patterns at the tower location
- Amount and thickness (gauge) of coaxial cabling or waveguide attached to the tower

- Maximum wind load (velocity) expected at the tower's location (most towers are designed to sustain maximum wind loading of around 100 miles per hour)

9.8 FAA Regulations

It is the responsibility of the Federal Aviation Administration (FAA) to regulate objects that project into or use navigable airspace. Communication towers, by FCC rules, must abide by FAA regulations. The FCC is the enforcer for the FAA with relation to tower issues.

9.8.1 The 7460 Forms

The 7460-1 FAA form must be filed to show site locations for towers higher than 200 feet or for towers that are in the glide path of any airport. The form must show the site's *polar coordinates* (latitude and longitude), the base elevation of the site above mean sea level (AMSL), and the overall height of the tower and antennas so the FAA can determine if the tower is a hazard to air navigation. Polar coordinates are designated using three sets of numbers: the first set designating *degrees*, the second set designating *minutes*, and the third set designating *seconds*.

Wireless carriers must also file a 7.5-minute USGS map of the cell site with 7460-1 forms delivered to the FAA. The FAA will study the application and notify the wireless carrier if the tower must be lit and/or marked. Once the FAA receives the 7460-1 form, it will notify the carrier if a 7460-2 form is also required.

The 7460-2 form is used to inform the FAA when a carrier will begin and end construction of a tower, so the FAA can notify regional airports. Carriers may need to obtain special clearance by the FAA before constructing towers, depending on their response to the filing of 7460-1 and 7460-2 forms.

9.8.2 Tower Lighting and Marking

As a rule, any objects taller than 200 feet are required by the FAA to have some type of marking and lighting. The various lighting and painting requirements are established by the FAA but enforced by the FCC. Tower marking is optional, depending on the type of lighting used. Marked towers have alternating red and white sections painted on the tower structure. Typical lighting options include red incandescent lights, white strobe lights, or a combination of both known as *dual lighting*. Of the two types of tower lighting, strobe lights are used in the daytime and incandescent red blinking lights are used at night. The FAA allows carriers to use both types of lighting, if they choose. If towers are marked with red and white paint, no lighting is required during the daytime. However, towers that are not marked with red and white paint are required to have a flashing strobe (or red incandescent light) operating during the daytime.

Key: Wireless carriers can be fined if the paint on marked towers is too faded. The FAA used to warn carriers it was preparing to come out to inspect the quality of the coloring of the towers, but the FAA now comes with no warning. Some carriers have swatches of the proper quality of the coloring for tower markings and occasionally use the color swatches to ensure they are within the legal scope of how brightly the tower should be marked.

All tower lighting controls must have alarms to indicate lamp failures (for the red incandescent lights), flash failures (for white strobe lights), or power outages. If the tower lighting fails for any reason, the FAA must be quickly notified so aircraft operators can be forewarned. In most cases, if a safety light on a tower goes out, a site manager has 15 minutes to submit a report on the defective light or risk a large FAA fine for endangering the flying public.

Because lighting is so critical, it is monitored throughout the construction process of a tower. Lights must be installed at intermediate levels on the tower as construction progresses and at the top. This is to ensure that the highest point on the tower is visible to aircraft at all times. A completed tower will have one beacon at the top and may have two or more between the ground and the top of the tower, depending on the overall height of the tower structure.

The FAA is very serious about enforcing tower lighting and placement. One wireless carrier failed to properly light a tower during the construction phase, causing a medical evacuation helicopter to crash into the tower. Everyone aboard was killed. The wireless carrier was fined millions of dollars for this infraction. Another wireless carrier erected its tower higher than stated on the 7460-1 form it filed and was fined millions of dollars because of the proximity of the tower to an airport.

9.9 Co-Siting and Tower Colocation

It is always an option for wireless carriers to lease space on an existing tower by approaching the owner of the tower. This is also known as *co-siting*, or colocation. Wireless carriers generally try to lease space whenever possible, to save the cost of building their own tower. Leasing tower space also appeases local zoning boards who have developed a negative attitude toward construction of new towers in their municipalities. Sometimes wireless carriers end up leasing tower space to their in-market competitors. This has become a prominent issue with the advent of PCS carriers building out their markets. Many municipalities today even require carriers to colocate, wherever possible, to avoid having additional towers constructed in their towns. But leasing tower space may or may not be less expensive than erecting a tower, depending on many factors. How long does the carrier want to lease the tower space? Would it prove to be cheaper to build their own tower, depending on how long they plan on keeping their license in a particular market?

Since the late 1990s, a tower consolidation movement has taken root, where it has become the norm for multiple carriers to share one tower. Along these lines, there are a number of companies that do nothing but build or buy towers and lease the space to wireless carriers. This includes cellular, PCS, Enhanced Specialized Mobile Radio (E/SMR), and paging carriers.

Some wireless carriers have sold their towers to management companies, offering space for other wireless operators to build out on. Since the Telecommunications Act of 1996 was signed, many companies have become innovative when it comes to leasing tower space. One company that began buying up radio stations across the country also saw the potential to gain revenue by leasing space on all the towers that came with the radio stations they purchased. Because so many wireless carriers have faced challenges from communities blocking additional tower construction, this company saw a market need and developed the idea. In each market, this company had four to eight towers that were usually 300 to 500 feet tall. Each of those towers offered opportunities not only for the company that owned the towers, but also for wireless carriers as well. Colocation is a very common occurrence today, simply because it serves the purposes of both communities and wireless carriers very well.

9.10   Tower Maintenance

Cellular carriers (850 MHz spectrum) who own some of the oldest towers in the industry, say the secret to a long-lasting tower is a strong maintenance program. Since the early 1980s, carrier towers have held up fine. But extra weight from ice, wind, other carriers' equipment, and additional antennas will weaken any tower over a period of time. If carriers do not catch the wear and tear early, it can eventually put a tower completely out of commission. A strict maintenance program can ensure that towers hold up for many years. Regular inspections every 1 to 3 years, ideally by companies who are not the original builders of the tower, are the key. An important practice for any tower inspection is to review the original tower drawings. The purpose of having tower drawings is to ensure factual information about the age, structure, and capacity of a tower. Even if a carrier is leasing tower space from another company or carrier, they should request the tower drawings or inquire about tower maintenance records. Many tower inspections are done visually. An inspector usually will check ground-level supports, bolts, and paint degradation but will not climb the tower.

Key: The best way to see what kind of maintenance a tower needs is just to get out there and climb it. If the tower is not visually inspected from the top down, corrosion at the top of the tower and eroding metal interiors will often go unnoticed.

Guyed towers require inspections more regularly, mainly due to stretching of the guy cables. For these towers, the basic items that should be checked during any tower inspection are the guy wires, cable tension, and the connections of the guy cables. A device known as a *tensiometer* is used to test the tautness of the cables that support a guyed tower. If the tensiometer reading determines that a tightening of the cables is necessary, that is maintenance that a wireless carrier (or tower owner) cannot ignore. Vandals, inappropriate installation, and weather are all determining factors in guy cable maintenance.

For all tower types, carriers should also check the foundation and the condition of the foundation bolts. There are ways of checking a tower's foundation with detailed accuracy. Stress-wave technology can ascertain the depth of a foundation, giving a carrier a better sense of its load limit without drilling into the foundation itself.

9.10.1 Weather, Corrosion, and Loading

Towers take constant abuse from Mother Nature, especially in coastal locations or areas susceptible to tornadoes and storms. Carriers may want to inspect a tower after a harsh storm because the tower may have been exposed to wind speeds higher than the tower's design load can handle. To help prevent corrosion in these areas, carriers should avoid using hollow-leg towers because it may be difficult to inspect the inside of these structures to verify that corrosion is not occurring inside the tower. Because towers can rust from the inside out, carriers should consider conducting an ultrasound test to determine if there is any interior corrosion. But other measures can be taken to protect towers from corrosion. One major carrier frequently repaints their towers in regions with intense sunlight, heat, or wind to maintain sufficient protection of the tower structure. Ice puts the most significant load on a tower, so carriers operating in the northeastern United States engineer their towers and tower foundations to handle the extra weight that ice adds to a tower.

Weather is not the only thing that can weaken towers. Equipment overloading causes most tower failures. Failures start in the legs of lattice towers, and monopoles deteriorate because of buckling or compression failure in the steel. If the steel is painted, not overstressed, and inspected on a routine basis, a tower can last for decades with good maintenance. Towers should also be reinforced if a carrier is adding new equipment or when inspections report damage to tower *members* (legs or cross arm sections).

9.10.2   Preventive Maintenance

As with any equipment, preventive maintenance can be the key to keeping a tower in good shape over a long period of time. There are four key areas that carriers should focus on to maintain a long-lasting tower:

- Towers should be painted on a regular basis to keep water from corroding them. Sealants can be used around joints to ensure protection against the elements. Towers should never be painted in the summertime because when the temperature goes up, the tower's metal expands, causing paint to shrink and crack when the metal contracts during cooler temperatures.

- Whether a carrier rents space on a tower or owns their own structures, coax cable needs to be maintained too. Coax should be changed out every 10 to 15 years. The thicker the cable (higher gauge), the longer it will last. Thinner cables should be checked on a more regular basis. Coax cables can also be the victim of target-practicing vandals. If a bullet grazes coax cable, it could result in a significant signal loss over time.

- Carriers should send their engineers and installers to tower maintenance certification programs, which are frequently offered by tower and consulting companies.

- Tower lights should be replaced regularly if any of them appear weak (red incandescent) or slow (strobes).

Test Questions

True or False?

1. _____ One of the advantages of monopole towers is that they are the most aesthetic of all the tower types.

2. _____ Failure to notify the FAA of a lighting failure at a tower can result in stiff fines.

Multiple Choice

1. Guyed towers:
   a. Are the tallest tower type
   b. Are more predominant in rural areas
   c. Are the same width at the base and the top
   d. Are the most expensive tower type to install
   e. a, c, and d only
   f. a, b, and c only

2. Which structural design for towers allows for rust and corrosion to occur internally from condensation or pooling of water?
   a. Solid-leg towers
   b. Angular-leg towers
   c. Tubular-leg towers
   d. None of the above

3. Tower loading describes which of the following?
   a. How many cross arm sections a guyed tower can support
   b. How deep the cement footing is placed at the base of a tower
   c. How much wind resistance a tower can withstand
   d. How much weight a tower can support
   e. All of the above
   f. c and d only

Base Station Equipment and Radio Frequency (RF) Signal Flow

10.1 Omnidirectional Antennas and Space Diversity

Today, omni base stations exist only in rural environments for the most part. This is because of the lower subscriber densities in rural areas—there is no requirement for the increased capacity that is afforded by using directional antennas and sectorized base stations.

Omnidirectional base stations are noted by their use of omni antennas, which are slender, long, and tubular. Omni base stations will have one of two appearances:

- A configuration where the two (base station receive) antennas are mounted on the underside of the tower antenna platform, and one (base station transmit) antenna is mounted on the top side of the antenna platform. This is a configuration that was seen more frequently in the 1980s and early 1990s.

- A configuration where two, or sometimes three, omni antennas are mounted only on the top side of the antenna platform. This is the configuration that is seen more often today. When only two antennas are mounted on the top of the platform, it is due to the use of duplexers, which will be explained later in this chapter.

See Figure 10-1 for an illustration of antenna mounts at an omni base station.

Key: There are always two receive antennas at every base station, which are known as *receive zero* and *receive one*, otherwise known as RX0 and RX1. The purpose of having two receive antennas at every base station is to provide for what is known as *space diversity*. Space diversity, also known as *receive diversity*, compensates for Rayleigh fading in the uplink to the base station. Space diversity is a tool used to optimize the signal received by a base station (transceiver); it counteracts the negative effects of Rayleigh fading. It ensures that the best possible receive signal is used to process all wireless calls.

Here is how space diversity works: When a wireless customer presses the send button on the phone to place a call, the signal travels through the atmosphere and enters *both* of the receive antennas. The signal then

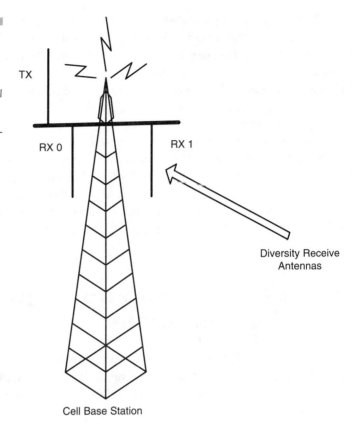

Figure 10-1
Horizontal view of tower mounting of omnidirectional base station antennas

TX

RX 0

RX 1

Diversity Receive Antennas

Cell Base Station

travels down the coax cable and into the base station transceiver that has been designated by the system to support the call. At that point, a device in the base station transceiver known as a *comparator* examines *both* of the received signals and selects the best signal of the two received. The comparator continues to *dynamically* select the better of the two receive signals for the duration of the wireless call. The call can change from being carried by one receive antenna (RX0) to being carried by the other receive antenna (RX1) dozens or hundreds of times during an average wireless call. Space diversity can provide anywhere from a 0.5 to 12 dB difference in received signal strength between one receive antenna (RX0) and the other receive antenna (RX1), at any given instant in time. This is due to the fact that the strength of the signal being received in either receive antenna can be dramatically altered by reflections in the environment, especially when the mobile is in a state of motion (i.e., talking while driving).

When there are three antennas mounted at an omni base station as described, two of them are the diversity receive antennas (RX0 and RX1), and one of them is the *transmit antenna*. The transmit antenna at a base station is used to transmit calls from the base station to the mobile phone; it transmits the downlink or *forward channel* signal. The receive antennas (RX0, RX1) receive the mobile-to-base signal, or *reverse channel* (uplink) signal.

In scenarios where there are only two omni antennas mounted at a base station as described, this means that one of the antennas is transmitting both the transmit signal and the RX0 signal; the second antenna is transmitting the RX1 signal. The act of putting both the transmit signal and the RX0 signal onto one antenna is known as *duplexing*, which will be explained in Section 10.2.2.3.

10.2   Base Station Equipment Configurations

This section contains an explanation of the various components that are required to generate and manage RF signal flow through any given base station, including both the uplink signal flow as well as the downlink signal flow.

10.2.1   Transceivers

Transceivers, or base station radios, are functionally the wireless network equivalent of the mobile phone. Transceivers contain the following elements:

- Two receive ports (for the two diverse receive signals)
- A comparator, whose function is described in Section 10.1
- One transmit port
- An *audio-in* channel
- An *audio-out* channel
- A data line (the data line runs the functions of the transceiver as it communicates to the other equipment in the base station and MSC)

See Figure 10-2 for a functional illustration of a typical base station transceiver.

10.2.2   RF Signal Flow Through a Cell Site: The Downlink (Forward Channel)

Whether a mobile subscriber originates a wireless call by dialing a phone number or receives an incoming call, this signal flow explanation applies in both cases. In other words, this signal flow applies to every wireless call. But the explanation in this section describes *only* the signal flow from the base station to the mobile phone—the end user. In other words, this section explains only the *downlink* signal processing.

Key: The MSC has already assigned a paired channel and frequencies to this call. The assigned channel and frequencies will remain the same for the duration of the call, unless a call handoff is required, such as in cases where the wireless subscriber is in motion.

10.2.2.1   Power Amps (PAs) and Linear Amps   A PA is used in the downlink signal path to boost the radio signal. Amplifier signal levels are measured in decibels. There is one power amplifier assigned to each

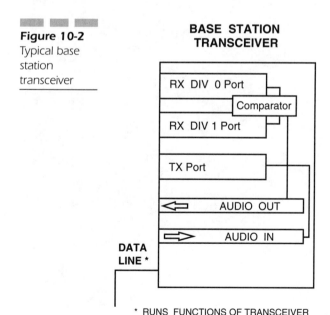

Figure 10-2
Typical base station transceiver

BASE STATION TRANSCEIVER

RX DIV 0 Port

Comparator

RX DIV 1 Port

TX Port

AUDIO OUT

AUDIO IN

DATA LINE *

* RUNS FUNCTIONS OF TRANSCEIVER

channel pair, or transceiver. Today many wireless carriers are using *linear amplifiers*, which have a maximum allowable total output. The power required for each respective mobile call is increased or decreased as necessary. The total amount of decibels supported by the linear amp, however, cannot be exceeded.

Here is an example of how a linear amp works: The total maximum power capacity of a linear amp is 100 watts. One mobile user may require 10 watts from the linear amp. A second user requires 15 watts, another user 20 watts, another user 25 watts, and another user 30 watts. All mobile subscribers have received sufficient power from the linear amp, but together their allotments do not exceed the capability of the linear amp: 100 watts. See Figure 10-3 for a picture of a typical power amplifier.

Figure 10-3
Typical power
amp (Photo
courtesy
Andrew
Corporation)

10.2.2.2 Combiners Combiners are essentially RF *multiplexers*, or *muxes*. Combiners fall into two main categories: cavity combiners and hybrid combiners. Cavity combiners are usually grouped together in fours and are more power efficient, while hybrid combiners are grouped together in increments of two, three, or five. The purpose of the combiner is to allow for the use of one antenna for multiple radio channels, rather than having one separate antenna dedicated to each channel at the base station. In other words, without a combiner, there could theoretically be one antenna for every radio channel at the cell. Combiners eliminate the cost and poor aesthetics of having to install a separate antenna and run coax cable down the tower for each and every radio channel. A by-product of the combiner's call processing is that it reduces signal levels. The strength of the signal when it exits a combiner can be reduced by up to 50 percent. However, today's modern combiners usually deliver less than 50 percent signal loss.

Each combiner port must be tuned to the frequency of the digital radio channels assigned to the cell by a wireless carrier technician. A device known as a *star connector* links combiners together. See Figure 10-4 for a photo of a combiner.

10.2.2.3 Duplexers The RF signal is now routed into a duplexer. The function of the duplexer is to enable both the transmit and receive signals to be routed through the same antenna. The call is then transmitted from the cell base station to the mobile phone at a base station transmit frequency. Duplexers are not always used by all carriers, but they avoid the cost of having to use two diversity receive antennas instead of one. A key benefit of using duplexers is that they can reduce the need to install antennas at the base station. These savings can add up quickly when dozens or hundreds of base stations are in play. Figure 10-5 shows a photo of a duplexer. Figure 10-6 provides an illustration of RF signal flow over the downlink.

10.2.3 RF Signal Flow Through a Cell Site: The Uplink (Reverse Channel)

Whether a mobile subscriber initiates a wireless call, or is receiving a call, this signal flow explanation applies in both cases. In other words, this signal flow applies to every wireless call. But the explanation in this section describes *only* the signal flow from the mobile phone (the wireless

Figure 10-4
Combiner
(Photo
courtesy
Andrew
Corporation)

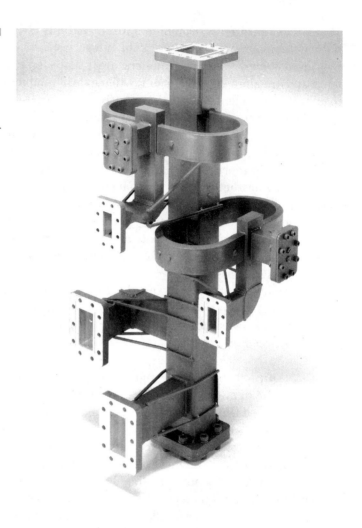

Figure 10-5
Duplexer
(Photo
courtesy
Andrew
Corporation)

subscriber) to the base station. In other words, this section explains only the *uplink* signal processing.

10.2.3.1   Bandpass Filters   Once an uplink signal is received into the RX0 and RX1 diversity antennas at the cell site, the first element it passes through as it is transmitted into the base station is a bandpass filter. The function of this filter is to filter out all frequencies except the receive frequency used to process any given wireless call. Remember that the MSC has assigned a paired channel to the call during the call setup phase, just like signal flow in the downlink. Figure 10-7 shows a picture of a bandpass filter.

Figure 10-6
RF signal flow
over downlink

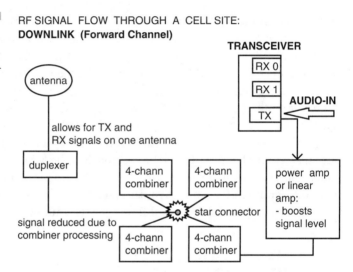

RF SIGNAL FLOW THROUGH A CELL SITE:
DOWNLINK (Forward Channel)

Figure 10-7
Bandpass filter
(Photo
courtesy
Andrew
Corporation)

10.2.3.2 Low-Noise Amplifier The signal is then routed to a low-noise amplifier (LNA), or *preamp*. The function of the preamp is to boost the received signal to a level that is strong enough to be split into multiple RF outputs—one for each radio channel. This is necessary because of loss due to impedance as the signal flows through the coax cable. The low-noise amplifier contributes very little noise to the received signal.

10.2.3.3 Multicouplers The call is then routed to a device called a *multicoupler*. The function of the multicoupler is to split the two received signals independently into multiple coaxial receive outputs, which are then routed to separate and distinct transceivers. The multicoupler's function is somewhat analogous to the combiner's function (but in the reverse direction), in that both pieces of equipment allow for the consolidation and interleaving of radio signals from multiple wireless calls.

Remember that each radio (transceiver) has two receivers: one connected to the RX0 diversity antenna and one connected to the RX1 diversity antenna. Thus, the multicoupler will receive and route to each transceiver its respective RXO and the RX1 signals. The transceiver constantly selects the better of the two received diversity signals via the comparator and continues to process the call. Once the transceiver selects the best receive signal, it then routes the call to the audio-out output and then to the MSC. Once the call reaches the MSC, the switch completes call processing. *Least-cost routing is taking place at this point* (for more information on least cost routing, see Chapter 14). Figure 10-8 provides an illustration of RF signal flow over the uplink.

Test Questions

1. Draw a diagram that shows the RF signal flow over the downlink into the base station. Include each piece of cell site equipment in your diagram, along with a brief explanation of the function of the equipment.

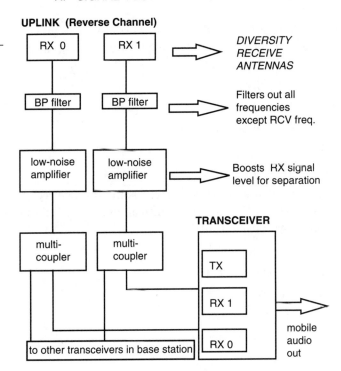

Figure 10-8

RF signal flow over uplink

RF SIGNAL FLOW THROUGH A CELL SITE:

True or False?

1. _____ A device in cellular base station transceivers known as a *complexitor* examines both of the received signals in the RX0 and RX1 antennas and constantly reselects the best signal of the two received during the processing of cellular telephone calls.

2. _____ There are three diversity receive antennas at each cell base station.

3. _____ The combiner essentially acts as an RF mux in the downlink signal flow.

Multiple Choice

1. The concept that describes the use of two antennas to receive one base-to-mobile signal, allowing for the best signal to be chosen for processing, is known as:
   a. Time diversity
   b. Space duplexity
   c. Space diversity
   d. Duplexing
   e. None of the above

2. Which cell site network element splits the two received signals from the RX0 and RX1 antennas into an average of 15 independent, coaxial-receive outputs?
   a. The multicoupler
   b. The combiner
   c. The transducer
   d. The star connector
   e. None of the above

Capacity Management, Propagation Models, and Drive Testing

The capacity of a wireless system is a function of the number of radios (channels) operating per base station in each wireless market. There are many approaches that a wireless carrier can employ to manage system capacity. In this chapter, we review the predominant options used to address expanded capacity requirements in wireless systems. What methods any particular carrier uses can depend on any number of factors: Capacity-engineering practices could be carrier-specific, the engineering management at certain carriers could favor certain practices or technologies, some practices could be more cost-effective than others, and some practices may be easier to implement using certain digital wireless technologies. Ultimately, the type of practices or technology that a wireless carrier will use depends on the carrier and the nature of their network.

11.1   Cell Splitting

Cell splitting is the process of dividing the geographic area occupied by two existing cells into three roughly equivalent areas, so a new, *third* base station can be deployed. The new cell is installed approximately in the center of the area that is the shared border of the two existing base stations. Cell splitting is done to increase system capacity, due to increased wireless traffic in a specific geographic area of a market.

Key: Cell splits are dictated when cells in a given geographic area are consistently at 80 percent of capacity or when all other avenues of increasing capacity have been exhausted. The existing frequency-reuse pattern must be closely adhered to when doing cell splits to reduce or eliminate the potential for interference.

The power levels of all surrounding base stations affected by a split must be adjusted downward to reduce expansive coverage overlap that can cause interference. Because splitting adds coverage and capacity to a system, the intended coverage area of all nearby cells will usually shrink, and existing cell coverage areas will become smaller. Most carriers try to avoid doing cell splits due to the vast amount of work and cost associated with constructing and deploying a brand-new base station, which would cost up to an average of $250,000. Most carriers will try to use other technologies to gain more capacity from their existing cell base stations. But in the end, at the rate wireless subscribership is growing, adding new base stations to a wireless system is likely the most customer-friendly thing to do. See Figure 11-1 for a depiction of cell splits.

11.2 Overlay/Underlay

Overlay/underlay, also sometimes known as *cell layering*, describes the process of virtually building a small cell inside a larger cell by manipulating the power levels of every other radio channel in a base station. Nokia calls this technique *super cell*, and Ericsson calls it a *hierarchical cell*. Lucent refers to it by its original name, overlay/underlay, because the concept was said to be developed by Bell Labs. This is accomplished by tuning radios at neighboring cell sites so that the decibel power levels of every other channel within the cell are vastly different from each other. This means that every other channel at neighbor base stations

Figure 11-1
Cell splits. Notice that the coverage area stays approximately the same, yet there are three base stations in place after the cell split.

Coverage Area of Two Base Stations Before Cell Split

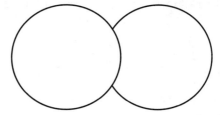

Coverage Area of Two Base Stations After Cell Split

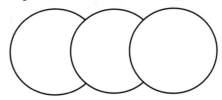

The new base station is placed in the center of the former coverage area. The original two base stations' coverage areas are reduced to accommodate the new base station and its intended coverage area.

would have vastly different power levels assigned to them. This provides for the reduction and/or elimination of adjacent channel interference between neighboring cell sites. Roughly half of the channels in the base station will have lower power assigned to them, and the other half will have much higher power assigned to them. The theory holds that adjacent channels could be assigned to the lower-power, interior channels in one base station, and the neighboring base station adjacent channels would have higher power assigned to the exterior channels. If there is good separation of frequencies between the lower-power channels and higher-power channels, a significant difference of power levels is not necessary. See Figure 11-2 for an illustration of the overlay/underlay concept.

11.3 Directed Retry

Directed retry defines the process where a given cell (cell A) reroutes its call-processing attempts to an adjacent cell (cell B) or to other nearby cells. *It can apply to both the origination of wireless calls as well as the termination of wireless calls.* In other words, it applies whether the mobile subscriber initiated the call him- or herself, or whether he or she is receiving a call. Directed retry is utilized if no channels are available to process a call in cell A. In some cases, a long list of cells could be on the

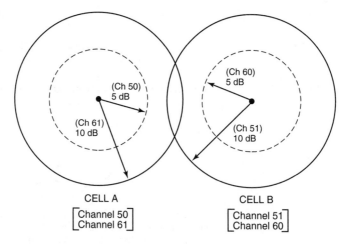

Figure 11-2
Overlay/underlay concept

(Ch 50)
5 dB

(Ch 61)
10 dB

(Ch 60)
5 dB

(Ch 51)
10 dB

CELL A

⎡Channel 50⎤
⎣Channel 61⎦

CELL B

⎡Channel 51⎤
⎣Channel 60⎦

retry list to ensure that the processing of a call is accomplished. These cells are usually also neighbor cells used for call-handoff purposes. This may account for the occurrence of long hang times (postdial delay) when wireless customers are attempting to place calls. Obviously, because calls generate revenues for the carrier and keep customers satisfied, the carriers will develop and use all means necessary to ensure that all call processing attempts reach the system.

> *Key:* In order for any cell or cells to support the directed retry process, the control channels at these cells must be able to detect the call-processing attempt, whether mobile-originated or mobile-terminating. The control channel at all the cells on a directed retry help list must be able to receive the signal from the subscriber's wireless phone.

The retry process is coordinated by the MSC and can be activated and deactivated on a cell-by-cell basis. The directed retry feature may be only a temporary measure to allow the customer to access the system until the capacity of cells in the affected area can be increased. This is why it is also known sometimes as *congestion relief*. See Figure 11-3.

Figure 11-3
Directed retry.
The mobile's
signal must be
"seen," or
detected, by all
adjacent base
stations in
order for
directed retry
to be feasible.

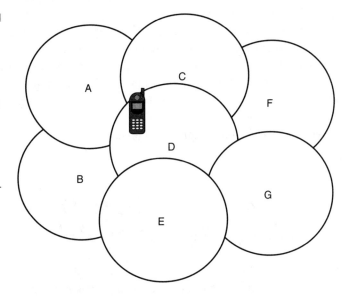

11.4   Propagation Modeling Tools

Software tools are available that wireless carriers use for predicting RF propagation. They use highly complex mathematical formulas, developed over years of field testing, to *try* to predict what signal levels will appear to be in any given terrain situation. These tools use many factors as input to the software program: RF power levels, antenna types (omni, directional, or downtilt), antenna gain, antenna F-B ratio, antenna height, and the gauge and type of coaxial cable, among other things.

The output of RF propagation models is usually a predictive graphical RF propagation that is merged (overlaid) with a mapping software program to show predicted RF coverage in a specific geographic area of a wireless market. The end result is a two-dimensional map of a county or a state, with multiple colors shown on the map to indicate varying coverage levels. These programs include depictions of towns, interstate highways, and secondary roads.

The real world will not fit neatly into any propagation models because the real world is constantly changing: Trees are planted, torn down, and lose their leaves for 5 months of the year in certain parts of the country; buildings are torn down or erected, trucks drive by, and highways are built. All these things—anything that sticks up above the ground, in fact —will affect wireless RF coverage. No software model could possibly keep up with changes in terrain features or predict RF propagation to 100 percent accuracy in a constantly changing world. There is simply insufficient time to constantly implement updated information into propagation models. Predictive modeling software programs can have only a given percentage of accuracy, depending on how many factors of terrain and clutter are taken into account when the programs are executed. Real-world printouts of propagation maps depict coverage levels in a tiered, hierarchical manner based on color.

11.4.1   The Hata-Okumura Propagation Model

The Hata model is an empirical model derived from the technical report made by an engineer named Okumura, so the results could be used in a software program. The Hata-Okumura propagation model is used for large, high-power, macrocell base stations that are located on towers, rooftops, and water tanks, among other things. The target parameters used for a base station design using the Hata-Okumura model are as follows:

- Range (the cell radius, or distance from the BTS itself to the outer edge of the base station coverage area, can be 1 to 12 miles, or up to 20 kilometers)

- Height of the mobile phone: 1 to 10 meters, or up to 30 feet

- Height of the base station antennas: 90 to 600 feet

- Carrier frequency (base station transmit frequency): 150 MHz to 1,500 MHz, or 1.5 GHz

    Therefore, the Hata model should *not* be employed when trying to predict path loss (RF propagation) of less than 1 kilometer from the cell site, or if the site is less than 90 feet in height.

11.4.2   The Cost-231 Walfisch/Ikegami Propagation Model

The Cost-231 Walfisch/Ikegami propagation model is mainly used to model RF propagation for a microcell or base stations that cover less than 1 mile of geography (radius is less than one-half mile). This model was not being used in a widespread fashion until the early to mid-1990s. The Cost 231 model was developed for use in GSM systems in Europe but is now also used in many wireless propagation models employed throughout the world.

   The Cost 231 propagation model is used for estimating the path loss in an urban environment for cellular communication. The Cost 231 model is a combination of empirical and deterministic modeling used to estimate the RF propagation in an urban environment in the frequency range of 850 MHz to 2,000 MHz (20 GHz).

11.5   Drive Tests

Drive tests are tests conducted through a wireless market usually twice per year, or whenever a new cell base station is deployed. The tests are performed when foliage is present (warmer months) and when foliage is not present (winter months) in climates that experience real winters.

Drive tests entail wireless carrier technicians (or engineers) *driving* a market to confirm RF power levels and call-handoff criteria. The information obtained from a drive test is used to tweak or retune base station power output, for one or more base stations in the market. It may even result in major reconfiguration of a particular base station, up to and including replacing one antenna type with another antenna type (i.e., higher gain). As the vehicle progresses through designated coverage areas from one cell to the next, wireless carrier employees use equipment such as spectrum analyzers to test coverage parameters in the market. RF signal measurements that determine when call-handoffs are dictated are one of the things that are recorded and turned in to RF engineers for analysis. Like propagation models, drive tests are limited in their effectiveness because they capture RF propagation characteristics in a moment of time. Drive tests are more accurate than propagation models but still not the complete answer for predicting how RF signals will propagate at any given time. This is because drive tests give an engineer only a snapshot in time of how RF is propagating from a specific point. It is a physical trait of RF that, at one spot, signal levels could vary greatly at any given moment. After powering up a wireless phone, move the phone around or just keep looking at the signal level indicator. The signal strength will constantly fluctuate. This is the type of dynamic physics of RF that drive tests attempt to quantify.

How often drive tests are performed depends on the ownership of the market in question. Many carriers will only perform drive tests when they do a major retune of a market. But usually wireless markets will be tested around two times per year.

█████ Test Questions

True or False?

1._____ Cell splits are dictated when cells in a given area are consistently at 30 percent of capacity, and all other avenues of increasing capacity have been attempted.

2. _____ Drive tests give a cellular engineer only a snapshot in time of how RF is propagating from a specific point.

3. _____ The real world fits easily into propagation models because the real world rarely changes.

4. _____ Drive tests are more accurate than a propagation model.

Multiple Choice

1. Directed retry

   a.   Defines the process of creating a smaller cell inside a larger cell by manipulating the power levels of certain radio channels

   b.   Defines the process of routing call origination attempts to nearby cells

   c.   Defines the process of subdividing an omni cell into three equal subcells

   d.   Defines the process of creating a new cell by splitting the distance between the centers of two existing (or planned) cells into two or more cells

2. What technique describes the process of (figuratively) building a small cell inside a larger cell by manipulating RF power levels of alternate channels at neighboring cell sites?

   a.   Triplexing

   b.   Franchising

   c.   Overlay/underlay

   d.   Directed retry

12

The Mobile Switching Center, the Network Operations Center, and the Backhaul Network

▓▓▓ 12.1   Overview

The mobile switching center (MSC) is responsible for connecting wireless calls together by switching digital voice data packets from one network path to another—a process known as *call routing*. MSCs also provide additional information to support mobile service subscribers, including user registration, authentication, and location updating. The MSC is the location of the switch and peripheral equipment that serves a wireless system. The switch itself is similar in function to a class 5 end-office switch in the PSTN. Major manufacturers of MSCs are Lucent Technologies, Nortel Networks, Seimens, Alcatel, Ericsson, and Nokia.

> *Key:* All cell base stations in a wireless network are, and must be, electronically connected to the MSC.

The primary purpose of the MSC is to provide a voice path connection between a mobile phone and a landline telephone, between two mobile phones, or between a mobile phone and the Internet. The MSC serves as the nerve center of any wireless system. Functionally, the MSC is supposed to appear as a seamless extension of the PSTN from the customer's perspective. The MSC is composed of a number of computer elements that control switching functions, call processing, channel/frequency assignments, data interfaces, and user databases. Wireless switches are some of the most sophisticated switches in the world today.

▓▓▓ 12.2   The Base Station Controller (BSC)

An integral element of all MSCs today is the existence of the base station controller (BSC). The BSC is a network element used in most wireless network implementations that is used to interconnect cell base stations to the MSC. The BSC facilitates call handoffs from one base station to another. Radio stations themselves do not have much built-in intelligence to manage and control the network, so they are connected to these controllers. In GSM systems these controllers are called BSCs, and in 3G/UMTS systems they are called radio network controllers (RNCs). The controllers manage the radio network and are responsible for call setup and call-handoffs.

> *Key:* BSCs are not always deployed or actually required elements in a wireless network; however, the GSM standard defines them as a required network element. Also, in those cases where the mobile switch itself is manufactured by one company and the base station equipment is manufactured by a different company, a BSC is a necessity to bridge the different models of equipment together. Even though they are not absolutely required in some systems, the current demands of mobile networks dictate that they *should* be used to offload management of base station activity from the MSC. The MSC then can attend to other critical tasks such as the actual switching of calls, managing interactions with the HLR/VLR, seizing PSTN trunks, and managing Internet access transport and related traffic.

BSCs essentially serve as front-end processors to the switches themselves. They can be located at the MSC site itself, or they could also be geographically distributed throughout a wireless system, or both. This depends on the actual number of base stations existing in the network, real estate owned by the carrier, and so forth. The purpose of the BSC is to offload some of the base station and call-management functions from the switch, so the switch can perform more switching-related and database-related functions. See Figure 12-1 for a depiction of a high-level MSC topology.

12.3   MSC High-Level Functions

High-level functions of the MSC and the BSC include the following (many of these functions require the operation of network peripherals, as noted here):

- *Switching of Mobile Calls and Least-Cost Routing.* The switch located at the MSC site *switches* all wireless system calls. The BSC does call setup and teardown. But the switch manages channel/-frequency assignment and call path assignment up to and including interconnection to the PSTN if the destination of the call is a landline telephone. In the case of calls that terminate to a landline phone, the MSC coordinates the seizure of trunks to the PSTN and

Figure 12-1
High-level MSC
topology

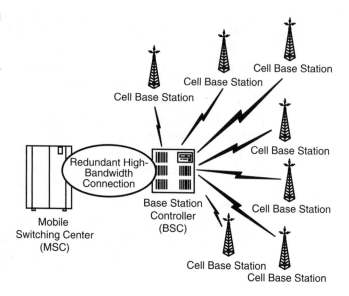

selects the least-cost route. Call processing by the MSC also includes mobile-terminated calls, where the MSC manages the paging function to locate mobile subscribers.

- *Tracking.* The MSC, through the BSC, tracks all wireless users in its system through a process known as *autonomous mobile registration* and constantly monitors the technical health of all cell base stations connected to it. If there is an equipment problem within a base station, the BSC will be notified through an alarm process. The BSC will then forward the alarm information directly to the network operations center (NOC) for processing and action. This transmission of alarms will usually take place over a carrier's own internal wide area network (WAN).

- *Paging.* The paging function is the means by which the MSC locates mobile subscribers for purposes of terminating calls to a mobile phone that were either originated by a landline phone or another mobile phone. For example, the MSC will attempt to locate the mobile by initiating a paging signal in the last 10 cell sites that the wireless user's signal was registered. If the user is not located in those 10 cells, the MSC may initiate a regional page to locate the mobile (e.g., 50 cell sites). If the user has still not been located, the MSC may initiate a global page, where it sends a paging signal to all base stations in its system (e.g., 200 to 300 cell sites). The details of this process are very carrier and equipment specific and would take only seconds to occur.

- *Call-handoff.* Call-handoff is the process where the BSC coordinates the transfer of a call in progress from one cell to another cell. Though today the handoff process is managed by BSCs in most cases, the MSC is still involved to some extent. Also, as noted in Chapter 6, the mobile phone itself can also play a key part in the call-handoff process, depending on the technology in use in the wireless system.

- *PSTN Interconnection.* The MSC manages the control of traffic between the wireless network and the PSTN. This includes the seizure of PSTN trunks and trunk group selection (i.e., least-cost routing).

- *Internet Connectivity.* The MSC manages connectivity to and *from* the Internet. This involves connections between the MSC and dedicated web servers based at the MSC location and the pass-through processing of Internet sessions between mobile subscribers and the Internet.

- *Billing.* Customer billing is controlled by the MSC. Upon the completion of every wireless call, the MSC generates an automatic message accounting (AMA) record, which contains the details of the call (e.g., customer mobile number or account number, the number of minutes the call lasted, and whether it was local, roaming, or long distance). The MSC will compile these records into batch files that are stored in a server collocated at the MSC site, and the batch files are then downloaded nightly to the actual billing system over the wireless carrier's own internal WAN.

- *Validation of Subscribers.* Through MSC databases known as the home location register (HLR) and visitor location register (VLR), the switch tracks which mobiles registering on the system are *home* subscribers and which subscribers are *visiting*, or roaming, subscribers. Home subscribers are those wireless customers who obtained service in a particular market and are making a call(s) in that same market. Visiting subscribers are those wireless customers who obtained service in another market and are making calls in a city that is not the town where they originally obtained their wireless service. Once the MSC has determined that a roamer is resident in its system and making calls, it knows to direct the billing information to a special data file, where ultimately the billing information is routed to a roaming clearinghouse (a third-party company). The roamer will then be charged accordingly on their monthly bill, according to the roaming rates that were agreed upon between the two carriers. For more information on this process and its details, see Chapter 15.

■ *Authentication of Subscribers.* This is a challenge-response, antifraud process where the mobile issues a secret code when attempting to place calls, and the MSC holds the key to this code. If the calculation that is inherent in the authentication process is validated by the MSC, the mobile is allowed to make calls. The authentication standard went into effect in the mid-1990s and has played a large part in eliminating technical wireless fraud. Today, fraud still occurs, but it is largely contained to only subscriber fraud. Subscriber fraud is defined as the type of fraud that occurs when dishonest people sign up for wireless service under false identities or pretenses, and they obviously have no intention of ever paying for the service. Authentication works hand in hand with the validation process described just previously and could take place in an MSC environment via the use of AAA servers. An AAA server is a server program that handles user requests for access to computer resources and, for an enterprise, provides authentication, authorization, and accounting (AAA) services. The AAA server typically interacts with network access and gateway servers and with databases and directories containing user information. The current standard by which devices or applications communicate with an AAA server is the Remote Authentication Dial-In User Service (RADIUS).

■ *Network Operations Statistics.* The MSC collects data on all system and call-processing functions, to include the following:

   ■ Traffic statistics on base station trunk groups (e.g., busy hour data)

   ■ Trunk groups that connect the wireless system to the PSTN

   ■ Call-handoff statistics

   ■ Amount of dropped calls (per BTS)

Maintenance statistics such as the amount and type of alarms are generated at all base stations and the MSC itself.

12.4   MSC Adjunct Processes and Network Peripherals

The MSC site also houses many adjunct functions and peripherals that reflect the multifaceted and growing applications that are supported by wireless carriers. Many of these peripherals are shown in Figure 12-2 and are as follows:.

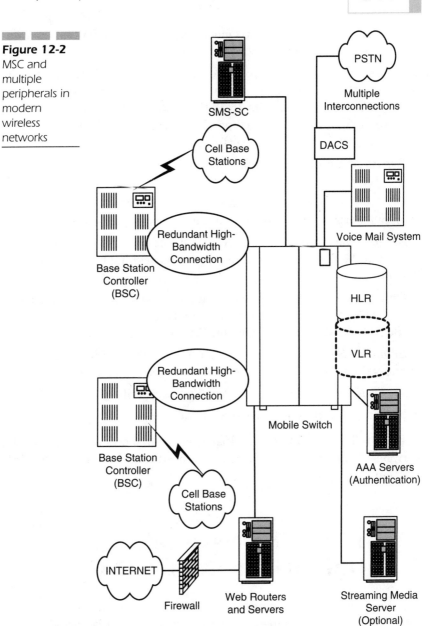

Figure 12-2
MSC and
multiple
peripherals in
modern
wireless
networks

- *Voice mail systems* are similar to any other typical voice mail system but likely much larger. Mobile subscribers' voice mail messages are stored and retrieved from these systems.

- *Short Message Service—Service Center* (SMS-SC) is an MSC adjunct server whose function is to manage the assimilation and transport of short messages originated by mobile subscribers per SMS standards (SMS will be covered in detail in Chapter 16).

- *Streaming media servers* could be located at the MSC site as well, to support streaming transport of music, news, or video services that mobile users may subscribe to. With the impending popularity of these types of services, this type of adjunct server may well become the norm at MSCs in the near future.

- *Web servers, routers, and firewalls* required to support Internet connectivity.

- *Digital Access Cross Connect Systems* (DACS), used to manage, groom, cross-connect, and optimize all trunking to the PSTN.

- *Asynchronous transfer mode* (ATM) *switches*, if ATM technology is used in the fixed network to manage and backhaul base station traffic.

- *Optical equipment* used to terminate optical-based interconnections to the PSTN, if applicable—that is, Synchronous Optical Network (SONET) multiplexers or Gigabit Ethernet.

- *Signaling System 7* (SS7) equipment/nodes. This equipment will many times be incorporated directly into the switch itself.

- *Authentication (AAA) servers* to validate subscribers.

- Servers to support locator-based services.

Refer to Figure 12-2 for an illustration that shows how many of the peripherals noted previously could be used (and housed) at an MSC location.

12.5   The Network Operations Center (NOC)

All telecommunication companies and network providers of all types maintain a Network Operations Center (NOC). This includes local exchange carriers (LECs), competitive LECs (CLECs), Internet service providers (ISPs), interexchange carriers (IXCs), voice over IP (VoIP) carriers, and so forth. Like all of these telecommunications companies, all wireless carriers maintain some type of NOC as well. NOCs are usually staffed 24 hours a day, 7 days a week, 365 days per year. The network data that is routed to a NOC by all MSCs is managed by a network management system such as HP Openview or a similar system that may even be proprietary to a wireless carrier based on the maker of its network infrastructure. The data used to provide network status at a NOC will be housed on an independent, fault-tolerant computer system that would operate on its own virtual local area network (VLAN). Today's NOCs are housed in large rooms, where the status of all network systems is projected onto a large, overhead screen. This screen will show the topology of all networks that the NOC is managing, in most cases in the form of a U.S. or world map, where the status of all nodes is shown as either green (operational) or red (alarm state). NOCs also have a TV in place with the Weather Channel on at all times, to monitor weather across the country because severe weather could have a major impact on system integrity in the carrier's market areas. NOCs will also have a TV with CNN on at all times, to monitor social events that may have an impact on system operations as well (e.g., earthquakes, social unrest). In the wireless world, all MSCs are connected back to the NOC for purposes of monitoring the technical health of a carrier's system. This monitoring functions in a tiered manner, going all the way down to the base station level: All base stations are connected to an MSC, and all MSCs are connected to the NOC. See Figure 12-3 for an illustration of the hierarchical nature of the NOC function in a wireless system.

The NOC performs all of the core network-management functions of any standard network: fault management (system alarms), configuration management, security management, accounting management (traffic data), and performance management. This will include alarms that are generated by equipment failures at any BTS or MSC. It also includes alarms generated by physical security breaches, such as a break-in at a base station shelter or an MSC location. A wireless carrier, like other telephone companies, can have one or more NOCs, depending on its size. If more than one NOC exists, they will be regionalized. They will be geographically distant but deployed symmetrically so they can support entire regions of the United States. In cases where carriers have multiple NOCs, the NOCs themselves will be connected together for redundancy purposes, to act as backups for each other in the event an entire NOC goes down for some reason. NOCs can also house large data centers where the wireless carrier's billing system resides. The data centers use completely redundant information systems or databases known as system *clusters*, where a mirrored copy of all billing data within the system is copied in real time, or at designated intervals, from one server or system to another.

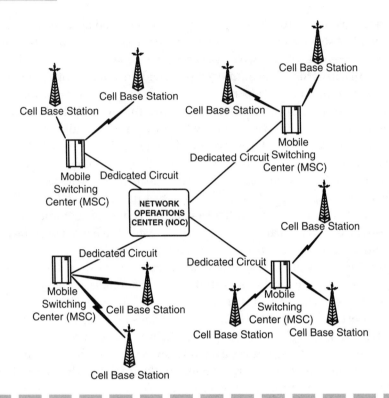

Figure 12-3
NOC monitoring topology. The NOC can monitor the entire wireless network all the way down to equipment in the base station due to the hierarchical nature of the monitoring topology.

12.6   The Backhaul Network: Cell-to-MSC Connections

All base stations in a wireless system must be electronically connected to the MSC. This is why a backhaul network, also known as a fixed network, is required in all wireless systems. Because of the need for call-handoff, the wireless system also requires that all cells must be able to communicate with all other cells, electronically (logically), through the MSC. This is how the MSC coordinates signal strength measurements that are used in the call-handoff process to determine when handoffs are required. The terms "fixed network" and "backhaul network" are synonymous and interchangeable. All cells must always be able to signal all other cells, and this is another reason why the hex metaphor was chosen as the ideal design and management tool for wireless systems. The hexagon design accounts for conceptual and physical overlapping between all cells because wireless coverage is depicted by circles from an engineering standpoint. The fixed network has to satisfy capacity demands and provide reliable service. This can be achieved if the following criteria are incorporated in the network's design:

- All routes and links between base stations and the BSC/MSC must be properly sized to meet traffic demands. Forecasted growth should be accounted for and incorporated into network build-out plans.

- Mobile-originated traffic must be routed in the most economical manner through the network, and on to the PSTN (if applicable).

- Survivability must be built into the network wherever possible.

It is critical that a transmission plan is established indicating current and potential future cell sites and that this plan is updated on a regular basis.

Key: The backhaul network can consist of connections based on leased facilities such as copper cable or fiber optics. The backhaul network may also consist of microwave radio links or a combination of any of these options.

All fixed network connectivity, regardless of the actual medium used (e.g., microwave or leased line), is accomplished via DS-n or OC-n transmission systems.

12.7   The Packet Transport Option in the Backhaul Network

The time is coming when some, or all, fixed network transmissions will be all TCP/IP-based, per the use of VoIP in a wireless network infrastructure. It is important to note that this cannot and will not occur until wireless IP handsets are developed, launched, and mass marketed. But that day is not far off (possibly 2008), considering the rapid pace at which VoIP is gaining acceptance in both the business and consumer worlds. If VoIP is used in the wireless network, these calls could still be carried over a DS-n (or OC-n) transmission system. Using G.711 codec compression, roughly 15 VoIP calls can be transported over each DS1 circuit in a backhaul network when frame relay is the transport system in use.[1] G.711 is the international standard for encoding telephone audio on a 64 Kbps channel. It is a pulse code modulation (PCM) scheme operating at an 8 KHz sample rate, with 8 bits per sample. According to the Nyquist theorem, which states that a signal must be sampled at twice its highest frequency component, G.711 can encode frequencies between 0 and 4 KHz. Telcos can select between two different variants of G.711: A-law and mu-law. A-law is the standard for international circuits.

VoIP technology could be used in the core of the backhaul network fairly easily today. What will be more challenging, but not impossible, is to eventually see VoIP transmissions occur *through the core network*, then all the way to the mobile handset. Although this is not impossible today, it is complicated by the fact that wireless carriers and handset manufacturers have enough on their hands trying to ensure that their networks and handsets can operate in multimode, multiband ways. What also complicates the issue is that more and more functionality is also being packed into today's cell phones, such as speakerphone functions, camera functions, personal digital assistant (PDA) functions, Windows functionality, and so forth. To also try inserting VoIP functionality is feasible but not a priority yet for the handset manufacturers. Today, wireless carriers are using ATM in the backhaul network in some cases, especially for intermarket switching. When ATM is used in a wireless fixed network, one real-time constant bit rate (rtCBR) permanent virtual circuit (PVC) will be required for each call.

Key: The process of transporting traffic across a wireless network back to the MSC for switching is known as backhauling. Historically, each wireless call has occupied one DS0 of bandwidth over the fixed network when traditional TDM-based transport is used such as DS-n or OC-n systems. Today, commonly used transcoding technologies now compress two, four, eight, or more calls per DS0. As carriers migrate toward TCP/IP-based transport and transcoding technologies, datagrams over the fixed network, will be spread across an entire DS1 of bandwidth.

---

[1] Verizon Wireless and Sprint PCS are said to have used VoIP in their efforts to simulate Nextel's push-to-talk feature, which did not take off in the marketplace.

▓▓▓ 12.8   Transmission Media

A number of transmission media are available for use by wireless carriers when implementing fixed networks. The most common media used are copper cable, optical fiber, microwave radio systems, and on rare occasions coaxial cable. The two prevailing options are copper cable and optical fiber when leasing facilities from an incumbent telephone company or CLEC.

> *Key:* When leasing DSn or OC-n facilities, the underlying transport media used is always defined by the telco, based on what is available in the area where leased circuits are requested by a wireless carrier. Usually, the only time optical fiber would be used is to support SONET (OC-n) installations or gigabit Ethernet installations. The exception to this case is, of course, if a wireless carrier specifically orders SONET facilities, which can be delivered only over optical fiber. The bottom line is that the type of media the telco uses to support a circuit order is driven by the nature of the service ordered. In other words, in an urban area, were a wireless carrier to order a major SONET ring to support their backhaul network, optical fiber would be required because it is part of the Synchronous Optical Network (SONET) standard.

Transmission media options may be limited due to the availability of facilities. For instance, in rural areas, it is very unlikely that optical fiber would be available. This is because, due to the thin density of population and business, it would be difficult or impossible for any landline carrier to recoup its investment in optical fiber. It should be noted that about 75 percent of the cost to install fiber or copper cable is labor charges (primarily trenching costs). In some rural areas, especially in the northwest United States, copper cable may not even be available from the local telco. Optical fiber facilities are ubiquitous in metropolitan areas of the United States. Along with SONET *ring* services, point-to-point OC-3, OC-6, and OC-12 circuits are also widely available in metro areas. When responding to major requests for proposal (RFPs), some telcos and CLECs may be willing to build out fiber systems, or SONET rings, on a contract basis in exchange for a long-term contract commitment and/or a committed amount of switched traffic over the rings over the period of the contract. The switched traffic relates to interconnection to the PSTN. Fixed network-related circuit purchases can be intertwined with interconnection issues when appropriate and agreed upon by both parties. The main disadvantage to leasing transmission facilities from local and competitive telephone companies is that a wireless carrier is completely dependent on the telephone company in terms of network reliability. This includes response time to outages (mean time to repair, or MTTR). This disadvantage can be minimized if the telco facilities have built-in redundancy features, such as SONET rings.

Microwave radio systems are also an option for use in the backhaul network. This option will be reviewed at length in the next chapter.

12.9   Network Configuration Options

Several methods exist by which base stations can be connected to the MSC, similar to the configuration options that exist for WANs. These options are a star formation, a ring formation, or a daisy-chain formation. The fixed network topology that is implemented may depend on the following:

- The nature of the wireless market and existing infrastructure. In other words, whether the market is in an urban area or a rural area.

- The network design and finance policies of the wireless carrier. Do they wish to spend capital on microwave radio systems that can be capitalized (and amortized)? Or do they wish to mainly use leased facilities, which are expense items?

- The facilities available from a telco. When operating in a rural area, for example, facilities options may be very narrow in scope, thus limiting or dictating fixed network options.

All interconnections to the PSTN must *also* ultimately connect into the MSC. These interconnections can traverse the same fixed network that the cell base stations use to connect to the MSC, if the interconnection to the PSTN occurs from a base station instead of the MSC. This may occur when a base station is closer to a telco central office than an MSC and would make business sense considering the telcos charge per airline mile for circuits used to connect the wireless network to the wireline network (PSTN). In these cases, it is important to ensure that an amount of bandwidth that is equal to the interconnection itself is reserved on the wireless carrier's fixed network, all the way back to the MSC. For example, if a DS3 interconnection is obtained and connects to Base Station X from a telco central office, a minimum of a DS3 of bandwidth from Base Station X back to the MSC must be reserved to support the interconnection traffic.

12.9.1   Star Configuration

Network topologies that use only star formations are rare today, in any type of network, because they do not allow for transport diversity. When star topologies are used, it is usually in microwave networks, where the star portion subtends a hubbed portion that contains path diversity between the major hub nodes. In effect, this topology is a combined star-ring design (see Chapter 13). Star formations that connect all cells to the MSC are almost unheard of, because of the potentially high cost of independently connecting every single cell base station to the MSC. This is especially true for rural cellular markets, where the geographic expanse of the market is much larger than that of an MSA. Economically, it would not make sense to have many long, stand-alone network links in place from each cell site to the MSC in a rural market. It might make more sense to install another MSC in the market. See Figure 12-4 for an illustration of a star topology.

12.9.2   Ring Configuration

When referring to ring configurations in the wireless fixed network, the rings can be implemented in one of three ways:

- Using SONET rings via facilities leased from either the incumbent telco (ILEC) or a competitive carrier (CLEC).
- Deploying SONET via microwave radio networks, which are usually designed, built, owned, and maintained by the wireless carriers themselves. This is possible because microwave systems have SONET characteristics that mirror those of optical fiber transport systems.
- Some wireless carriers deploy and manage their own SONET networks by leasing dark fiber from any number of wholesale dark fiber providers and directly purchasing the optical multiplexing nodes themselves. This option must be weighed against the first option in the preceding list by means of a business case: By

Figure 12-4
Fixed network
star topology

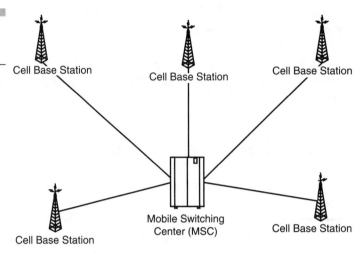

purchasing and operating the SONET systems themselves, wireless
carriers would also have to account for compensation and benefits
for personnel needed to manage the SONET network. This function
could be outsourced by the wireless carrier, but nonetheless it would
impact whether or not this option would economically be a
worthwhile pursuit compared to leasing the facilities from another
carrier as described in the first bulleted item. SONET systems will
have transmission capacities that can range from OC-12, OC-24,
OC-36, OC-48, and even OC-192 (remember that each OC number
equates to the equivalent of a DS3 of bandwidth). See Figure 12-5
for an illustration of a bidirectional line-switched ring (BLSR)
SONET system, used in a wireless fixed network.

BLSR SONET systems are deployed in oval ring formations to achieve
diversity through what is known as *path switching*. This means that the
system always has two diversely routed transmission paths available to
it, and it is constantly checking these paths to ensure that traffic is
routed over the path with the best overall signal. If (voice) traffic is
routed over path A, and path A's signal degrades, then traffic will be
routed in the reverse direction over path B; it will be *path switched*. Obviously, the most glaring example of signal degradation on any given path
is a complete fiber cut, also known in the industry as *backhoe fade*
because the use of backhoes is what accounts for major fiber-optic cable
cuts in many cases. This redundant capability of SONET rings is also
known as fault tolerance, or survivability. The SONET ring option is
widely available in urban areas, so it is therefore best suited as an option
in an urban wireless market. It is not widely available in most rural

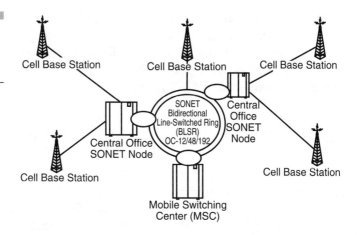

Figure 12-5
Fixed network ring topology using SONET

areas because the demand does not exist to make the economic case for telcos or CLECs to build SONET systems in a rural area.

> *Key:* When wireless carriers do employ the SONET ring option in the fixed network, the MSC itself is implemented as a node on the ring, as are multiple ILEC or CLEC central offices and/or wire centers.

When wireless carriers use SONET rings in their fixed networks, it is important to note that although the ring itself is fault tolerant (survivable), the links from each base station to the entry point of the ring (telco or CLEC central offices or wire centers) are in most cases not survivable. This is what is known as the *single point of failure* in a (fixed) network. Yet this still means that in the event of a major failure of leased facilities on the ring itself or in any base station's connection into the ring, only one base station connection would be lost instead of many base station connections. This was shown in Figure 12-5. The microwave radio option in the fixed network is used more often in rural environments for the following reasons:

- It is more likely that there is a possibility the telcos (or CLECs) would not have (copper) facilities in some rural areas.
- The LOS required to support microwave radio is easier to achieve in flat, rural areas.

Zoning in rural areas makes it easier to erect tall towers. Towers will be taller in rural areas because base station coverage areas are wider and more expansive. Taller towers equate to potentially longer microwave shots, which meshes well with the concept of a larger coverage area per base station and longer distances between base stations. Microwave radio systems can also be designed and deployed in ring configurations, similar to SONET rings.

12.9.3   Daisy-Chain Configuration

In larger, rural wireless markets, a daisy-chain architecture may be a good deployment option because of the large expanse of land that must be traversed for cells to connect to the MSC. In a daisy-chain fixed network architecture, all cells throughout a wireless market connect to each other as they home in toward the MSC. See Figure 12-6.

Key: In a daisy-chain configuration, the cumulative amount of network capacity (i.e., trunks or bandwidth) on any link between two cells gradually becomes greater as the links move toward the MSC. This is due to the incremental effect of adding more and more base-station traffic to the network as it approaches the MSC. For instance, the link with the largest capacity (and therefore the largest overall amount of trunks/bandwidth) will be the last link in the network chain between the final cell in the chain and the MSC itself.

A stand-alone link that runs off the main backbone of a daisy-chain network is known as a *spur*, as shown in Figure 12-6. It is critical to plan for growth when the daisy-chain method is used for the fixed network. If planning is not done intelligently—with an eye toward expansion—the capacity on any one of the cell-to-cell links could be maximized as base stations are added in any given market area. As the wireless network grows, depending on where new cell sites are deployed, lack of good planning could require a major reconstruction of entire sections of the daisy chain, or even the whole chain itself. A common approach is to use DS3 circuits for the core transport of the network, with DS1 circuits used to feed sites further downstream. Another option would be for the wireless carrier to add another strategically placed MSC at a point toward the distant (opposite) end of the main backbone of the daisy chain. One nationwide 850 MHz carrier with a large base of rural markets uses microwave radio links, laid out in a daisy-chain fashion, to connect base stations to

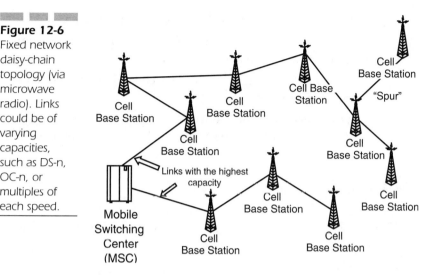

Figure 12-6
Fixed network daisy-chain topology (via microwave radio). Links could be of varying capacities, such as DS-n, OC-n, or multiples of each speed.

the MSC in most of its markets. The daisy-chain architecture has been very effective and optimizes the carrier's network infrastructure.

> *Key:* The daisy-chain and ring configurations also allow a wireless carrier to attain economies of scale and network optimization. No need to maintain duplicate networks exists, where one is for fixed network cell-to-MSC connectivity and one is for interconnection connectivity (MSC-to-PSTN). Both cell base station radio trunks and PSTN interconnection trunks can traverse the same network in a wireless system if necessary.

Test Questions

True or False?

1. _____ The monitoring of the wireless network by the NOC is accomplished via a hierarchical method of monitoring, from the NOC all the way down to the base station level.

2. _____ Interconnection to the PSTN includes the performance of least-cost routing by the MSC.

3. _____ All base stations in a wireless network do not have to be electronically connected to the MSC.

4._____ All BSCs must be physically located at the MSC site.

5._____ A stand-alone link that runs off the main backbone of a daisy-chain network is known as a hatch.

6._____ All cell base stations must be electronically connected to the MSC in a wireless network.

7._____ The wireless backhaul network cannot be used to transport both cellular radio traffic as well as interconnection traffic.

Multiple Choice

1. Which network element manages the connection of all base stations to the MSC?
   a. The DACs
   b. The NOC
   c. The BSC
   d. The AAA server
   e. None of the above

2. Which network element manages the grooming and cross connect of interconnections to the PSTN?
   a. The NOC
   b. The DACS
   c. The BSC
   d. Web servers

3. The process of transporting cellular system traffic across the fixed network back to the MSC for switching is known as:
   a. Demultiplexing
   b. Downlinking
   c. Cell relay
   d. Backhauling
   e. None of the above

4. Which transmission system listed below is not a viable option for use in a wireless backhaul network?
   a. Copper
   b. Optical fiber
   c. Microwave radio
   d. Satellite

CHAPTER 13

Microwave Radio Systems

13.1   Overview

Today's digital microwave radio systems provide a feasible backhaul solution for wireless transmission at distances up to 80 km (approximately 48 miles) point to point. Planning and developing a microwave system in a wireless carrier environment for the backhaul network is a dynamic and continuous process.

A typical microwave radio node—one end of an actual system, or shot —consists of three main components: an indoor mounted baseband shelf, an indoor or outdoor mounted RF transceiver, and a parabolic antenna. Each node transmits and receives information to and from the opposite node simultaneously, providing full duplex operation. The objective for any microwave system is to provide the best distortion-free and interference-free service from point A to point Z. Overall, the reliability of a microwave system depends on equipment failure rates, power failure rates, and propagation performance over any individual path.

The main advantage to deploying a large privately owned and operated fixed network using microwave radio is that a wireless carrier has ultimate control over that network in terms of the reliability of the system, system support and maintenance, and the nature of the hardware components that are purchased and deployed.

Key: A (point-to-point) microwave radio system is also referred to as a *hop* or a *shot.* This is due to the daisy-chain, linked nature of most microwave networks. Microwave radio signals are "shot" from one node (A) to the next (Z), all the way through the chain.

13.2   Microwave System Development and Design

Microwave network designers need to be mindful of many things when embarking on the design of a microwave radio system. Some items can be relatively mundane, yet common sense dictates they be incorporated into initial and ongoing network design (i.e., network documentation, site studies, and path studies). Other items are more technical in nature yet just as important (i.e., impact of Fresnel zones on network design). All these issues are explored in this section.

13.2.1   Network Documentation

At any given moment, a design engineer should be able to view the evolution of a microwave system. Therefore, a preliminary, documented network design is required in the form of a large diagram map to establish all the nodes in a network (cell base stations) that require transmission links between them. This map can become the main reference document for network planning and implementation. Standard symbology for the map should be adopted, agreed upon internally, and strictly adhered to. This symbol structure should permit illustration of the different types of elements in the network, along with the varying traffic capacities that are used and available on the different links. Once a network diagram is established, it can aid in evaluating the impact of future network growth, and forecasts for the future can be superimposed accordingly onto the network map.

Although modern digital microwave products are designed to be modular, providing minimum-disruption upgrade routes, for a number of logistical reasons it is beneficial to have an allowance for capacity growth inherent in the network from the outset. Future growth in the number of transmission links can have significant consequences for site selection, and the network diagram can be used as a vehicle to highlight these areas. A new diagram should be produced every 3 to 6 months, as it is important to look at the consequences of growth both in terms of capacity and the overall number of microwave links.

13.2.2   Network Design

The network design should be drawn to plan and illustrate the network topology that is to be adopted. Typically, two types of microwave-network topologies are used: namely, star networks and ring networks. Such topologies will contain one or more hub sites at strategic locations that serve *spurs*, or chains of subordinate sites from the centralized hub. In many cases today, the star and ring topologies will be combined to form a hybrid star-ring topology. The hub sites in these networks should be limited to serving a maximum of six or seven cell sites to maintain good network reliability. The completed star-ring network design is accomplished by implementing transmission loops, or rings, between the hub sites in the network. The advantage is that the rings can be used to provide path diversity and integrity to the network, removing the need for duplication of single links. See Figure 13-1 for a depiction of a hybrid star-ring network.

Key: Ring architectures can be successfully achieved only if the necessary routing and grooming intelligence exist at all appropriate points in the network to allow for path-protection switching to occur when necessary. Further, the capacity of each link in any one ring has to be sufficient to support all sites in the loop in the event protection switching takes place. This capability is shown in Figure 13-1, at the sites labeled *hub site*.

13.2.3   Site Selection

Once a prospective microwave node site is accepted by all functional groups (i.e., engineering, operations, real estate, regulatory), the final microwave design can be completed. This will involve the following activities:

- A final path study to confirm a line of sight (LOS) exists from A to Z
- Urban/rural area considerations
- Frequency selection—ensuring interference with other systems will be avoided
- Contacting regulators, if necessary
- A review of available frequency bands, frequency approval, and finally decisions involving weather and frequency band versus path distance considerations

Minimizing the number of required sites in the network will bring logistical benefits and control real estate investments (or site leasing costs). Therefore, it is always critical when selecting sites that no specific network element is considered in isolation. A number of specific items are important when designing microwave radio links:

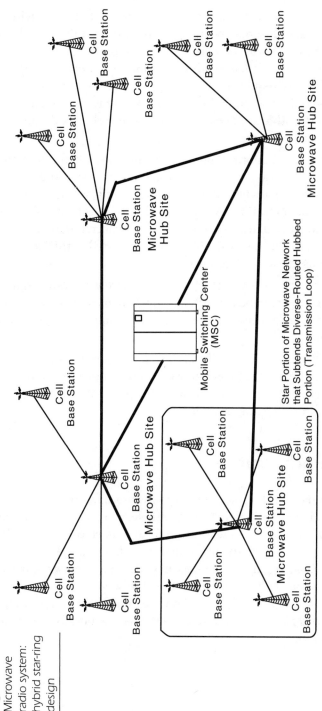

Figure 13-1
Microwave
radio system:
hybrid star-ring
design

- Good microwave sites, particularly in relation to hub sites, will be relatively high geographic points to give maximum LOS availability. Whenever possible, the cell base station equipment and the microwave system outdoor equipment (antennas and cabling) should share any required towers or poles. Likewise, indoor microwave equipment can be housed in the same equipment shelters as the other radio equipment (e.g., cellular, PCS) and should be planned accordingly.

- Microwave-related information must be included when calculating tower loading, if new tower installations are proposed, and the loading calculations must take antennas, cabling, wind, and ice loading into account. If a new microwave system is being added to an existing tower, calculations are required to ensure the incremental loading of microwave equipment can be supported within the specifications of the *existing* tower.

- If a site is in a rural or remote area, the service access should always be considered, particularly in times of inclement weather such as heavy snowfall. This is the same concept that applies to base station site selection.

- Attention should be given to future growth requirements in all areas, especially if the site is likely to develop into a future hub. It is a good practice to inform landowners of any potential future growth to prevent problems at a later date.

13.2.4   Path Studies

A microwave path is the geographic span comprising the two ends of a link, the A end and the Z end. A final path study must include propagation analysis and take into account reflective surfaces such as lakes, rivers, drainage fields, sandy areas, marshland, and large flat roofs. All these things could significantly affect how RF in a microwave path will propagate from one end of the link to the other end.

If the microwave path is in an urban area and the corresponding site can be seen with binoculars or a telescope, this is generally sufficient for LOS confirmation.

In rural areas the microwave paths are usually longer (greater than 9 miles) and use tower structures instead of buildings. It is difficult to verify the LOS if the tower is not yet built. In these instances the microwave path must be plotted on a map and an analysis performed using software tools to determine antenna heights, taking into consideration clearance and reflection criteria (see Section 13.2.4.2).

13.2.4.1 LOS Requirement LOS is fundamental to microwave radio, with a clear transmission path between the two nodes of a link, the A and Z ends. This means there can be no natural or man-made obstructions in the proposed path between the two ends of a microwave system. Establishing LOS can be accomplished in one of two ways: by creating a path profile or by surveying the actual path physically.

A path profile is established by using topographical maps and translating the contours of the map into an elevation profile of the land between the two sites in the path. Earth curvature and known objects can be added. The Fresnel zone calculation (see Section 13.2.4.2) can then be applied, and an indication of any clearance problems can become known. There are various software tools available to assist with this exercise.

A path survey can be undertaken by visiting sites and observing that the path is clear of obstruction. It is important to make note of potential *future* interruptions to the path such as tree or foliage growth, future building plans, nearby airports, subsequent flight-path traffic, and any other transient traffic considerations.

How a particular carrier establishes LOS feasibility is a matter of that carrier's standard engineering practices. It will be dependent upon factors such as link length, site locations, availability of topographical information, and availability of software tools. It is not uncommon to use both techniques—path profile and path survey—for certain links.

13.2.4.2 Fresnel Zones When transmitting from one end of a microwave path to the other, the electromagnetic signal disperses as it moves away from the source, and therefore the LOS clearance must take this dispersion into account. Particular attention should be paid to objects near the direct signal path to ensure that the required signal levels reach the receiving antennas. This is referred to as the *Fresnel zone clearance*.

There are *even*-numbered Fresnel zones and *odd*-numbered Fresnel zones. Again, they exist in layers under and over the direct microwave signal. These different Fresnel zones are determined by the respective degree of phase reversal of the indirect radio signal that occurs along the route between the transmitter and the receiver.

Odd-numbered Fresnel zones incur a half-wavelength phase reversal (180 degrees) between the transmitter and the receiver, but the indirect radio energy arrives at the receiver in phase with the direct radio signal. Therefore, odd Fresnel zones add to, or complement, the total composite radio signal at the receiver because they arrive at the receiver *in phase* with the direct signal. All odd Fresnel zones (e.g., first Fresnel, third Fresnel) are half-wavelength multiples of the direct radio beam. For example, first Fresnel is a half-wavelength phase reversal of the direct beam, and third Fresnel is a one-and-a-half-wavelength phase reversal of the direct radio beam.

Even-numbered Fresnel zones also incur a half-wavelength phase reversal (180 degrees) between the transmitter and the receiver, but the indirect radio energy arrives at the receiver out of phase with the direct radio signal. Therefore, even Fresnel zones lessen, or detract from, the total composite radio signal because they arrive at the receiver *out of phase* with the direct signal. All even Fresnel zones (e.g., second Fresnel, fourth Fresnel) are full-wavelength multiples of the direct radio beam. For example, second Fresnel would be a one full-wavelength phase reversal of the direct radio beam. Fourth Fresnel would be a two full-wavelength phase reversal of the direct radio beam, and so on. See Figure 13-2 for an illustration of microwave Fresnel zones.

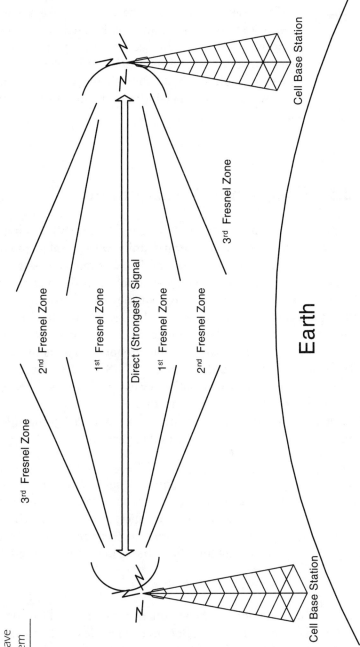

Figure 13-2
Fresnel zones
in microwave
radio system

Key: The goal in designing a microwave system is to ensure that no more than first Fresnel zone is obtained between the direct radio beam and the terrain in order to avoid unwanted signal reflections. In other words, the only reflected, indirect signal that should be allowed into the microwave propagation design is the first Fresnel zone, and no more.

13.2.5 Frequency Management

Many factors determine which frequency band will be used in a microwave system. Early microwave links were implemented using lower frequencies, such as 2 GHz and below. Frequencies were easily obtained and equipment was readily available. But these bands are now badly congested. Typical microwave frequency bands now in use in wireless backhaul networks are 8, 10, 13, 15, 18, 23, and 38 GHz, though some of these may not even be available in some countries. Wireless operators are usually assigned three or four frequency bands for use in the fixed network, the most common being 8, 15, 18, and 23 GHz. Unlicensed microwave systems are also an option for use in a wireless backhaul network and will be reviewed in Section 13.3.

Weather is an important consideration when designing microwave systems, because it can impact their efficiency.

Key: Raindrops attenuate higher frequencies, and this must be taken into account when working with frequencies of 10 GHz and higher in microwave systems. Due to the physics of RF propagation at frequencies of 10 GHz and higher, the potential for what is known as *rain fade* is more prevalent.

Along with rain fade, higher-frequency bands are also attenuated more by the atmosphere (higher free-space loss) than are lower-frequency bands. Ideally, frequency bands should be matched to a given microwave path as follows:

- Higher frequencies should be assigned to shorter paths to account for the higher degree of free-space loss.

- Lower frequencies should be assigned to longer paths. Although one disadvantage of using higher-frequency bands on longer paths is rain fade, another drawback of using lower frequencies on shorter paths is frequency congestion. For example, if 15 GHz systems are used for path distances of 1 mile, this will use up the 15 GHz frequency band quickly, making it unavailable for future path distances of 4 to 9 miles, which is better suited to 15 GHz propagation. Practical design calls for using the 18 or 23 GHz band for 1 kilometer paths and the 15 GHz band for longer paths. Table 13-1 shows the ideal path distance to frequency-band association in microwave systems.

Once the frequency band has been chosen, the proper frequency channel must be assigned to the microwave link. This should be selected so that it will not interfere with the operation of other radio systems (e.g., PCS base stations transmitting at 1.9 GHz in the United States).

With the preliminary network design in place, a clear picture is available of the different path lengths and capacities of the links required. It is then necessary to determine if this is achievable within the bounds of local regulations governing frequency availability and management. As stated previously, the propagation characteristics of electromagnetic waves dictate that the higher the frequency, the greater the free-space loss, or attenuation, due to the atmosphere. As a result, using lower-frequency bands for longer paths and higher-frequency bands for shorter paths, as shown in Table 13-1, can make more efficient use of the frequency spectrum. The majority of national frequency management administrations (e.g., the FCC) will also have some form of link-length policy in adherence with this philosophy.

Table 13-1

Microwave Frequency Band and Path Use Recommendations

Path Distance	Ideal Frequency Band
Less than 4 mi	18 or 23 GHz
4 to 12 mi	13 or 15 GHz
Greater than 12 mi	2, 8, or 10 GHz

There are six frequency bands where private microwave operators can apply for FCC licenses to implement microwave systems. These frequencies are in the common carrier bands: L, S, X, Ku, K, and Kn bands. See Figure 13-3.

13.2.6 Diversity and Protection Schemes

Microwave systems are available in unprotected and protected configurations. Several protection schemes are available, including monitored hot standby (MHSB), frequency diversity, and space diversity.

From an equipment perspective, a protected terminal provides full duplication of all active elements, such as both the RF transceiver and the baseband components. This is an example of MHSB.

Both space diversity and frequency diversity provide protection against path fading due to multipath propagation (i.e., Fresnel zones) in addition to providing protection against equipment failure. These techniques are typically required only in frequency bands below 10 GHz, specifically for long paths over flat terrain or over areas subject to atmospheric temperature inversion layers (e.g., bodies of water or high-humidity areas).

FREQUENCY BANDS

Figure 13-3

Microwave frequency bands

HF	3-30 MHZ
VHF	30-300 MHZ
UHF	300-1000 MHZ
L-BAND	1.0 - 2.0 GHZ
S-BAND	2.0 - 4.0 GHZ
C-BAND	4.0 - 8.0 GHZ
X-BAND	8.0 - 12.0 GHZ
K_U BAND	12.0 - 18.0 GHZ
K-BAND	18.0 - 27.0 GHZ
K_A BAND	27.0 - 40.0 GHZ
MILLIMETER	40 - 300 GHZ

Space diversity requires additional antennas, which must be separated *vertically* in line with engineering calculations. Frequency diversity can be achieved with one antenna per terminal, configured with a dual-pole feed. However, it should be noted that this will complicate frequency management in the long run.

Key: If any microwave link carries traffic from more than one site, it needs some sort of protection mechanism. As the network's number of cell base stations increases, a system of transmission loops should be established between major hub sites and the MSC (or MSCs) to increase survivability and reliability in the network. Transmission loops in a network infrastructure provide diverse routing, thereby increasing the transport system's reliability.

One option is to keep a redundant route available with the capacity to carry all the traffic of the loop in case the main route fails. Another option is to use the diverse route at all times so that traffic flow is split 50-50 between the two routes. This design reflects what is known as *load balancing*. If unprotected radios must be used, they should be deployed only to serve single-end sites, known as spurs. Refer to Figure 13-1 for a depiction of transmission loops in a wireless carrier backhaul network. Also, the photo shown in Figure 13-4 shows a major microwave hub site that illustrates this route diversity/protection configuration. Also note the use of horn reflector microwave antennas at the top of the tower in the photo (these antennas are covered in Section 13.5).

Microwave radio systems today employ digital carrier systems using time-division mutiplexing (TDM). There are very few (if any) analog microwave systems existing in the field today. The older analog systems used frequency-division multiplexing (FDM). Some utilities may still be using the older analog microwave systems.

13.2.7 Microwave System Capacity Options

Capacity requirements are an important consideration when designing and deploying a microwave system. Microwave radios can be configured to carry a certain amount of traffic in a specific frequency range. Capacities range from DS1 all the way up to OC-48 (48 DS3s). For example, a carrier could select a 16-DS1 capacity radio operating at 28 GHz to carry

Figure 13-4
Microwave
radio hub site
(Photo
courtesy Telus
Corporation)

a significant amount of traffic over a path distance of 3 to 5 miles. A carrier could also select a 12-DS1 capacity radio operating at 2 GHz to carry traffic over a path distance of about 23 miles.

13.2.8 Microwave System Reliability: Index of Refractivity

The reliability of the propagation of a microwave radio path is impacted by what is known as the *index of refractivity* of the microwave radio beam. Three atmospheric conditions have the most effect on the index of refractivity: barometric pressure, humidity, and temperature.

Temperature inversion is a condition that has an adverse impact on microwave radio systems and may cause the systems to become inoperative at times. About 99 percent of the time, the air closer to the ground (and closer to the microwave radio system) is warmer than the air in the lower atmosphere. When the opposite occurs and the air closer to the ground is cooler than the air high above, temperature inversion occurs. Temperature inversion distorts the microwave radio beam. As a rule, microwave beams normally bend toward the earth. When temperature inversion occurs, the beam bends *away* from the earth, causing the microwave system to become inoperative because the signal is not received at one end of the shot. Temperature inversion occurs more frequently at higher radio frequencies, especially in the 11 to 18 GHz frequency range. See Figure 13-5.

13.3   Unlicensed Microwave Systems

Unlicensed microwave systems are also an option for wireless carriers, especially when the bandwidth requirements are lower. This option is easier to implement in rural areas because much less potential exists for interference with other radio-based services. What must be remembered is that anytime an unlicensed radio technology is used, the user of that unlicensed spectrum runs the risk of either causing interference to other users of the spectrum or having his or her own system's signal integrity negatively influenced by interferors in the same unlicensed spectrum (other entities using the same spectrum).

13.3.1   The Basics

Along with all the microwave radio options that require formal licensing by the FCC via the 494 form, other unlicensed, point-to-point microwave options are available for wireless carriers. Like anything, both benefits and disadvantages exist for pursuing an unlicensed microwave radio option.

The *benefits* of using unlicensed microwave systems are as follows:

■ No Form 494 requirement exists with unlicensed systems as it does with licensed systems. This could take weeks or possibly a month or more off the timeline required to install and turn up a microwave system.

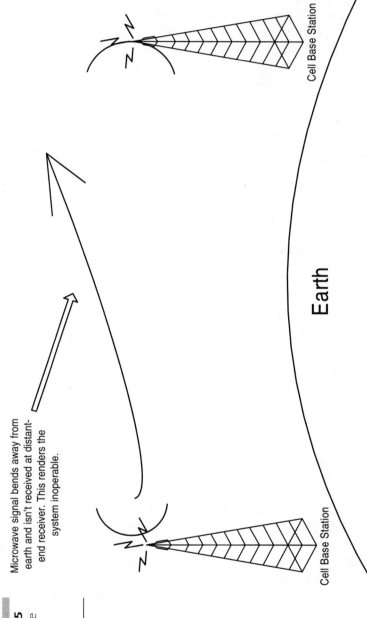

Figure 13-5
Temperature inversion in microwave systems

Microwave signal bends away from earth and isn't received at distant-end receiver. This renders the system inoperable.

■ Unlicensed systems' hardware could be less expensive to deploy than licensed systems'.

■ Unlicensed systems provide a viable alternative to leased facilities provided by incumbent or competitive carriers. The wireless carrier has sole management and control over the microwave system. If the system has problems, they are also not reliant on another party to troubleshoot and repair the system.

■ Unlicensed microwave systems provide reliability that is on par with other licensed systems (e.g., five 9's reliability, which is considered carrier class and equates to only approximately 5.3 minutes of downtime per year).

■ Unlicensed systems are more economical than licensed facilities due to more rapid payback.

■ Unlicensed systems have a considerably smaller physical footprint compared to many large, licensed, high-capacity systems.

The *disadvantages* of using unlicensed microwave systems are as follows:

■ Because these systems are sharing and using unlicensed spectrum with many other services, the possibility of interference is a reality. Unlike licensed microwave systems, frequency studies are not required (and therefore not usually conducted) with unlicensed systems. This opens up the possibility for interference by other radio-based systems that share the same unlicensed spectrum, which is usually the industrial, scientific, and medical (ISM) spectrum of 2.4 GHz in many cases. It should be noted that Wi-Fi hot spots use this same spectrum.

■ Most unlicensed microwave systems will not have some of the capabilities that licensed systems have, such as routing and grooming intelligence. Most unlicensed systems also will not have the capability to offer very high bandwidth such as many licensed systems.

13.3.2 Specifications

Many unlicensed microwave systems contain proprietary encryption capabilities to ensure transmissions stay secure. They are also available in multiple capacities, but obviously they only operate using unlicensed frequencies, such as the 5,725 MHz (5.7 GHz) to 5,850 MHz (5.8 GHz) range or the 2,400 MHz (2.40 GHz) to 2,483 MHz (2.48 GHz) range. These systems provide autosensing 10/100 BaseT, RJ-45 Ethernet interfaces for network management purposes. They are also SNMP-capable.

One of the industry leaders in this space is a company named Proxim Wireless Networks, makers of the Lynx unlicensed microwave family. Unlicensed microwave radio options provided by Proxim include the following:

- A 1 to 2 DS1 capacity system that operates in both unlicensed bands (2.4 GHz or 5.7 GHz). Distance can be less than 1 mile to more than 50 miles.

- A 4 or 8 DS1 capacity system operating at 5,725 to 5,850 MHz at distances up to 36 miles (4 DS1) or 32 miles (8 DS1).

- A 4 or 8 DS1 spread-spectrum system that operates in both unlicensed bands (2.4 GHz or 5.7 GHz). Distance can be less than 1 mile or greater than 40 miles.

- A DS3 capacity system using either 5.3 GHz or 5.8 GHz unlicensed frequencies, at distances from less than 1 mile to more than 15 miles.

Harris Corporation, a manufacturer of many microwave systems that operate in the licensed spectrum, also makes many unlicensed microwave systems as well. The Harris systems have many of the very same features of the Proxim systems, including many of the same low-capacity options.

13.4   Coax and Waveguide

There are two types of cabling used to support the installation of microwave antennas: coaxial cable or waveguide.

Coaxial cable was invented in 1929 and first used commercially in 1941. Coaxial cable, also known simply as coax, is the kind of copper cable used by cable TV companies, and sometimes also by telephone companies from their central office to the telephone poles near customers.

Coaxial cable is called *coaxial* because it includes one physical channel that carries the signal, surrounded (after a layer of insulation) by another concentric physical channel, both running along the same axis.

The outer channel serves as a ground. In microwave radio systems, coax cable is usually used for microwave radios that employ frequencies of 3 GHz or *less*. See Figure 13-6.

Waveguide is a hollow, elliptical metal cable that connects the RF equipment to the microwave antenna and is used in systems of 3 GHz or *higher*. A waveguide is an electromagnetic *feed line* used in microwave radio systems, as well as broadcasting and radar installations. A waveguide consists of a hollow rectangular or cylindrical metal tube or pipe and is used in microwave systems to support systems transmitting at frequencies of 3 GHz or *higher*. Waveguide cable is bulkier than coax cable, and its size (and sometimes its weight) must be taken into account carefully for tower-loading purposes.

The electromagnetic field in waveguide propagates lengthwise. To function properly, a waveguide must have a certain minimum diameter relative to the wavelength (frequency) of the signal being used. If the waveguide is too narrow or the frequency is too low (i.e., the wavelength is too long), the electromagnetic fields cannot propagate properly. At any frequency above the cutoff (the lowest frequency at which the waveguide is large enough), the feed line will work well, although certain operating characteristics vary depending on the number of wavelengths in the cross section of the waveguide itself. See Figure 13-7 for a photo of waveguide.

Figure 13-6
Coaxial cable (Figure courtesy www.whatis.com)

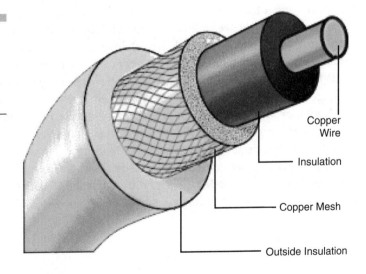

Copper Wire

Insulation

Copper Mesh

Outside Insulation

Figure 13-7
Waveguide
cable: note
elliptical shape
(Photo
courtesy
Andrew
Corporation)

13.5   Microwave Radio Antennas

Today's microwave antennas come in two predominant forms. The most common form is a type that appears as a concave dish on communication towers: the parabolic reflector.

13.5.1   Parabolic Reflectors

Parabolic reflectors are the most common type of microwave radio antennas in the common carrier band today. They appear as dishes (on towers), and the dishes come in diameters of either 6, 8, or 10 feet. This type of antenna has a feed horn mounted inside of the dish. Both the transmit and receive signals are reflected off the dish into and out of the feed horn. From the feed horn, the signals travel down coax cable or waveguide into the microwave receiver. About 90 percent of new microwave systems use this type of antenna. Figure 13-8 shows how these systems operate, and Figure 13-9 is a photo of an actual parabolic reflector.

Radomes are available in the form of either dome-shaped convex equipment or flat-shaped fiberglass covers. Their main purpose is to reduce wind resistance and thus tower loading when mounted onto parabolic reflector antennas. Radomes also protect the antenna from the elements. See Figure 13-10 for a photo of parabolic reflector microwave antennas that use both flat and convex radomes.

Figure 13-8
Parabolic
reflector
microwave
antenna
function

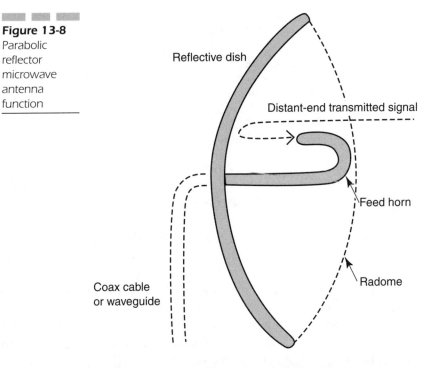

Figure 13-9
Parabolic
reflector
microwave
antenna
(Photo
courtesy
Andrew
Corporation)

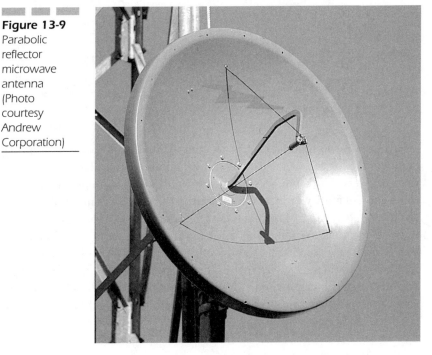

Figure 13-10
Radomes used with parabolic reflectors (Photo courtesy Andrew Corporation)

13.5.2 Horn Reflectors

Horn reflectors are microwave antennas that look similar to an inverted, angular funnel or a horn from a distance. With horn reflectors, a radio signal is beamed off a reflector inside the horn. In effect, the radio energy is funneled into and out of the *horn* with this type of antenna. This antenna type has a direct signal feed via waveguide, as opposed to the parabolic reflector antenna, where the feed horn reflects the radio energy into and out of the system via the reflection of the antenna (dish). Figure 13-11 shows a photo of horn reflectors.

13.6 Microwave Radio System Software Modeling Tools

Microwave radio engineering is as complicated and specialized as RF engineering for cell base stations. As with base station RF engineering, microwave system engineers can use special software programs to determine the feasibility of installing microwave radio systems in any given situation, no matter if LOS exists in the first place between the intended A and Z locations. When using the software programs, a microwave system engineer can change many of the parameters of a simulated microwave radio path in order to determine what factors, together, would optimize the radio shot. Some of the parameters of simulated microwave radio systems that can be changed to determine feasibility of a specific path are as follows:

Figure 13-11
Horn reflector microwave antennas (Photo courtesy Terry Michaels/AT&T)

- The microwave radio frequency
- The size of the antenna (dish)
- The height of the dish on the tower
- The diameter of the cable (if coax is being used)
- The transmitter output power
- The receiver sensitivity threshold, or how well it accepts the transmitted signal

Two of the most prevalent microwave system software modeling programs are *Rocky Mountain* and *Pathloss*.

The reliability of a microwave radio system is a company-specific issue and depends on the actual grade of service that the carrier wants to provide. Most companies engineer microwave radio systems for a reliability factor of 99.99 percent uptime per year; that is, the system will be in use and active 99.99 percent of the time.

The FCC 494 form must be filed with the FCC to obtain permission to activate a microwave radio system. The 494 form confirms the frequency assignment for the microwave system to include that a frequency study has been completed, which will ensure that no interference will exist with the proposed new system and any existing microwave systems within a specific geographic radius.

Test Questions

True or False?

1. _____ Coax cable is used in microwave systems that employ frequencies of 8 GHz or less.

2. _____ The two most popular forms of microwave antennas used today are the parabolic reflector and the horn reflector.

3. _____ For microwave systems of less than 4 miles, the ideal frequency to use is 2, 8, or 10 GHz.

4. _____ There is more rapid fading of microwave signals at higher frequencies.

Multiple Choice

1. Three atmospheric conditions have the most control over the microwave radio system index of refractivity. Which item below is *not* one of those three conditions?
   a. Temperature
   b. Sunlight
   c. Humidity
   d. Barometric pressure

2. The oval- or flat-shaped fiberglass covers used with parabolic reflector microwave antennas are known as:
   a. Wind domes
   b. Enhancers
   c. Equalizers
   d. Radomes

Interconnection to the Public Switched Telephone Network (PSTN) and the Internet

▣ 14.1   Overview

Interconnection is defined as the point where any two carriers' networks are physically linked together. It can involve a wireless carrier's connection to a landline phone company or local exchange carrier (LEC), an Internet service provider's (ISP) connection to an LEC, an LEC's connection into an interexchange carrier (IXC), or a wireless carrier's direct connection to an ISP.

About 60 percent of all wireless phone calls made today are directed and terminated to a landline telephone. This number has been steadily changing to the point where more wireless traffic is being originated from landline phones and terminating on wireless phones. Because of the large volume of land-to-wireless traffic, it is imperative that a wireless carrier develop and maintain a cost-effective series of interconnections to the public switched telephone network (PSTN), or landline network.

 Key: The PSTN is the world's collection of interconnected voice-oriented public telephone networks, both commercial and government-owned. It is also referred to as the plain old telephone service (POTS). It is the aggregation of circuit-switching telephone networks that has evolved from the days of Alexander Graham Bell's first patent applied for in 1876. Today, it is almost entirely digital in technology except for the final link from the central (local) telephone office to the user. In relation to the Internet, the PSTN actually furnishes much of the Internet's long-distance infrastructure.

To develop the interconnections, a wireless carrier must first examine *where* to place interconnections and what *type* of interconnections should be ordered and implemented.

This decision can be based on several factors. First, a logical location for connection to the public telephone network is to place an interconnection, also known as a point of presence (POP), where the competition has an interconnection. (The term POP can be another word for interconnection and actually refers to a point where any two telecommunication carriers link their networks together, whether they are incumbent local telephone companies, long-distance carriers, ISPs, or competitive LECs [CLECs].) That way a wireless carrier can stay abreast of the competition, providing local service and a local telephone number where the competition is doing the same. Second, cost factors must be taken into account to justify the placement of an interconnection. Marketing projections should prove that the forecasted amount of sales (and, in turn, call volume) will outweigh the installation and monthly charges inherent to interconnection. In some cases special construction charges may also apply to install interconnection circuits, especially in rural areas where a wireless carrier's cell site may be miles away from the local telephone company's nearest cable terminal or central office. Special construction charges reflect the costs to lay cable through a right of way (ROW) where current cable capacity is exhausted. Third, the population density of a given geographic region must be examined to determine whether a local interconnection will suffice or whether a connection that offers low cost-per-minute charges over a wider area is more appropriate.

There are many facets to interconnection:

- The elements that compose the interconnection itself, such as circuits, trunks, and phone numbers
- The different types of interconnection available
- The cost structures of interconnection and how they can vary from one phone company to the next
- Legal issues relating to how interconnection agreements are formalized

14.2   The Structure of the PSTN

The local telephone company, anywhere, is known as the LEC or the telco. Usually, the incumbent phone company in a particular region is one of the former Bell companies—several of them have even merged into larger entities. Today, the dominant, former Bell companies—also known as *incumbent telcos*—are SBC, Verizon, Quest, and Bellsouth.

SBC is comprised of the former Southwestern Bell states and now includes the five-state footprint of the company formerly known as Ameritech (which SBC purchased in 1998), as well as the two-state footprint of Pacific Telesis (which was purchased by SBC in 1997). SBC also purchased Southern New England Telephone (SNET). SBC's total footprint includes 13 of the 50 United States.

Verizon is the largest of the dominant phone companies today and is a combination of the former Bell companies, known as Nynex and Bell

Atlantic, and the huge, former independent phone company with a nationwide footprint, GTE.

Quest is the new name of the former U.S. West and still incorporates the same states where U.S. West operated. But Quest also includes the assets of the former company known as Global Crossing.

Bellsouth is still Bellsouth, having somehow escaped consolidation of some sort over the years (up to 2005, at least). They are also still the only remaining dominant phone company with the word "Bell" in their name. In the early twenty-first century, widespread rumors existed that AT&T would buy Bellsouth, or vice versa, but these rumors never morphed into reality.

The remainder of the local telephone company landscape, nationwide, is made up of independent telephone companies and large CLECs. Some of these independent companies are very large, at least in terms of the size of the parent company that holds the reins. For example, many local phone companies nationwide are owned by Sprint. These larger, independent telcos have distributed footprints throughout the United States, in just about every region of the country. Others are mom and pop telcos, usually named according to the geographic areas where they operate. Examples are Lufkin-Conroe Telephone Company of Lufkin, Texas, and Mud Lake Telephone Company in Mud Lake, Idaho.

Examples of large CLECs still operating today, after the telecom meltdown of 2001, are Level3, XO Communications, and a restructured Focal Communications. All three of these companies declared bankruptcy at some point since the late 1990s, but again, they have reemerged and are some of the major nationwide CLEC players still left standing at this time. There are also many smaller, regional CLECs operating across the United States.

 Key: The dominant local carriers in every state are more heavily regulated than the independent telephone companies. This is a result of a regulatory mandate of ensuring a level playing field or competitive landscape among CLECs, the independent telcos, and the dominant telcos.

When a wireless carrier obtains interconnection to the public network, it is obtained from either the dominant carrier, a CLEC, or an independent telephone company. This depends on which type of phone company has been granted the right to operate in a specific geographic area by the respective state public utility commission (PUC) as well as price points offered by the competing carriers.

14.2.1 Local Access and Transport Areas

When AT&T was divested in 1983, there was more to the breakup than dividing the country up into seven regional Bell companies. As part of the divestiture, the modified final judgment (MFJ) called for the separation of *exchange* and *interexchange* telecommunications functions. This meant that exchange services were provided by the regional Bell operating companies or independent telephone companies, and interexchange services were to be provided by IXCs. The United States was thus broken up into 197 geographical areas known as local access and transport areas (LATAs). LATAs serve the following two basic purposes:

- They provide a method for delineating the area within which the LECs may offer services.
- They provided a basis for determining how the assets of the former Bell System were to be divided between the Regional Bell Operating Companies (RBOCs) and AT&T at divestiture. This included outside cable and supporting infrastructure (outside plant) as well as wire centers, switching offices, and the equipment housed therein (inside plant).

As the name implies, the basic principle underlying the structure of LATAs is that within an LATA, only local telephone companies (LECs) could provide exchange services, carry local telephone traffic, and provide a myriad of intra-LATA connectivity services. LATA sizes vary and are usually based on population densities of a given area. For example, in the western United States, because of the sparse population of some states, an entire state may be one LATA (e.g., Wyoming). In other states, such as Illinois, about a dozen or more LATAs exist.

Key: Each LATA has a three-digit number assigned to it for identification purposes. For example, Chicago is in LATA number 358. In LATAs, where an independent telco is the dominant local exchange carrier, the LATA designation number begins with the number 9. For instance, LATA number 949 is in North Carolina, and the dominant LEC there is the independent telco Sprint Mid-Atlantic Telephone Company. For the most part, LATAs do not encompass more than one state, although they can cross state boundaries in some cases.

LATAs were designed not only to delineate exchange areas, but also to keep intact geographic areas that were socially, economically, and culturally tied together. This is why Chicagoland and northwest Indiana (#209 in two different states) are one LATA (#358).

14.3 Elements of Interconnection

The technical reference for wireless interconnection to the PSTN is found in Telcordia General Reference (GR) 000-145. This document provides the technical definitions associated with wireless interconnection to the PSTN.

There are two basic elements to wireless interconnection to the PSTN: (1) transport via transmission systems, such as DS-n, OC-n, and trunks (if interconnection is based on time-division multiplexing [TDM]), or TCP/IP, and (2) telephone numbers for resale to wireless customers. When interconnection is TDM-based, each DS0 acts as a trunk in this context. In the majority of cases today, telephone numbers are assigned to wireless carriers through a company called NeuStar, the North American Numbering Plan Administrator (NANPA). Just several years ago, telcos obtained phone numbers through the NANPA, which at that time was Bellcore and is now Telcordia. At that time, most telcos charged wireless carriers when they assigned them ranges of telephone numbers. The highest charges were applied when wireless carriers obtained entire NXX codes, and the charges could be in the tens of thousands of dollars.

14.3.1   The Circuit

DS-n or OC-n circuits comprise the underlying transport for wireless interconnection to the PSTN, similar to how these types of transmission systems are used heavily in the backhaul network. When TDM-based transport is used, each channel (DS0) is used to transport one wireless call.

Interconnection to the Internet could occur differently. Internet connectivity could take the form of a high-speed gigabit Ethernet pipe to the wireless carrier's designated ISP. It could also take the form of a frame-relay circuit to the ISP's POP from the wireless mobile switching center (MSC). The traditional TDM method could also be used, where unchannelized DS1, DS3, or OC-n circuits could be used.

No matter what is transported from wireless to landline networks—voice, data, image, or video—the underlying foundation of the transport will be a circuit of some type. It should be noted that compression techniques such as Adaptive Differential Pulse Code Modulation (ADPCM) are typically *not* used in PSTN interconnections.

14.3.2   The Trunk and Bandwidth

It is important to understand the difference between a trunk and a line. A line is that which connects a piece of equipment to a switching system or a computer system. For example, a person's home telephone is connected to the telephone company central office by a line. In contrast, a trunk connects two switching systems together or a switching system to a computer system. For example, in the case of wireless interconnection, DS0s running over DS1/3 circuits function as trunks that connect the MSC with the switch that is housed at a telephone company central office. Each DS-n channel—each DS0—functions as a separate trunk.

Bandwidth is what is procured to support Internet connections originated by mobile subscribers that terminate to the Internet. This bandwidth can come in many forms, but most always the nature of the traffic running over these connections will be TCP/IP (IP). From the wireless carrier to the Internet, the connections could be frame relay, switched Ethernet, point-to-point gigabit Ethernet, unchannelized DS3 circuits, or even unchannelized Synchronous Optical Network (SONET).

14.3.3 Telephone Numbers (For Resale)

Inherent with most types of interconnection to the PSTN is the ability of wireless carriers to obtain blocks of phone numbers, which are then sold (actually resold) to wireless customers when they initially purchase service from wireless carriers.

Phone numbers can be purchased in blocks of 100 or 1,000, in what are called *line ranges*. Some telcos will sell blocks only in certain increments, in other words only in blocks of 100 or only in blocks of 1,000. The increments the numbers are sold in is telco-dependent. Remember that these line ranges are the information that is shared with other wireless carriers when roaming agreements are put into place.

14.4 Interconnection Operations

Interconnection is still a large expense for wireless carriers, even though the cost-per-minute rates charged by the landline telcos to wireless carriers have trended steadily downward since 1996. It is very important for wireless carriers to ensure that they are always implementing the best type of interconnection throughout their markets. It is also wise to employ a small staff of billing analysts who can decipher the interconnection bills to determine if incorrect rates are being charged to the wireless carrier. In most instances, the staff of billing analysts can pay for their own salaries several times over by catching costly errors on interconnection invoices.

Calls originated by a mobile phone will be switched at the MSC and routed to their destination using standard least-cost routing methods, which are developed by the network engineers of any given wireless carrier.

Key: When wireless carriers obtain interconnection service from a telco, they also obtain the phone numbers that they will resell to their customers. These phone numbers are obtained from North American Numbering Plan Administration (NANPA). The NANPA administers all phone number allocations for the entire United States. Calls originating from a landline telephone and terminating to a wireless subscriber (land-to-wireless calls) are routed through the PSTN to the telco central office where the wireless carrier's interconnection was obtained. This routing is determined by the NPA-NXX (area code and exchange) or the mobile subscriber who is being called. Landline telcos know how to route incoming calls dialed from the PSTN (or other mobile phones) because the issuing authority for telephone numbers, NeuStar, will send out a notification document to all telephone companies and wireless carriers in the United States (and the world), notifying them when a new line range or NXX code is activated. The Common Language Location Identifier (CLLI) code of the particular central office where the wireless carrier obtained interconnection—and the line range of phone numbers—will be included in the notification to all telcos and other wireless carriers. The telco stores the line ranges owned by the wireless carrier in its central office (CO) switch that connects the wireless carrier to the telco. In other words, the telephone numbers (TNs) are stored in the telco switch where interconnection actually occurs. This is how all telecom carriers know to route calls to a specific (mobile) phone number. Land-to-mobile calls are routed to that particular telco central office based on the called person's phone number—their NPA-NXX—and from the telco central office to the wireless carrier's MSC over the actual interconnection circuits.

Wireless interconnections to the PSTN can occur from either a cell site location or an MSC location. If a desired interconnection can be accomplished from a cell site that is closer to a telco CO than the MSC, it makes more sense from a cost perspective to place the interconnection into the cell site. This is because telcos charge per airline mile for (interconnection) circuits, and if a wireless carrier base station is closer to a telco CO than the wireless carrier's MSC, it might make more sense to do the interconnection into the base station. The interconnection traffic is then backhauled to the MSC for switching.

> *Key*: If a wireless carrier installs a PSTN interconnection circuit from a base station, as opposed to directly from the MSC, it is imperative that network engineers reserve capacity on the fixed network back to the MSC that is equivalent to the bandwidth of the interconnection obtained into the base station. Also, in these situations, the wireless carrier may want to do a net present value-based (NPC-based) cost analysis to compare the two situations to see which makes more financial sense: doing the interconnection from the BTS or doing it from the MSC.

14.5   Types of Interconnection

Multiple types of interconnection are available to wireless carriers. Each type of interconnection has its own functional characteristics, and each type is also ordered for specific marketing or operational reasons, or for both. The underlying principle when ordering any type of interconnection used for carrying wireless traffic to the PSTN or Internet is to obtain the largest available geographic footprint at each point of interconnection for the lowest cost per minute across that footprint to terminate wireless traffic. The larger the footprint, the more cost effective the interconnection. Each type of interconnection will also have its own calling area.

14.5.1   Type 2A Interconnection

A Type 2A interconnection consists of a circuit connection (and associated trunk group) from a wireless carrier's cell or MSC to a telco access tandem, known in traditional Bell System nomenclature as a Class 4 CO. Access tandems serve as hubbing centers for multiple end offices in an LATA. They will also serve as the access point for toll tandems, known in Bell System nomenclature as Class 3 COs. An end office is also known as a CO or a Class 5 CO in traditional Bell nomenclature. End offices are the lowest link in the PSTN hierarchy, providing last-mile connectivity, dial tone, and Internet access to residences and businesses.

> *Key*: Access tandems exist in the PSTN hierarchy because it is not always practical or cost effective for a LEC to link every end office in a given LATA to every other end office within that LATA. With access tandems in place, PSTN switching can become a hierarchical operation, which is more efficient and cost effective than linking all end offices to each other. There will also be very high capacity trunking systems known as intermachine trunks (IMTs) connecting access tandems to other access tandems. This entire switching and routing hierarchy is also where the origins of good wide area network (WAN) designs originated. In most instances, the local calling area for Type 2A interconnections is the entire LATA.

It is the objective of wireless carrier network engineers to obtain the largest possible LATA-wide footprint when obtaining Type 2A interconnections. Wireless carriers must obtain an entire exchange code through the NANPA when procuring a Type 2A interconnection. An entire exchange code consists of 10,000 phone numbers (e.g., 607-689-0000-9999).

Type 2A interconnections represent the procurement of interconnections at a wholesale rate, because this type of interconnection gives wireless carriers the lowest cost for the biggest footprint. Therefore, because this type of interconnection is wholesale in nature, the MSC is viewed functionally as the next level down in the PSTN hierarchy—a telco Class 5 end office switch—in terms of call processing and tariffs. See Figure 14-1.

14.5.2 Type 2T Interconnection

A Type 2T interconnection is a circuit connection (trunk group) to a telco access tandem, which is used to route equal access traffic from a wireless carrier's network to an IXC's point of presence. Equal access provides wireless customers the opportunity to presubscribe to an interexchange carrier of their choice. Type 2T interconnection was required of RBOC-based wireless carriers (e.g., Bell Atlantic Mobile Systems, Ameritech Mobile) in 1983, as a result of the MFJ ruling by Judge Harold Greene.

This type of interconnection provides wireless customers access to all long-distance carriers via the following:

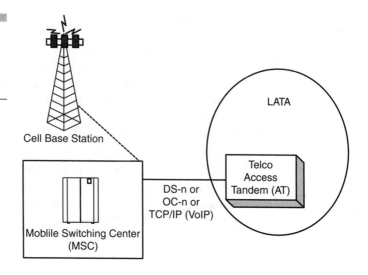

Figure 14-1
Type 2A
interconnec-
tion to the
PSTN

- A Feature Group D (FGD) trunk group, which carries the automatic number identification (ANI) of the calling party
- The carrier identification code (CIC) of the appropriate IXC (10-XXX)
- The dialed number

> *Key:* Type 2T interconnections likely are not used much anymore, as the concepts of long-distance calling and selecting your own long-distance carrier are becoming a quaint notion in today's telecom world. Long distance itself is becoming a concept whose days are numbered, with the advent of voice over IP (VoIP), low-cost transport, and the commoditization of roaming and long distance as services.

14.5.3 Type 1 Interconnection

A Type 1 interconnection consists of a circuit connection (trunk group) between a wireless carrier's cell or MSC to an LEC end office. This interconnection has more functions than any other type of interconnection.

Type 1 interconnections could be ordered for marketing purposes, simply to obtain local access telephone numbers or to present a local presence in a specific geographic area. This strategy applies especially to wireless carriers operating in rural areas.

Key: In some rural areas, it might be necessary to order a Type 1 because it is not cost effective to order other types of interconnection, such as Type 2A. For example, in a very remote but relatively large community, a wireless carrier would definitely want to order local telephone numbers. That way, calls from mobile phones to landline telephones (and vice versa) would be terminated and rated at the cheapest possible rate, while simultaneously appearing as local numbers to the local population. Also, it is possible that in rural states there may only be two access tandems in the entire state—one per LATA. It would be inefficient and too costly to use a Type 2A interconnection in the scenario described here, because all calls either originated by or terminating to wireless subscribers would have to be tromboned up to the access tandem and then back to the wireless carrier's MSC.

As a rule, calls that terminate outside the local calling area of Type 1 interconnections are rated as *toll traffic* by the local telephone company and might therefore be rated as toll traffic by the wireless carrier as well. Sometimes wireless carriers will assume payment for these toll charges themselves and not pass the charges on to their customers. This practice is known as *reverse billing* and is done for marketing and competitive purposes.

Along with local traffic, Type 1 interconnections also carry 911 traffic, 411 traffic (directory assistance), 0+ traffic (credit card calls), 0- traffic (calls to the telco operator), and toll-free traffic (800, 866, 877 numbers). Type 1 interconnections are also capable of carrying traffic to any destination LATA-wide, but Type 2A interconnections are the preferred method for transporting these calls because the rates are usually lower to terminate a call to a landline phone via a Type 2A interconnection. 911 calls from wireless customers will be routed to a public safety access point (PSAP) for processing.

Type 1 interconnections can also be used as a pipeline for traffic that needs to be routed to an interexchange carrier when the destination of the call is inter-LATA or interstate. When the amount of traffic in a given wireless market is not great enough to justify a dedicated DS1 circuit that ties the wireless carrier's network directly to an IXC's POP, wireless traffic is routed to the IXC through a Type 1 interconnection in switched access (SWAC) form. This means calls are processed on a call-by-call basis. *Access* to the IXC's network occurs when calls are *switched* over a Type 1 interconnection *through* the local telephone company, hence the term "switched access."

Directory assistance or 411 traffic can be routed directly to a telephone company call center over a Type 1 interconnection. Wireless carriers can connect to the call center directly from the MSC. Today wireless carriers also offer directory assistance call completion, where they will automatically route the call to the number requested for no charge when wireless customers press 1. This service is great for customers, because they do not have to stop and write down the number. Obviously, this is very helpful when driving. (Of course, none of us should be using mobile phones when driving, right?) Today, wireless carriers also have the option to route 411 traffic to independent call centers over dedicated trunks from the MSC directly to the independent call center. These call centers are run by companies who are pure entrepreneurs. They are competing with the telcos for DA business. The appeal of this service architecture is that the ICC will charge the wireless carrier a wholesale rate per call that is cheaper than what the local telco will charge for the same DA service. This fee structure is what provides wireless carriers incentive to use these companies in the first place, because it offers wireless carriers a potential revenue stream. Wireless carriers will then charge their subscribers at a retail rate. In other words, the independent call center company may charge the wireless carrier $0.75 per call, while the telco DA service charges the wireless carrier $0.90 per call—a charge that is passed through to the wireless customer with no margin tacked on. So, when using the independent call center company, the wireless carrier can charge the retail rate of $0.90 because of the familiarity this fee has with end users. But the wireless carrier makes $0.15 profit per call. This scenario represents a viable revenue stream for the wireless carrier. The only problem with this situation is that to keep their costs down, the independent call center companies will not update their directory listing databases as often as the telco DA centers. This could result in bad information being given to wireless subscribers who call 411. Another problem is, again, to keep costs down the independent call center may not staff its call center as fully as it should, which will result in longer hold times before a wireless customer actually reaches an agent. These downsides have to be weighed against the upside of the revenue stream potential.

Type 2A rates, making the Type 2B essentially the least expensive type of interconnection from a wireless carrier perspective. Put simply, Type 2B interconnections are used for least-cost routing purposes. This type of interconnection is becoming more and more prevalent, and is cheaper than Type 2A rates, because there is no tandem-switching element involved in the rates charged to wireless carriers.

14.5.5   Dedicated Interconnection to an IXC

When the volume of mobile-originated long-distance traffic in a particular market is low, a wireless carrier would send all inter-LATA and inter-state long-distance traffic out a Type 1 interconnection as SWAC long-distance traffic. In this scenario, the local exchange carrier routes long-distance calls to the wireless carrier's preferred IXC on a call-by-call basis through the LEC end office and on to the IXC's POP.

But when the amount of long-distance traffic in a wireless market is large, the wireless carrier can justify the installation of a dedicated interconnection to its long-distance carrier. This type of justification can be made by obtaining a quote for the monthly cost of a dedicated circuit from the wireless carrier's MSC to the IXC's nearest POP. If the quote for this link is equal to or less than the monthly cost of sending switched access traffic to the IXC over a Type 1 interconnection (the number of switched access minutes of use × the cost per minute charged by the IXC), the interconnection is justified.

Key: The rationale for this type of interconnection is that wireless carriers are charged lower cost-per-minute rates by the IXC, versus sending traffic to the IXC through the LEC via switched access over a Type 1 interconnection. This is because access charges are assessed to the IXCs by the LECs when long-distance traffic is sent to the IXC through LEC central offices (over Type 1 interconnections), which really means IXCs charge wireless carriers a higher cost per minute. These additional costs may be several cents per minute and are passed on to the wireless carriers. In turn, the wireless carriers will pass these costs on to their customers.

This type of interconnection is a win-win situation for both wireless carriers and their subscribers.

14.5.6   Intramarket Wireless Carrier-to-Carrier Interconnection

In some cases, it makes sense to do a direct, wireless carrier-to-wireless carrier interconnection within a wireless market. This applies especially to high-traffic areas such as urban markets.

Wireless carriers can obtain records from their MSC that will tell them how much customer-originated mobile traffic in a given market terminates to given NXX codes in a market. A database provided by Telcordia known as the Local Exchange Routing Guide (LERG) lists all exchange codes within each area code. In the LERG, a column next to each exchange code denotes the carrier that owns the exchange code. So through a combination of the switch records and the LERG, a wireless carrier can determine how much of its mobile traffic terminates to its in-market wireless competitor's telephone numbers.

Through nondisclosure and confidentiality agreements, two wireless carriers can determine how much of their respective customer traffic terminates to the other wireless carrier's NXX codes (mobile-to-mobile calls) in that market. These carriers can then get a quote for the cost of a point-to-point circuit that would directly link their two networks together, MSC to MSC. If the total cost to send the traffic to each other's NXX codes through interconnections to the PSTN (the number of minutes × the cost per minute) exceeds the cost of the point-to-point circuit, then it would behoove these two carriers to implement a direct connection between their two networks to transport traffic between each other's subscribers. The carriers would split the monthly cost of the circuit evenly between themselves.

The intercarrier link is essentially an IMT between the two carriers' networks. These trunks support customer-originated mobile-to-mobile traffic between the two wireless carriers. The bandwidth of the circuit connecting the two carriers' MSCs would be traffic engineered and based on the volume of calls between their mobile subscribers. This is the key information that is obtained from the switch records that each carrier obtains, which spurred them to pursue this interconnection in the first place.

The main benefit of this type of interconnection is that the wireless carriers are essentially bypassing the LEC to terminate calls to other mobile subscribers, thus saving money on interconnection costs.

14.5.7   Point-to-Point Circuits

Point-to-point circuits, also known as *leased lines* or *private lines*, are used to link two wireless carrier *locations* together. In the wireless world, this usually means one of the following three things:

- Linking two cell sites together
- Linking a cell site to the MSC
- Linking sales offices to MSCs

These circuits are called leased lines because they are leased from a telco; they are not a facility that is owned by the wireless carrier. These types of circuits are installed when a wireless carrier does not have other means to link two locations together. One scenario requiring a wireless carrier to lease a circuit from a telco is where there is no line of sight (LOS) for a microwave radio connection between two locations, presuming the wireless carrier prefers microwave facilities in its backhaul network. This means that natural or man-made obstructions exist between the two points, which would block the signal of a microwave radio system. This could be a mountain, mountainous terrain, or a building in the LOS between the two locations that rises to a point where it blocks the radio signal. Another situation requiring a point-to-point circuit is when a cell site is so far away from another cell that a microwave radio link is not an option, because the range would exceed the physical capabilities of microwave radio.

Which type of landline carrier the wireless carrier approaches to provide a point-to-point circuit depends on whether the link between the two locations is intra-LATA or inter-LATA. If the two locations are intra-LATA, the wireless carrier approaches the local carrier. If the locations are inter-LATA, the wireless carrier approaches an interexchange carrier. When a wireless carrier implements point-to-point circuits, the two ends of the circuit are known as the A and Z locations.

When the locations are in the same LATA, the circuit passes through a LEC CO at both ends before arriving at its final destination. The part of the circuit between the wireless carrier's network (cell site or MSC) location and the local exchange carrier central office is known as the *local loop*. See Figure 14-3.

When the locations are in two different LATAs, the connections at each end are made first to a LEC end office or wire center and then to an IXC's POP. The part of the circuit that runs between the IXC POPs is known as the interoffice channel (IOC). When ordering inter-LATA point-to-point circuits through an IXC, there are times when the costs for

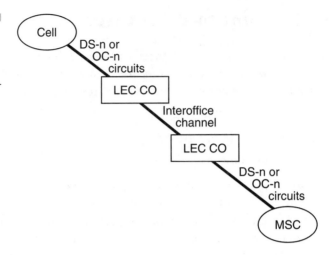

Figure 14-3
Intra-LATA
point-to-point
circuit structure

the interoffice channel will be very high. This is because regardless of what the wireless carrier's A and Z locations are, the locations of some IXC POPs at each end could be extremely distant from the two locations the wireless carrier is trying to link together. Remember that circuit costs are rated based on airline miles between the two end points of the circuit. If the circuit run is inter-LATA, the wireless carrier has no choice but to go through an IXC for service. The end result is that the total cost for the circuit may be very expensive, but if the wireless carrier needs the circuit, the cost may need to be overlooked. See Figure 14-4.

14.6   Cost Structures and Rate Elements

Most of the types of interconnection that have been reviewed in this chapter have some common cost structures and rate elements. Each interconnection is achieved by installing a circuit to connect locations between two networks—the wireless carrier's network and the landline network—regardless of the carriers involved. Some common elements of interconnection are given in the following list:

- For each circuit that is installed, there will be a nonrecurring charge (NRC) to install the circuit itself.
- For each circuit installed, there will be a monthly recurring charge (MRC) for the circuit itself.

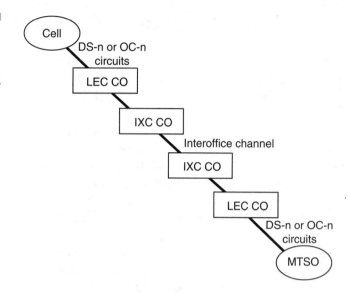

Figure 14-4
*Inter-LATA
point-to-point
circuit structure*

- For certain types of interconnection, there may be a charge to activate elements of the circuit that act as trunks. This would be an NRC when it exists, based on the telephone company involved. An example of this activity would be the charge to activate DS0s for a DS-n-based interconnection. For Type 2A interconnections, this charge no longer applies. However, it may apply for Type 1, Type 2B, or Type 2T interconnections.

- There may also be a recurring monthly charge for the trunks running over an interconnection circuit, based on the previous information. This charge is very carrier-dependent and usually applies to smaller, independent telcos.

For point-to-point leased circuits, only two charges apply:

- An NRC for installation of the circuit
- An MRC for use of the circuit

Key: Because point-to-point circuits are purchased as dedicated circuits, not circuits where the telco role is to switch traffic of any kind, no trunk charges apply to these types of circuits. Even if wireless traffic does get switched or routed over these circuits, when they are purchased by a wireless carrier, that carrier can use the circuit for whatever purpose they desire.

14.7 Cost-per-Minute Charges

Prior to the passage of the Telecom Act of 1996, rates for all types of interconnections ranged anywhere from $0.02 per minute (i.e., Type 2B interconnections) to more than $0.05 per minute (i.e., Type 2A interconnections).

Since passage of the Telecom Act of 1996, rates for all types of interconnections have dropped to a range of less than $0.01 per minute to $0.03 per minute. Independent telephone companies usually charge higher rates per minute, because they have to compensate for the fact that they have much less volume to work with than, say, the companies that grew out of the former Bell company (Verizon, SBC, Quest, Bellsouth).

For dedicated interconnections to IXCs, wireless carriers can obtain rates from around $0.01 to $0.04 per minute. The core issue, which drives the exact rate per minute charged to wireless carriers for this type of interconnection, is total billable minutes per year. At the same time, wireless carriers may be required to sign long-term contracts of up to 3 years in order to obtain lower rates. Volume (minutes of use) and term (years) commitments are what drive lower rates for interconnection contracts.

14.7.1 Rating Structures

There are two structures used to determine the cost-per-minute rate that telcos charge to wireless carriers for interconnection: *banded rate* structures or *flat-rated* structures. Either method can apply to tariffs or custom contracts (see Sections 14.9.1 and 14.9.2).

14.7.1.1 Banded Rates Banded tariff structures are still used by many landline carriers to assess charges to end users, whether they are residential consumer customers or business customers. Here is how this structure works.

A landline telco designates anywhere from three to eight concentric circular geographic bands whose center is a particular central office. In this context, the central office is called the *rate center*. The rate center is the central office where the interconnect exists and is used for rating and measuring mobile-to-landline wireless calls.

Each band encompasses a specific mileage range outward from the rate center. Calls that terminate to landline destinations within mileage bands that are closer to the rate center are cheaper. Conversely, calls to destinations within the farthest mileage band away from the rate center are rated as the most expensive. As the mileage bands progress outward from the rate center, the cost to carry traffic to destinations within mileage bands also increases correspondingly on a sliding scale. For example, if a telco uses three mileage bands, the first band could encompass 0 to 8 miles from the rate center. This is known as Band 1. The second band could represent a circular band that is 8 to 25 miles away from the rate center; it would be known as Band 2. The third mileage band represents a concentric circular band that is 25 or more miles from the rate center. Calls to someone living in Band 1 are rated the cheapest; calls to persons in Band 3 are rated the most expensive. Banded rate structures are more difficult to administer and deal with for all parties involved (see Figure 14-5).

14.7.1.2 Flat Rates The more progressive telcos, including long-distance carriers, have implemented what are known as flat-rate pricing structures. Because of intense competition in today's marketplace, more and more landline carriers are using flat-rate pricing. It is easier to administer for telcos, and it is easier for everyone to understand.

Flat-rate pricing reflects a very simple pricing mechanism: Telcos charge carriers one flat rate (e.g., $0.02 per minute) to terminate calls across an entire region, usually a LATA. All mobile-to-landline wireless calls are rated exactly the same from the rate center to any destination (LATA-wide in the case of a wireless interconnection). It does not matter how far away from the interconnection point, or the rate center, a mobile call terminates to the landline.

This rating structure has several advantages. First, it is much easier to administer and deal with for both the telco and the wireless carrier.

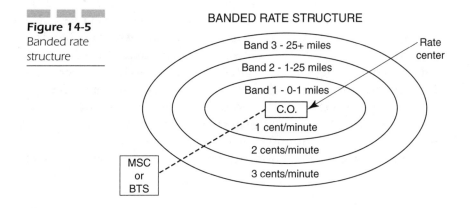

Figure 14-5

Banded rate structure

Another advantage that flat-rate tariffs provide is that, because they offer LATA-wide service, theoretically a wireless carrier does not have to install more than one interconnection in any given LATA. This would substantially decrease the monthly recurring costs for circuits and associated trunks over interconnections by reducing the net amount of interconnections required.

14.8 Special Construction Charges

When a wireless carrier orders any type of interconnection, special construction charges may apply. This usually occurs when the wireless carrier requires an interconnection or point-to-point circuit to be implemented in a very remote rural or mountainous area. If the wireless carrier places an order for service, and the telco's cable facilities are a long distance away from the wireless carrier's location, where no cable facilities are available, then the telco will have to make what is known as a *reusability assessment*. This assessment revolves around whether or not other entities (residences or businesses) will be able to make use of a newly required cable facility the telco will have to install to support the wireless carrier's order within a given period of time (e.g., 5 years). If no other entities will be able to make use of the cable, the cost to trench the cable must be borne by the wireless carrier, and they are then charged accordingly if they still wish to pursue the service order and pay the fee. This fee is known as a *special construction charge*. Complex formulas are used to assess whether or not the telco can pass on these costs to the wireless carrier. They are known as the *reusability factor*.

Depending on the amount of special construction that is required, the cost could be several thousand dollars, or even in the tens of thousands of dollars. If special construction charges are prohibitively high, the wireless carrier could seek other options for connectivity. These options could include seeking out other sites for interconnection or more creative options such as working with the telco to implement a microwave radio link between the two networks. This would involve the wireless carrier purchasing a low-capacity microwave radio system, placing one node or antenna at its site, and placing another node or antenna at the telco's nearest CO. The microwave antenna could be placed on the roof of the CO.

14.9   Interconnection Agreements

Two methods are used by landline carriers to develop interconnection agreements with wireless carriers: tariffs and custom contracts. This is how the two parties formalize an interconnection purchase. Sometimes a combination of the two is used when custom contracts are developed based on reduced tariff rates (see Section 14.9.2). Both tariffs and custom contracts contain common sections designed to protect both parties involved in the interconnection purchase: the telco and the wireless carrier.

One of these sections is the terms definition section where acronyms and other terms used in the documents are defined clearly. This is a legal way of protecting both parties from the potential to misconstrue meaning anywhere within the tariff or contract. Another section defines the roles and responsibilities of both parties. This is mainly designed to protect the telco, so if an interconnection is turned up after the due date given to the wireless carrier, the telco is covered if the tardy turnup is due to the wireless carrier not meeting its responsibilities that are stated in the tariff. These requirements could include not supplying power for circuit termination equipment, not supplying space for the circuit termination equipment, or even not supplying cable from its own telco/data room to the premises demarcation point, or *demark*. The demarc marks the physical separation of responsibilities between the telco and its customers, which in this case is a wireless carrier.

14.9.1   Tariffs

The cost-per-minute rates that the landline carriers charge to terminate (connect) calls to the PSTN are usually stated in the form of tariffs, which must be filed with and approved by the respective state PUCs. Tariffs state the rates charged for services, as well as the responsibilities of both the telephone company and the party that is purchasing service. In this case, that party is a wireless carrier. Although it may seem odd that the party purchasing service has responsibilities, tariffs are designed in this manner to minimize the legal risk to telephone companies in case there are service problems.

When a wireless carrier purchases an interconnection from a landline carrier, the tariff rates apply for all charges. Tariffs state the cost-per-minute rates for carrying mobile-to-land calls and rates charged for leasing the circuits.

 Key: No option exists to negotiate tariffs when ordering interconnections from a landline telco. Tariffs are used as a regulating measure by the states to ensure that the big boy on the block does not throw his weight around, so to speak. Rate stability for the largest telcos ensures that the smaller telcos and CLECs can compete on a more level playing field. Nonetheless, the landline telcos also have the option of negotiating custom contracts with their wireless interconnection customers via rules and special tariff options instituted by the state PUCs. Custom contract rates are usually derived from tariff rates.

Every few years, the incumbent telcos will gather their interconnection customers together to discuss rate restructuring (e.g., rate decreases). Once a proposal for tariff restructuring is presented to the incumbent telco, they review it internally. Once approved internally, the telco then submits the new tariff to the state PUC for approval. This process can take anywhere from 6 months to a year to complete. *Tariffs are public documents.*

Prior to passage of the Telecom Act of 1996 , one of the largest stumbling blocks that existed for wireless carriers in obtaining equitable and fair interconnection rates and terms was the fact that LECs did not grant wireless carriers the cocarrier status in accordance with FCC orders. Cocarrier status dictates that when LECs determine interconnection rates and terms they should treat wireless carriers just as they would treat another LEC or IXC in terms of how those carriers interconnect to the LEC's network and what rates they are charged per minute to transport calls into the LEC's network. This could apply to any number of relationships: incumbent telco to LEC interconnection, CLEC to LEC interconnection, independent LEC to independent LEC interconnection, or IXC to LEC interconnection. Technically, it is illegal for any LEC to charge a wireless carrier a higher rate to terminate calls than it charges other carriers to terminate *similar* calls into the LEC's network.

14.9.2   Custom Contracts

If interconnection rates charged to wireless carriers are not derived from a tariff, they are formalized through a written interconnection agreement known as a custom contract. Custom contracts are very similar to tariff agreements, but the rates contained in a custom contract are lower than tariffed rates. Custom contracts are used by all the telcos but sometimes under different rules. As usual, the rules that apply to an incumbent telco (usually a former Bell company) are more stringent than the rules that apply to other telcos such as CLECs and other smaller independents. *Contracts are private documents.*

14.10   Least-Cost Routing

Wireless carriers must ensure that all mobile-originated calls are routed to their destinations as inexpensively as possible. This process is known as least-cost routing.

Key: The wireless carrier's own network should always be used whenever possible as the first route to terminate mobile-originated traffic to the landline network or to other mobile subscribers within wireless carrier clusters. Remember, a cluster is a geographically contiguous group of wireless markets all owned by the same carrier. Ideally, mobile-originated wireless traffic should always be directed over the wireless carrier's own network to the PSTN interconnect that is *closest to the destination* of the wireless call. So, the bottom line is that least-cost routing can involve the use of an array of telco interconnects, along with the wireless carrier's own backhaul network.

For calls within a wireless carrier's own system, in a market cluster where the carrier has deployed its own signal transfer points (STPs), inter-MSC SS7/IS-41 links can and should be used for least-cost routing purposes. These links are known as IMTs, using SS7 ISDN User Part (ISUP) for call management.

Standard least-cost routing methodology usually employs three possible routes, cheapest to most expensive, for directing mobile-originated traffic from point A (the wireless network) to point Z (any landline destination or another wireless carrier's network). If the operational status of the first trunk group (the cheapest option) is *all-trunks-busy*, calls will automatically be directed to the second trunk group (the second cheapest option), and then to a third trunk group if necessary (the most expensive option) if the second group is also all-trunks-busy. This type of routing is accomplished by having what is known as a *route pattern* designated for every NPA-NXX combination in the United States, and every international country code in the world. The route pattern points to the three possible trunk groups that are available to route traffic. The route patterns exist in the switch—the MSC. All switches, whether PBXs, access tandems or toll tandems, have route patterns for all NPA-NXX combinations.

Key: Obviously, the goal with least-cost routing is to ensure that the first trunk group is always properly engineered to handle most, if not all, calls, thereby avoiding overflow to the most expensive routes. Usually, where there is overflow from the first trunk group to the subsequent trunk groups, it is due to a spike in network traffic that is tied to some type of public disaster or emergency.

14.11   Internet Connections

Wireless carrier interconnections to the Internet would take place ideally directly from the MSC to the ISP's POP. This is the ideal connection scenario because the wireless carrier requires adjunct server and router equipment in place to manage Internet traffic to and from its customers.

This connectivity could take many forms, simply because there are many options available today from incumbent LECs (ILECs) and CLECs. The exact amount of bandwidth required would be a reflection of wireless carrier traffic studies that spelled out how much bandwidth is needed. This projection should always take projected growth into account. The options wireless carriers have for connecting their networks to the Internet are as follows:

- The wireless carrier could implement a gigabit Ethernet connection directly to the ISP, as many ISPs today would accept this type of connection and plenty of carriers are out there who supply metro Ethernet connections. A great example is SBC's GigaMAN® service.
- Wireless carriers could connect to the Internet using OC-n circuits.
- Unchannelized DS3 circuits could be used.
- DS1 or multiple DS1 circuits could be used to support less populated, rural markets.

Key: As the number of Internet-based applications and services provided by wireless carriers continues to surge, this connectivity will become critical to wireless carrier infrastructures. That said, it is *always* wise to implement dual-homed Internet connections over separate and diverse physical routes to two distinct and physically separate ISP POPs. This also means that there should be diverse access facilities into the MSC site itself. This should have already been the case, to support sound backhaul network and PSTN interconnections. This type of design should be implemented even if it means obtaining Internet access service from two separate ISPs. See Figure 14-6.

14.12 Wireless Local Number Portability (WLNP)

Local number portability (LNP) describes the ability of customers to switch service from one carrier to another. In the wireless world, *wireless LNP* (WLNP) means changing wireless carriers but keeping your wireless phone number. It can also mean porting, or changing, your landline home phone number to become your mobile phone number. WLNP went into effect on November 24, 2003. It was an FCC mandate that required all carriers in the industry to prepare for interconnecting into a national database that managed the porting of numbers between wireless carriers. Naturally, portability applies to residential customers, business customers, and wireless carriers as well (for numbers they obtained through interconnection to the PSTN).

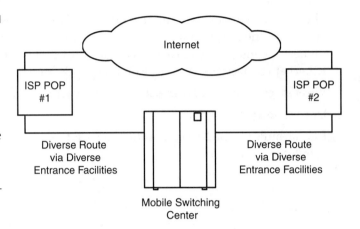

▬▬ ▬▬ ▬▬ ▬▬ ▬▬ ▬▬ ▬▬ ▬▬ ▬▬ ▬▬ ▬▬ ▬▬ ▬▬ ▬▬ ▬▬ ▬▬

Key: The portability rules that went into effect in November 2003 included service portability, in other words, the ability to port telephone numbers from landline telcos to wireless, as well as from one wireless carrier to another wireless carrier.

▬▬ ▬▬ ▬▬ ▬▬ ▬▬ ▬▬ ▬▬ ▬▬ ▬▬ ▬▬ ▬▬ ▬▬ ▬▬ ▬▬ ▬▬

For years, mobile subscribers complained about the inability to retain their mobile phone number, and how that has virtually held them hostage to their current wireless service provider. The FCC, along with consumer advocates, believed that WLNP would enhance competition and allow subscribers to choose the carrier they want.

After opposing the measure, carriers won a one-year delay of what they called an unfunded government mandate because they needed time to train personnel and coordinate WLNP with public safety agencies. The CTIA said it would cost $1 billion to initiate number portability and another $500 million *each year* to support the new service. The CTIA claims the cost will drain resources away from needed network upgrades, filling gaps in coverage, and delivering customer advanced services such as 3G. Churn, or service cancellation, was also a concern of wireless carriers. The concern of wireless carriers was echoed by a report released In-Stat/MDR. Analysts said that where portability was implemented overseas, the rate of customer churn increased by as much as 50 percent. The research firm estimated that over 22 million wireless subscribers would change carriers during the first full year that WLNP was in effect. That rate would "fall to 30 percent by 2004, finally leveling off at 10 percent annual churn." In reality, after one full year of being in place, only 8.5 million subscribers changed wireless carriers while keeping their wireless phone numbers. In any case, WLNP is a good thing for consumers, which means that long term, it will be good for the industry itself.

14.13 Reciprocal Compensation

Mutual compensation described a concept that existed in the early 1990s, where wireless carriers were trying in vain to obtain agreement with the landline telcos to have the telcos compensate the wireless carriers for calls that originated on a landline phone and terminated to the wireless network. The thinking was that it was only fair for compensation to be a two-way street, because wireless carriers also spent billions to develop their networks. They just did it in a shorter period of time.

This issue heated up in the early to mid-1990s because the proliferation of wireless service resulted in a large increase in landline-to-mobile phone traffic. The wireless carriers came to the point where they felt it was only fair that they also be compensated for calls that terminate onto their networks. When a wireless carrier transports a mobile-to-land telephone call, this call originates within the wireless network and terminates to the PSTN (the landline network). Wireless carriers are required to pay a specified cost per minute for each of these calls to the landline telco. However, when landline customers called a mobile telephone number (land-to-mobile calls), the landline telephone company did not pay the wireless carrier any monies for terminating any of these calls onto the wireless network. The LECs got away with not paying mutual compensation to wireless carriers for years. Part of the reason for this was because the LECs—the former Bell companies—had more political heft, which they used to keep the cellular carriers off balance and always at a disadvantage.

On December 15, 1995, the FCC issued a Notice of Proposed Rulemaking dictating that a concept known as *reciprocal compensation* should govern interconnections between commercial mobile radio service (CMRS) providers (cellular, PCS, paging, enhanced specialized mobile radio carriers, or EMSR) and LECs. Reciprocal compensation simply takes mutual compensation one step further by formalizing it, saying that under many (if not all) circumstances, landline and wireless carriers should bill each other for calls terminating to each other's networks. Reciprocal compensation is the ultimate maturing and resolution of the mutual compensation and cocarrier status issues.

The Telecom Act of 1996 formalized the reciprocal compensation issue by making it law. It stated that if a wireless carrier is being charged a certain cost-per-minute rate for calls terminating to the landline network, then the landline carriers have to pay the wireless carrier for calls that terminate to the wireless network as well. Reciprocal compensation was the government's way of ensuring that the financial playing field between landline and wireless carriers was finally leveled off; fairness is now ensured by the force of law.

14.14 Enhanced 911 Service (E-911) and Locator Technology

911 service becomes Enhanced 911 service when automatic number identification (ANI) and automatic location information (ALI) from a wireless phone is provided to the 911 operator—the Public Safety Answering Point (PSAP). In other words, the name and location of the caller is provided.

In the United States in the year 2003, 198,729 911 calls were made *daily*. Because there are now 175 million wireless subscribers in the United States, it is likely that many 911 calls will be originated by mobile subscribers.

The U.S. regulation known as the FCC E-911 mandate was passed in 1996, requiring all wireless service operators to provide an emergency 911 caller's cell phone location to the PSAP to within 100 meters 67 percent of the time. The initial deadline for the commencement of implementation was October 1, 2001, and no major wireless carrier met the deadline. The FCC granted temporary waivers and extension of time. However, noncompliant phone companies have received server fines (AT&T was fined $2.2 million in May 2002) and the FCC is now increasing the pressure for compliance. The primary difference between Phase I and Phase II is the degree of caller location accuracy. Phase I requires accuracy to the cell/sector level, and Phase II requires accuracy within 125 meters 67 percent of the time.

Sprint, Verizon Wireless, Cingular, T-Mobile, and AT&T Wireless (now part of Cingular) all filed requests for waivers to the October 1 deadline.

The biggest hurdles have been rolling out the required network software and infrastructure and coordinating with emergency response agencies.

The Phase 2 requirements of wireless E-911 were to be implemented by October 31, 2001. Phase 2 for network-based implementations includes the ability to identify the latitude and longitude of a mobile phone within a radius of less than 125 meters 67 percent of the time and within 300 meters 95 percent of the time. The handset-based implementation requires 100 percent compliance of new or updated mobile phones by the end of 2004. Full Phase II compliance is now required by December 31, 2005. As usual, the Phase II deadline had been December 31, 2004, but wireless carriers requested a postponement from the FCC (until December 31, 2005) and they received it.

The fundamental operational needs to support Phase I and II 911 include the following:

- A mapping and geolocation database for PSAP call routing
- Geographical-boundary provisioning
- Out-of-band signaling for real-time location updates
- A large processing capacity for location calculation
- Interactive call control
- High reliability and manageability

The intensive processing demands, as well as the need for large database support and management, are best served by an adjunct computer-based solution. Additionally, location technology offers myriad vertical service opportunities beyond wireless 911, which could help shoulder the anticipated Phase II cost burden.

The FCC proposed the implementation of regional (state and local) cost mechanisms to cover the cost of implementing wireless E-911. The FCC also stipulates that wireless service providers have the same legal liabilities as the landline providers to successfully deliver E-911 services to their customers.

14.14.1 Locator Technologies

There are two approaches wireless carriers can select to determine the location of a mobile subscriber for E-911 purposes: network-based solutions or handset-based solutions.

Network-based solutions employ triangulation and timing, using precise measurements of the wavelength differentials between base stations to reveal the location of a mobile user. In other words, network-based solutions rely on triangulation technology. This approach has been determined to cost a lot more than a handset-based solution (e.g., GPS receivers in mobile phones). One network-based solution uses time difference of arrival (TDOA) technology. With this approach, at least three base stations must be detected by the mobile handset over the control channel to resolve ambiguities in location. They must also be tightly synchronized. The base stations calculate the time it takes for the signals to travel from the stations to the handset, and back to the stations. This yields the distance of the handset from the base stations. The wireless systems will then map the three calculated distances, and wherever these locations meet reveals the position of the handset.

Another network-based technology option is the direction of arrival (DOA) technique. DOA requires only two base stations, and the base stations must have an array of antennas. These base stations also calculate the distance of the handset from the base stations like TDOA. In addition, DOA also measures the *angle* of the signals being sent back to the base stations. The distances and angles measured together can then be used to calculate the location of the handset.

These network-based technologies are the most promising triangulation techniques and the easiest to implement. No change in standards is required, and, best of all, they work with unmodified phones. However, one drawback is that the mobile position significantly degrades in the urban environment due to signal reflection, and accuracy decreases when the mobile phone is close to base stations(s) and/or in hilly areas. Service providers are discovering that implementing these new services can be costly and time-consuming. Some estimates say the average cost to implement a network-based solution—per base station—is $500,000.

Handset-based solutions use GPS and a GPS receiver in the handset to communicate user locations to the PSAP and the wireless service provider.

Key: GPS technology can pinpoint handsets within 15 feet outdoors and 100 feet inside buildings, which exceeds the FCC's requirements. GPS units are placed throughout a wireless network. As these units keep track of GPS satellites orbiting the earth, they pass along key satellite information, including estimated time of the signal's arrival to nearby wireless handsets, which are equipped with scaled-down GPS receivers. Then, based on time differences between when the network's GPS units and handset's receive signals from the satellites, it is possible to precisely pinpoint the caller's location.

One downside to this option is that because GPS receiver units must be included in the mobile handsets, the handsets will then consume a lot of power. But the upside is that the handset-based solution is more accurate and provides universal coverage. To attain widespread use of this option, new GPS-enabled handsets would have to be purchased by many wireless customers over a relatively short period of time. Considering the persistent level of promotions offered by wireless carriers to provide incentives for their customers to upgrade their wireless handsets, this objective does not appear difficult to achieve.

The two technology approaches to locator technology are based on micromeasurement of transmission time for wireless signals to travel through space and require the widespread deployment of specialized hardware units and network overlays.

The downside to locator technology is that it effectively turns the mobile into a de facto tracking device. Users can be located anytime, anywhere—especially with the GPS handset-based solution.

Locator technology, although Big Brother–like in nature, can be applied in many application scenarios as listed here:

- E-911 emergency location
- Personal child security
- Asset tracking and people finder
- Fleet management
- Telematics
- Driving directions (includes integration with software such as MapQuest)
- Wireless gaming
- Location-based billing
- Information directory service
- Push/pull advertising
- Electronic house arrest
- Enterprise applications
- Field force effectiveness
- Mobile worker management

It would be an understatement to say that the industry is lukewarm about wireless location-based services. The main reason: Carriers invested billions of dollars into getting their networks E-911 compliant and are now trying to recover those costs from a location-based services (LBS) market that many analyst firms project will not be generating more than tens of millions in the next few years. In spite of this, carriers are deploying location-based applications with regular frequency, and although pure-play LBS applications, like car navigation systems and telematics, may not be the hottest sectors, consumer and business applications with location-based elements are becoming (or will become) crucial to most carriers' service offers.

Many wireless carriers now offer location services for a fixed monthly fee, where tracking can be monitored via an Internet site accessed by customers. This is how many parents today are keeping track of their teenage kids. Locator technology is also being used to track wayward spouses and employees who tend to slack off. Intrusive? Yes. Effective? Yes. Scary? Yes.

The essential difficulty for the wireless carriers to deploying locator technology in their networks is that the current time measurement location technologies (based on triangulation) have marginal accuracy and require an investment of billions of dollars. The wireless industry does not have this capital to invest. Another concern is that this financial hurdle has the potential to set back the urgently needed enhancement to public safety and stall much needed revenue generation from premium location-based services. This is becoming a major concern due to the fact that the U.S. subscriber base is now over 175 million subscribers (as of late 2004).

The following information presents just a snippet of what some enterprising companies are developing to support the locator marketplace.

A company known as Landmat has developed a variety of location-based products. PalTrak does what its name implies: It locates friends connected to the wireless network and arranges meetings, identifies meeting places, and even keeps track of the children. DateTrak is a similar service with advanced messaging and more controlled tracking features built in. And its cobranded Time Out Mobile service links Time Out's reams of listing and entertainment content to Landmat's navigation system.

A firm called Xora has built the ultimate Big Brother technology for wireless devices, allowing companies to track their employees' time sheets, jobs, and physical locations using GPS-enabled phones.

MapQuest, one of the most recognizable brands on the Internet, has brought its mapping functionality to the wireless network, adding location-based features and even voice-recognition technology.

▰ Test Questions

True or False?

1. _____ Telephone numbers are not purchased for sale to cellular customers when a Type 2B interconnection is installed.

2. _____ Type 2A interconnections are direct links to telco end offices.

3. _____ Wireless carriers are required to purchase an entire NXX code when obtaining a Type 1 interconnection.

Multiple Choice

1. Type 1 interconnections to the PSTN:
   a. Terminate directly to a telco access tandem
   b. Terminate directly to a telco end office
   c. Carry local/EAS traffic
   d. Carry 911, 411, and operator-directed traffic
   e. a and c only
   f. b, c, and d only

2. The Telcordia Technical Reference for Interconnection to the PSTN is:
   a. TR-000-240
   b. TR-140-000
   c. Reciprocal termination
   d. GR-000-145

3. Which type of interconnection displays the ultimate case of least-cost routing?
   a. Type 1
   b. Type 2B
   c. Type 2A
   d. Type 2T

4. What is the difference between a banded interconnection structure and a flat-rated interconnection structure?

    a.   A banded rate structure charges one rate for calls terminating anywhere in a LATA; flat rate does not.

    b.   A flat-rate interconnection structure charges higher rates depending on how far away from the originating end office (or access tandem) the call terminates (connects). A banded-rate structure charges different rates for different states.

    c.   There is no real difference between banded interconnection structures and flat-rate interconnection structures.

    d.   Banded interconnection structures charge higher rates, depending on how far away from the originating end office (or access tandem) the call terminates (connects). Flat-rated interconnection structures charge one rate, LATA-wide.

5. The reason that wireless customers are charged for land-to-mobile calls, even though they did not originate the call themselves, is because:

    a.   Wireless carriers are greedy and have all engaged in price-fixing practices.

    b.   Land-to-mobile calls are still using extensive resources of the cellular carrier's infrastructure.

    c.   Wireless carriers are always charged for land-to-mobile calls also.

    d.   All of the above.

    e.   None of the above.

6. Dedicated interconnections to IXCs are justified by doing a cost analysis to verify:

    a.   That there is always going to be one million minutes per month directed over the interconnection to the IXC.

    b.   That a Type 2A interconnection is not necessary.

    c.   That the monthly cost for the DS1 circuit that allows for dedicated access to the IXC is equal to or less than the total cost of all mobile-originated calls that are routed to the IXC as switched access through a Type 1 interconnection to the LEC.

    d.   All of the above.

7. Wireless interconnections to the PSTN are also known as:

    a.   Localities

    b.   CLECs

    c.   POPs

    d.   IXCs

Roaming
and
Intercarrier
Networking

15.1   Overview

Roaming defines the use of wireless service outside a wireless subscriber's home service area. In other words, wireless customers are in a roaming state when they use their mobile phone outside of the general area where they originally purchased their wireless service. Each wireless carrier determines its own home service area on the basis of the specific market where they are operating. In each market, one carrier may try to maintain a larger home area footprint (coverage area) in order to gain a competitive edge over its in-market competitor. The home service area may be designated on a large wall map in sales offices where service is initially obtained by wireless subscribers.

Key: Note that due to the size of the 1900 MHz personal communication service (PCS) metropolitan trading area (MTA) markets, carriers operating in that frequency band will by definition have larger home areas because these markets are huge compared to the traditional 850 MHz cellular markets.

Once the mobile phone is powered on in a roaming market, customers become aware they are in a roaming state through the autonomous mobile registration process. When the mobile phone is powered on in a roaming service area, the phone registers with the system where it is currently located. Key elements of autonomous mobile registration process are the transmission of the mobile's mobile identification number (MIN), otherwise known as the 10-digit telephone number; the electronic serial number (ESN) of the phone, which is essentially the phone's fingerprint; and the System Identification (SID). The SID is a three- or four-digit number that uniquely identifies a mobile user as belonging to a specific home market. A roaming indicator will then appear on the liquid-crystal display (LCD) screen of the mobile phone, in the form of the letter *R* or the word "roam."

Wireless carriers' definition of home service areas is continually expanding due to aggressive cluster expansions by all major wireless carriers, even to the point of nationwide footprints, which are most common today. This is contributing to the commoditization of roaming as an operational and revenue concept, which is leading to its position as a differentiator in marketing efforts. In other words, many wireless carriers today offer calling packages where roaming is free. This, in turn, is leading to the slow death of roaming charges (see Chapter 18).

Special rates apply to wireless customers who make calls while roaming in those cases where carriers *do* charge for roaming. These rates are very carrier-specific and may fluctuate greatly between carriers across the United States. The rates are very high compared to regular call rates, but roaming rates are continuing to decrease as more advanced intercarrier networking systems have been implemented nationwide. These systems based on Signaling Systems 7 (SS7) ultimately reduce costs and increase efficiency.

Key: All wireless carriers impose higher fees than normal for customers who place calls while roaming. This is especially true for wireless carriers who have a large market presence in rural areas and who may rely heavily on roaming revenues by virtue of the fact that there are interstate highways prevalent in rural areas, even if the potential subscriber base is small. When mobile subscribers are traveling for business or vacation by driving, they have no choice but to register with and use roaming systems. In some cases roaming fees are a large part of these carrier's overall revenue streams.

15.2   The Early Days of Roaming

In the first 10 years of the wireless industry, the validation of roaming customers occurred by having all cellular carriers directly network their mobile switching centers (MSCs) together with private line circuits. These circuits validated customers for both antifraud purposes as well as billing purposes. This setup existed because at that time there were not any large, nationwide ANSI-41/SS7 networks in place supporting wireless carriers, which allow for carriers to more efficiently validate and bill their customers.

Today, when wireless carriers do charge for roaming, they simply charge a higher per-minute rate for calls made by customers who are roaming. But in the old days in the wireless industry (from 1983 to the mid-1990s), a typical rate structure for a roaming call could be as follows:

1. A daily fee was assessed, whether a person who was roaming made one call or 20 calls per day. This fee could be as high as $3 per day.

2. A cost-per-minute fee of anywhere from $0.50 to $1.50 was applied.

3. A per-call fee might have been assessed. This fee could be $0.50 to $1.00.

These higher rates were deemed necessary to cover the higher cost of intercarrier networking in that time period, which usually involved dedicated circuits either directly between carrier's MSCs, or to roaming clearinghouses, or to both. Roaming clearinghouses are third-party business entities that accept roaming data from wireless carriers via dedicated circuits to the clearinghouse database location. The clearinghouse then manages and allocates the roamer billing monies amongst all wireless carriers who subscribe to its service. Today the roaming clearinghouses are still used, but more for billing and monies exchange instead of precall validation because of the now-widespread base of SS7 systems in the wireless world.

From 1983 to the mid-1990s, in most cases someone who wanted to call a person who was roaming would have to follow a cumbersome multistep process in order to reach the person. This process was so customer-unfriendly that it inhibited roaming activity. The process was actually a deterrent to increasing roaming revenues because it was so clumsy:

1. A caller who wanted to reach a roaming customer would have to know the exact destinations of the person who would be roaming. When wireless customers in those days signed up for service, they were given a small catalogue that looked like a miniature telephone book, which had a nationwide listing of roamer access numbers for all cellular carriers (A band/B band).

2. The caller would then have to dial the special roamer access telephone number from the catalogue described in the preceding passage. This call would terminate into the switch (MSC) of the market where the wireless customer was roaming to a roamer access port.

3. Then the caller had to dial the roamer's 10-digit MIN once he or she accessed the roamer MSC.

This roaming structure is essentially extinct today for the most part, because of the deployment of more advanced intercarrier networking systems. These systems enable callers to transparently reach roaming customers anywhere, by simply dialing their 10-digit mobile phone number. These advanced systems are enabled via the networking standards known as ANSI/TIA-41 (American National Standards Institute/Telecommunications Industry Association) and SS7-signaling systems.

15.2.1 Roaming Agreements

All wireless carriers develop roaming agreements with each other, if they want to ensure their customers can roam anywhere in the United States (or world) and be able to place a wireless call. Ideally, they should also be able to use the features they are accustomed to in their home markets.

Roaming agreements are business agreements developed between wireless carriers to outline the terms under which mobile customers of each carrier can roam into each other's territories and still be able to place and receive wireless calls. The most important part of the roaming agreement is to iron out rates that are acceptable to both parties, in terms of what wireless customers will pay to place calls while roaming. Also, in order to operationalize a roaming agreement, the carriers must exchange all their active line ranges with each other so that roamers can be recognized when visiting nonhome wireless systems. Line ranges are defined by the bucket of NPA-NXXs that carriers sell to their wireless customers when they obtain wireless service (NPA = area code; NXX = exchange code). Along with the rates charged to roamers and line ranges, carriers also must exchange SID information—the system ID that uniquely identifies each carrier in the markets where they operate. Wireless carriers maintain entire departments of personnel dedicated to nothing but managing roaming and all its intricacies.

15.3 Modern Roaming Systems

15.3.1 Roaming and Digital Interoperability

A major roaming interoperability challenge exists with so many digital wireless technologies being fielded by 850 MHz cellular and 1,900 MHz PCS carriers. As of 2005, there are five types of digital wireless technologies in the U.S. marketplace:

- IS-136, fielded by AT&T Wireless until its network is fully converted to GSM and integrated into by Cingular Wireless. There are likely other smaller, regional wireless carriers who may be using IS-136 as well.

- GSM, fielded by Cingular Wireless, Western Wireless, and T-Mobile.

- CDMAOne, fielded by Verizon Wireless and Sprint PCS.

- CDMA 1XEVDO, a 3G technology fielded by Verizon Wireless and AT&T Wireless (now Cingular) in several cities as of late 2004.

- IDen technology, a proprietary TDMA technology fielded by Nextel Communications.

Motorola and other wireless handset manufacturers have addressed this situation with the development of a trimode, multiband handset that operates using most of the technologies listed in the preceding list. A quadmode handset is in the works. This technology will continue to expand and evolve, and in some form, mobile subscribers should always be able to place a call wherever they travel, regardless of where they obtained their service.

15.3.2   Signaling System 7 (SS7) Overview

Up until the early 1980s, the signaling for a telephone call used the same voice circuit that the telephone call traveled on—this was known as *in-band signaling*. This method of signaling used the same physical path for both the call-control signaling and the actual connected call. This inefficient method has been replaced by out-of-band or "common-channel" signaling techniques such as SS7. SS7 is an internationally standardized protocol used by telecom carriers of all types for interoffice signaling, call establishment, billing, routing, and information-exchange functions. In the public switched telephone network (PSTN) and today's wireless networks, SS7 is a system that puts the information required to set up and manage telephone calls in a separate network rather than within the same network used to make the call. SS7 signaling travels on a separate, dedicated overlay network. Using SS7, telephone calls can be set up more efficiently and with greater security. Because control signals travel in a separate network from the call itself, it is more difficult to violate the security of the system. Special services such as call forwarding and wireless roaming are easier to manage as well.

A network utilizing common-channel signaling such as SS7 is actually two networks in one:

- First there is the *circuit-switched user network,* which actually carries the user voice and data traffic. It provides a physical path between the source and destination.

- The second is the *signaling network,* which carries the call control traffic—the SS7 network. It is a packet-switched network using a common channel-switching protocol. SS7 has become the de facto signaling standard in the telecom world—both wireline and wireless.

SS7 has been in use in the wireline networks since the late 1970s and early 1980s. Throughout the 1980s, competing wireless carriers had their MSCs connected together directly in a meshed architecture to facilitate roaming activity with each other. As wireless network growth exploded, a hubbed signaling architecture became necessary to increase efficiencies and cut costs. SS7 was the answer. Wireless carriers have been implementing SS7 on a large scale since the early 1990s to facilitate nationwide roaming for mobile customers.

There are multiple functions of the SS7 protocol in today's networks, as follows:

- Setting up and managing the connection for a call

- Tearing down the connection when the call is complete

- Billing

- Managing signaling in support of vertical services such as call forwarding, calling party name and number display, three-way calling, and other intelligent network services

- Managing the routing for toll-free (800, 888, 866, 877) and toll (900) calls

- Wireless and wireline call service, including mobile subscriber authentication and roaming

An example of information that may be contained in SS7 messages:

1. "What's the best way to route a call to a specific NPA-NXX (e.g., 708-431)?"

2. "The route to network point 587 is crowded. Use this route only for calls of priority 2 or higher."

3. "John Doe is a valid wireless subscriber. Continue with setting up his wireless call."

In today's wireless networks, SS7 infrastructures are used to transport ANSI-41D messaging to other wireless carriers in support of key features that ANSI-41D allows for, namely (automatic) call delivery and intersystem handoff. The Mobile Application Part (MAP) of the SS7 standard is what addresses the registration of roamers and the intersystem hand-off procedure in wireless mobile telephony.

> *Key:* ANSI-41D messages containing subscriber profiles are encapsulated into SS7 messages and transported across SS7 backbone networks to authenticate roamers and support seamless roaming (see Sections 15.5 and 15.6).

The primary function of SS7 is to provide call control, remote network management, and maintenance capabilities for interoffice telecommunications networks. SS7 performs these functions by exchanging control messages between reserved or dedicated channels known as signaling links and the network points that they interconnect. There are three kinds of network points, which are called signaling points: service switching points (SSPs), signal transfer points (STPs), and service control points (SCPs). SSPs are PSTN or wireless network switches used to originate or terminate calls. They are also used to communicate on the SS7 network with SCPs to determine how to route a call or set up and manage some special feature. SCPs are databases used to support SS7 networks and can contain routing data and information on calling cards. Traffic on the SS7 network is routed by STPs.

15.3.3   Signal Transfer Points (STPs)

STPs are data switches that relay SS7 messages between network switches and databases in an SS7 network. Their main function is to route SS7 messages to the correct outgoing signaling link, based on SS7 message address fields known as signaling point codes (SPC), or point codes for short.

Key: An SPC, or signaling point code, is a numeric address that uniquely identifies each signaling point in the SS7 network. In this case, the STPs are network nodes, as are the MSCs. Point codes are used to direct messages to the appropriate network destination. Each switch, STP, and Intelligent Peripheral (IP) has a unique point code in an SS7 network. All wireless carrier STPs are identified by unique point code addresses. Point codes are equivalent to the IP addressing of the SS7 world, and their numbering convention is xxx.xxx.xxx. Point codes consist of nine digits. The first three digits represent the network, the second three the cluster, and the third three identify the member (the node or device). Any device that processes SS7 traffic in any way must have a point code assigned to it.

STPs have a dual functionality: They act as both network hubs and SS7 packet switches. To facilitate SS7 functionality in the wireless world, wireless carrier MSCs are connected to STPs. Physically, STPs can be built-in adjuncts to the MSCs, or they can be stand-alone systems such as Tekelec's Eagle family of STPs. This configuration depends on the maker of the MSC as well as the engineering and operations practices of the wireless carrier. Larger wireless providers procure their own STPs, and then connect their STPs to the MSCs, usually in a regional architecture. These links that connect MSCs to the wireless carriers' mated STP pair are implemented using 56 Kbps or DS1 circuits, and are known as A (access) links. This is how the STP acts as a hub: Multiple MSCs will connect to one mated STP pair. Smaller, regional wireless carriers sometimes lease STP space from the larger carriers to avoid the cost of procuring their own STPs.

Just as the wireline telephone companies deploy their STPs in mated pairs in order to achieve equipment redundancy, wireless carriers also deploy their STPs in mated pairs. All telephone companies need the redundancy that is afforded by having a backup STP. The STPs must be connected to each other for redundancy purposes, and these links can be implemented by using either 56 Kbps data circuits, multiples of 56 Kbps circuits, or full DS1 circuits. The link that connects a mated pair of STPs to each other is known as a C (cross) link.

> *Key*: A mated STP pair will *not* be housed within the same location (building). They will be geographically separated within a region, by at least 50 miles or so. This deployment strategy is used to ensure availability and to ensure that at least one STP is functioning in case of regional disasters such as fire, flood, earthquake, and so forth.

15.3.4 The Home Location Register (HLR)

The HLR is the main database that houses identity information on valid "home" subscribers in a wireless network. A "valid" customer is one who pays his or her bills on time, and whose MIN/ESN has not been declared illegitimate due to fraud (which is becoming a rare occurrence). The HLR is an integral component of wireless networks, necessary to acknowledge and maintain subscriber information relating to the identity of valid customers in the switch's home area, or home subscribers. When home subscribers power on their phones and register with the system over the control channel, their registration is posted as a record in the HLR. The HLR will maintain information on every active mobile in its system until they power off their mobile or move out of their home service area.

In the earlier days of the industry, from 1983 to the mid-1990s, most wireless carriers (850 MHz cellular carriers) deployed HLRs as physical adjuncts to their MSCs in every market. As SS7 technology was deployed, and as wireless markets expanded and became more sophisticated with the deployment of digital wireless technology and increased customer expectations, wireless carriers considered modifying their networks so only one HLR was used—a centralized HLR. Many large wireless carriers today have deployed centralized HLRs. This represents one HLR—one database—for all the carrier's market holdings. If a wireless carrier owns multiple wireless markets, the MSCs from all these markets would all be connected over the carrier's own WAN backbone to the centralized HLR. In these cases, the HLR will usually be at a geographic location that is central to all the carrier's market holdings. The centralized HLR could also be located at the carriers' network operations center (NOC), or wherever the carrier's billing system is housed.

Key: In some cases, the NOC, the billing system, and the HLR may all be housed within the same facility so the carrier can maintain all its critical systems in one spot. It should be noted that sound network planning and common sense would dictate that these systems all have mirrored route backups operating 24 hours a day, 7 days a week, ideally in a completely separate geographic location. Diverse route connectivity into this location would also be mandatory.

The advantage of having a centralized HLR compared to having an HLR operating at every MSC location is that it is easier and more efficient to maintain one HLR, rather than a multitude of HLRs. Specifically, the key benefits are as follows:

- Less capital required to purchase the HLRs themselves—only one required versus many. Even a huge HLR would be less expensive to purchase and maintain than many HLRs.
- The MSC has more processing power to use for other purposes, such as switching calls—its most important function.
- Only one HLR to maintain versus multiple systems. This translates into fewer man hours dedicated to maintenance.
- Less floor space is taken up at MSC locations.
- Increased consistency with how the HLR is programmed and maintained.

The only possible drawback to having one centralized HLR is that a carrier would have to closely monitor the utilization on its internal WAN backbone to ensure that the HLR-related validation and signaling traffic is not overwhelming the capacity of the WAN. Regular monitoring of the utilization of the WAN backbone should preclude capacity problems from becoming an urgent issue. In a worst-case scenario, the wireless carrier would have to expand its WAN backbone to increase its capacity. Also, if a centralized HLR is used, some type of hot standby should be deployed at a geographically diverse location, because the HLR is a critical network element.

15.3.5 The Visitor Location Register (VLR)

The VLR is a database maintained by wireless carriers to track mobile users who are roaming in the provider's home territory. The VLR is resident within every wireless switch, or its functionality can also exist within a centralized HLR as a partition of the HLR itself. Its purpose is to acknowledge and obtain subscriber information relating to the identity of valid roaming customers in a switch's home area.

Once a mobile phone is powered up in a nonhome (roaming) system, it will immediately perform a mobile registration. The mobile will transmit its SID, ESN, and MIN during the registration process. Once the phone registers into the visiting system, the roamer MSC will validate the subscriber and obtain feature information from the customer's home switch (via SS7 signaling). All information pertinent to that roaming subscriber will be stored in the VLR until he or she leaves the roaming area.

15.3.6 Third-Party SS7 Roaming Networks

What is happening today is not a new phenomenon. Throughout history, there have been many new technologies and infrastructures that have dramatically advanced communications and commerce, such as the railroad, air travel, electric grids, telephones, and so forth. But all of these systems were unable to reach their full potential until they deployed an overlay of coordination and control provided by intelligent infrastructures.

The late 1980s saw North American cellular becoming standardized as network growth and complexity accelerated. As cellular technology itself surged in popularity by around 1990, it became clear that an economical, efficient method was required to allow for precall validation and call control amongst wireless carriers.

A sizable portion of roaming today still occurs in markets that are adjacent to subscriber's home markets. But as the use of wireless phones has expanded over time, it is no longer feasible to rely on direct carrier-to-carrier connections to validate users and manage call setups, teardowns, and so forth. The solution is to deploy SS7 hubs and signaling architectures to facilitate roaming amongst wireless carriers. The first carrier to build such an intelligent cellular network was McCaw Cellular Communications (owned by Craig McCaw), with the launch of

North American Cellular Network (NACN) in 1991. NACN was a standing regional (then national) SS7 network provider that provided interconnections for wireless carriers to allow for precall validation, billing and roaming clearinghouse functions, and transparent roaming, also known as automatic call delivery. NACN served mainly A-side cellular band carriers. At its peak, NACN covered over 98 percent of A side cellular carriers in the United States, enabling A-side cellular customers to roam throughout the United States, Canada, Mexico, Puerto Rico, New Zealand, and Hong Kong. The NACN was also the first network to offer automatic call delivery while roaming in GSM networks.

Another nationwide SS7 network provider serving wireless carriers was a company originally known as Independent Telecommunications Network (ITN), which then changed its name to Illuminet. Illuminet operated in the same manner as NACN, offering SS7 interconnection services to mainly B-side wireless carriers throughout the United States. Both of these companies no longer exist, but they took the wireless industry a long way forward in terms of demonstrating how wireless networks could become more intelligent, seamless, and functional. NACN folded, and Illuminet was purchased in 2001 by Verisign.

Verisign's Intelligent Infrastructure Services enable businesses (and people) around the world to connect across today's nationwide and worldwide wireless networks. Verisign operates one of the largest telecom signaling networks in the world, enabling services such as wireless roaming, text messaging, caller ID, and multimedia messaging. On a daily basis Verisign enables over 14 billion Internet interactions, and 3 billion telephony interactions (think SS7). As next-generation networks are emerging, Verisign is now deploying intelligent infrastructures required for everything from radio frequency identification (RFID) supply chains to voice over IP (VoIP) to mobile content.

Verisign's ANSI-41 and GSM-MAP Transport Services allow wireless carriers to offer subscribers seamless roaming with the same features and security they have in their home areas. Their seamless roaming services provide real-time information for troubleshooting, fraud detection, reports for monitoring network integrity, coordination of network setup, and data administration.

Independent SS7 networks like those offered by Verisign (and its predecessors) allow wireless carriers to operate more efficiently and compete more effectively because of their access to any wireless VLR or HLR, via direct connectivity to major wireless carriers and gateway access to other networks. International roaming to Europe, Asia, Africa, and most of the GSM world is available with Verisign's SS7 network infrastructure and experience with ANSI-41 and GSM technologies. Interstandard roaming supports GSM, CDMA, or TDMA networks. Verisign's international roaming service is designed to support automated (seamless) roaming throughout participating wireless carriers' markets worldwide. A central (SS7) signaling node supports both serving and home carriers across 25 countries. Advanced systems are used to manage and log all subscribers' IS-41 and GSM messages across all subscribing carriers through one central link. This service also offers technical updates, multilingual support team services, and online access to roaming coverage queries.

Roamer validation and automatic call delivery are provided through transmission, monitoring, and intelligence-gathering services (i.e., profiling systems that detect usage anomalies that indicate possibility of fraud). Interswitch message transport uses the ANSI-41 industry standard, sending validation traffic from the serving (roamer) switch to the home switch.

Verisign consolidates roaming agreements for carriers and provides a cross-protocol roaming gateway and interstandard mediation platform to perform authentication and conversions from cellular intercarrier billing exchange record (CIBER) to Transferred Account Procedure. CIBER was the first cellular wholesale billing protocol.

Verisign performs standard clearinghouse functions also, as previously described in this chapter (Section 15.2). Verisign consolidates all roamer calling data and facilitates payment amongst wireless carriers. Settlement and exchange service is available both domestically and internationally, as Verisign enables carriers and operators to reconcile financially using CIBERNET, or to conduct financial settlement directly with roaming partners using a secure web-based option. CIBERNET engineered CIBER and introduced multilateral financial settlement to the wireless industry.

Daily reports and monthly totals are available on a secure Web site within 24 hours of data receipt. Wireless carriers have access to standard reports to support financial, marketing, and fraud management activities, as well as point-and-click functionality to drill down into data details.

Verisign's settlement exchange service for GSM uses the industry standard TAP. Roaming data is transferred to both GSM and non-GSM customers and their clearinghouses, or GSM conversion agents to enable clearance and settlement.

15.4 The ANSI-41 Signaling Standard

In 1988, the analog networking cellular standard called TIA-IS-41 was published. This interim standard is now formally known as ANSI-41. The objective of ANSI-41 in wireless networks is to unify how network elements operate: the way various databases (e.g., HLR, VLR) and mobile switches communicate with each other and with the landline telephone network.

ANSI-41 is the wireless industry standard for interswitch (inter-MSC) signaling that allows for precall validation, call completion in nonhome markets, and access to advanced databases such as Wireless Calling Name, E-911, and Wireless Number Portability. The ANSI-41 standard allows for transaction-based services that support applications such as short message service, and access to HLR/VLR in wireless networks.

Despite ownership or competitive influences, for the betterment of the industry all wireless systems across America (and really the world) need to act as one larger wireless system. It is in this manner that roamers can travel from system to system without having a call dropped. It is also via this philosophy that calls can be validated to check against fraud, subscriber features can be supported in any location, and so on. All of this activity relies on network elements cooperating in a uniform, elegant manner. This is what the ANSI-41 standard brings to the table in the wireless industry.

Seamless networks enable wireless subscribers to place calls from virtually any place and maintain uninterrupted wireless service when subscribers travel from one region or market to another. Prior to the mid-1990s, seamless roaming capabilities were available only by relying on switching equipment from a single manufacturer because the equipment communicated in a proprietary messaging format. In other words, if two different wireless carriers wanted to implement networking capabilities between themselves to support seamless roaming, both carriers had to use wireless switches from the same vendor.

So the demand for multivendor, seamless networks led to the development of an industry standard for intersystem wireless networking: ANSI-41(D). Prior to being ratified as an official ANSI standard, ANSI-41 was known as IS-41 (for interim standard). The ANSI-41D standard has evolved through multiple revisions since the early 1990s, as an interim standard (IS-41 Revisions 0, A, B, and C). It was formally standardized as TIA/ANSI-41D in late 1997.

IS-41 Revision 0 defined the methods for call-handoff interactions and precall validation between separate and distinct wireless systems serving different regions. Precall validation was and is still a means to thwart wireless fraud. It confirms that a mobile identity has not been illegitimately duplicated for nefarious use. To complete intersystem handoff when neighboring wireless markets were involved, a point-to-point DS1 link was required between the MSCs of the two providers.

IS-41 Revision A (IS-41A) utilized HLRs and VLRs to house and manage the transfer of subscriber profiles, providing the foundation to standardize the methods for automatic roaming, automatic call delivery, and intersystem handoff. This marked the beginning of a subscriber's ability, while roaming, to use the same system features that he or she normally uses in his or her home systems.

IS-41 Revision B (IS-41B) defined additional enhancements for roaming, including Global Title Translations (GTT), handoffs for IS54/136 digital wireless systems, and new messages, which supported the use of three-way calling and call waiting for calls that underwent an intersystem handoff.

IS-41 Revision C (IS-41C) offered major enhancements in intersystem signaling. It built on IS-41B by standardizing services that were projected to be commonly used by subscribers while roaming, such as short message service, authentication, and calling number identification presentation and restriction. IS-41C supported cellular systems (850 MHz carriers), as well as 1900 MHz PCS network operators, with a great deal of flexibility.

ANSI-41D, published in 1997, incorporates all of IS-41C functionality into a formal standard and is the latest evolution of the ANSI-41 standard. ANSI-41D defines support for authentication, short message service, automatic call delivery in border cells, message waiting notification, and caller identification (ID), to name a few. Today, the majority of wireless carriers use SS7 transport capability to provide connectivity to other wireless carriers across the country and the world. ANSI-41D messages are encapsulated into SS7 Transaction Capabilities Application Part (TCAP) facilities, which provide a reliable, high-capacity message transport medium to support intercarrier networking.

Key: The ANSI-41 standard for wireless interswitch signaling is not restricted to links between *different* wireless carriers. Wireless carriers use ANSI-41 for inter-MSC signaling within their own networks too. If a wireless carrier owns a market cluster with more than one MSC in the cluster, these MSCs can also be connected via ANSI-41 signaling over the carrier's own internal WAN. This use of intrasystem ANSI-41 links from MSC to MSC allows for intracluster calls to be sent from one MSC to another seamlessly. These clusters could include the expanse of an entire state, multiple states, or even the entire United States. This use of ANSI-41 signaling enables a carrier to avoid using the PSTN for intrasystem call transport, thereby reducing its interconnection costs. This is a fundamental example of least-cost routing.

15.5   Automatic Call Delivery (ACD)

ACD defines the process of seamlessly routing and terminating phone calls to wireless subscribers who are not in their home market. ACD allows wireless subscribers to receive calls regardless of where they travel when roaming.

15.5.1   The ACD Process

The ACD process works as follows:

1. John Doe is the person who is roaming out of his home service area. His mobile ID number is 708-422-3743. His home market is Chicago and he is roaming in Los Angeles.

2. When John Doe powers up his mobile phone, it immediately performs an autonomous mobile registration with the visiting wireless system in Los Angeles. As part of the registration process, John Doe's system identifier is downloaded to the HLR/VLR in the roamer switch. The SID is a three- or four-digit number that

identifies him as a subscriber of another carrier's system (Chicago) and reflects a unique identification for John Doe's home market. The roamer MSC then sends an ANSI-41 message encapsulated in an SS7 packet to John Doe's home MSC containing the SID, John Doe's MIN, and his phone's electronic serial number (the "fingerprint" of the phone) to inquire if he is a valid customer. Via HLR records, the home MSC then performs an authentication check, informs the roaming MSC in Los Angeles that John Doe is a legitimate customer, and forwards a copy of his feature set to the roamer MSC.

At this point, the roaming system's MSC knows that John Doe with MIN 708-422-3743 and ESN X is now resident in its system. If someone from John Doe's home market attempts to call him by dialing his mobile phone number, his home MSC will send an ANSI-41 message over the nationwide SS7 network, notifying the visiting MSC that there is an incoming call for John Doe. The roamer MSC then assigns John Doe a temporary listed directory number (TLDN) of 713-555-1212. The TLDN is obtained from a pool of phone numbers that is set aside from a line range that the roaming system's carrier obtained via interconnection in its market.

3. The roaming system then signals John Doe's home system (using ANSI-41 over SS7 signaling), informing the home system's MSC that John Doe has a temporary number in its system, and the number is 713-555-1212.

4. At that point, because John Doe's home MSC knows he is in the visiting system with a TLDN of 713-555-1212, so the home MSC routes the call to the roaming system using a combination of the PSTN (bearer path) and the nationwide SS7 network (signaling path). The home MSC routes the call to the roamer system via its dedicated IXC trunks, while an SS7 message is sent to notify the roaming system that the call is on the way.

5. Once the call reaches the roaming MSC, the paging process begins to find John Doe in the cell where he is located, and the call is routed directly to John Doe's mobile phone (to the TLDN number), as if he were a home subscriber in that market.

The TLDN's purpose is to act as a pointer—a temporary MIN—to John Doe's phone while he is roaming. It acts as an alias phone number while John is roaming in that foreign wireless network. That way, calls to John's wireless number are pointed directly to the roaming carrier's MSC via the TLDN, routed over the landline PSTN, and are eventually terminated to John's phone in the market where he is roaming. This is all made possible because John Doe's home MSC has all nationwide NPA-NXX combinations in its routing tables. Once the roamer MSC informs John Doe's home MSC what the TLDN is, the home MSC knows to route the call over its long-distance trunks because of the least-cost routing directions in its switch routing tables.

 Key: If someone attempts to call John Doe, that person does not need to know that John is roaming or even where he is located.

Automatic call delivery, also known simply as call delivery, is the process of seamlessly transporting a wireless call to a roaming customer who is in a nonhome wireless market. The market where the customer is roaming could be anywhere in the United States, or in many cases, the world, where wireless coverage exists and the carrier has SS7 connections into an SS7 network. The call is transported seamlessly because callers do not need to know the location of the mobile customer they are calling. Callers attempting to reach the wireless subscriber also do not need to dial a special roamer access telephone number.

Most large wireless carriers have massive market footprints today, in many cases stretching from sea to shining sea. In the past, cellular (and PCS) carriers relied heavily on third-party SS7 network providers such as NACN or Illuminet to support their seamless roaming needs—to support call delivery operations. These three options are described in the proceeding section. But today, wireless carriers have several options available to them when it comes to nationwide SS7 signaling and seamless call delivery operations.

15.5.2 SS7 Interconnection (ACD) Between Wireless Carriers and Third-Party SS7 Networks

First, in rare cases today, a wireless carrier could use a third-party SS7 network provider such as Verisign, if they so choose. This might be the case in a remote, rural market where the wireless carrier has no STP/SS7 presence and a low level of roamer traffic. It could be more economical to rely on a third party such as Verisign for SS7 transport, assuming Verisign has an SS7 presence in the area.

In this scenario, wireless carriers would connect *their own* mated STP pairs to a Verisign mated STP pair to allow for the termination of calls to *their* wireless subscribers in the Verisign-supported area.

Key: The connections from a wireless carrier's mated STP pair to a mated STP pair at a different (higher) network level are known as *D links*. They are called D links because they represent the diagonal linking of SS7 network elements at different levels of the SS7 network hierarchy: a regional STP pair (i.e., wireless carrier) to a nationwide STP pair. They are usually implemented using 56 Kbps data circuits, or in some cases DS1 circuits. In this scenario, the D link connections are known as a D-link quad because four STPs are actually all connected together, in a fully meshed, dual-homed configuration. This configuration allows for full redundancy of the STP pairs between the wireless carrier and the nationwide SS7 network provider.

See Figure 15-1 for an illustration of a D-link quad between a wireless carrier, mated STP pair and a nationwide SS7 provider's mated STP pair (Verisign). See Figure 15-2 for an illustration of this type of SS7 network topology between wireless carriers and third-party SS7 network providers.

Figure 15-1
D link quad connects mated STP pairs at different hierarchical levels of the SS7 network, such as a wireless carrier's mated STP pair to a Verisign mated STP pair.

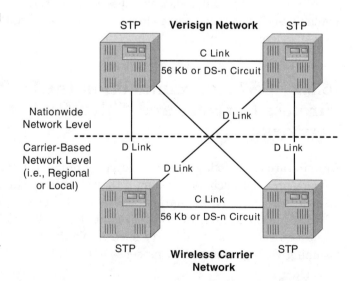

Figure 15-2
ACD to a roaming mobile subscriber using third-party SS7 network connections (e.g., Verisign). The actual call to a roamer is transported over landline PSTN. Note all links depicted (A, C, D) are landline-based circuits.

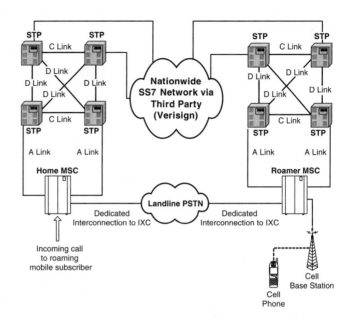

15.5.3 SS7 Interconnection (ACD) Between Different (Competing) Wireless Carriers

A second option available to wireless carriers for seamless roaming transport and SS7 connectivity is to connect *their* mated STP pair(s) to another wireless carrier's mated STP pair(s). This scenario could be mutually beneficial in a market where both carriers transport a high volume of wireless traffic *to each other's mobile customers* in that particular area. In this scenario, each carrier would connect its mated STP pairs to each other, in what is known as a B Link Quad.

A B link quad is *functionally* the same as a D-link quad. But again, the naming convention that describes the linking of mated STP pairs and the associated transport of traffic between carriers reflects a hierarchical architecture. The "B" in B link stands for "bridging." These links connect mated STP pairs at the same hierarchical network level. They are called *bridging links* because they are building a connection/signaling bridge from one type of carrier to another that operates at the same hierarchical network level. See Figure 15-3 for an illustration of a B Link Quad. See Figure 15-4 for an illustration of this type of SS7 network topology.

Figure 15-3

B link quad

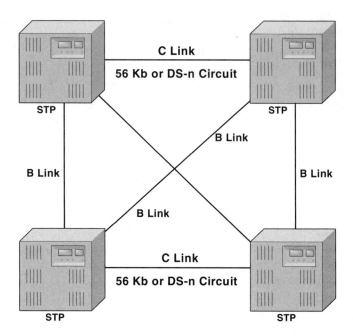

Figure 15-3
B link quad

Key: A, B, C and D links are all implemented using land-line, private, leased line circuits. Many times these links will be a part of the wireless carrier's own internal WAN. All these links in the SS7 world, including the (redundant) A links that connect wireless switches to STPs, have redundancy built into their operations as well as their architectures. Each network element has two connections into other network elements. This concept is known in the telecom industry as *dual homing* and is used extensively in sound WAN, MAN, and LAN designs. All of these links are also load balanced, meaning that the amount of traffic flowing over each link of a dual-homed connection is split roughly 50-50 onto each link. Therefore, each link operates at only 40 percent capacity. That way, if one of the A links goes down for some reason, the link that is still operational can sustain the traffic generated toward both STPs, with 20 percent excess capacity for traffic spikes.

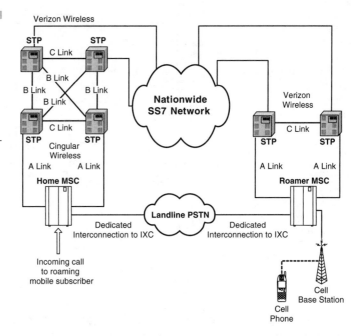

Figure 15-4
ACD: one wireless carrier to another wireless carrier. Carrier names are for sample purposes only.

15.5.4 Intracarrier SS7 Interconnection (ACD)

Naturally, because the major wireless carriers have nationwide market and coverage footprints, they also have nationwide SS7 networks. In most cases wireless carriers can and will simply use their own internal SS7 networks to support the signaling of wireless calls from one end of the country to the other. This applies mainly to transporting calls only between their own customers (i.e., Verizon customer to Verizon customer; Cingular customer to Cingular customer). See Figure 15-5 for an illustration of this setup.

15.6 Intersystem Handoff (IHO)

Intersystem handoff (IHO) is also made possible through the use of ANSI-41 signaling. Intersystem handoff is the process of transparently handing off a call in-progress from one wireless system to a geographically neighboring wireless system. This process is made possible because of an

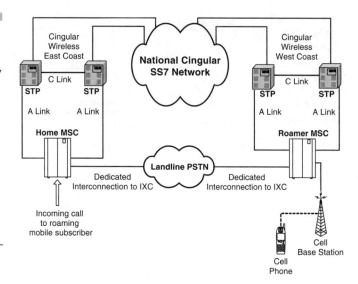

Figure 15-5
Intracarrier
ACD:
nationwide SS7
auto call
delivery
network. Note
that Cingular
Wireless was
used for this
example, but
any wireless
carrier could
apply in this
diagram.

MSC-to-MSC link between the neighboring markets that is used as the
bearer path for wireless conversations handed off over the border. This
link is known as an intermachine trunk (IMT). Any direct switch-to-
switch links are known as IMTs.

The intersystem handoff process is very similar to the regular (intramarket) call-handoff process and works as follows:

1. Engineers from wireless carriers whose markets are next to each
   other must select certain cells on their market borders that will be
   used for handing off calls between the markets.

2. A point-to-point DS1 circuit must then be installed between the
   MSCs in the two markets. An intersystem handoff requires a
   minimum of a DS1 circuit, because this IMT not only will have a
   signaling channel, but the conversation bearer channels also will use
   channels of the DS1 circuit. DS3 circuits (or higher) could also be
   used for an intersystem handoff. This depends on the amount of
   traffic over a given market border. The neighboring wireless carriers
   could also use SS7 signaling between their two MSCs to manage the
   IHO, versus dedicating a channel from the dedicated trunk group
   that acts as an IMT for the bearer path of the calls that traverse the
   border.

3. If John Doe is on the phone in his home system and he is traveling
   toward the neighboring wireless system, the cell that is carrying
   his call sends a message to the home MSC that John Doe's signal is

getting weaker. Over the control channel of that cell, what is also signaled to John Doe's home MSC is that he's about to enter a market boundary area and that an intersystem handoff may be required. This is all determined via received signal strength indicator (RSSI), just as it is with an intrasystem handoff.

4. One of the cells that is identified as a handoff candidate will be a border cell site in the neighboring system. This will be a cell that was selected for handoff by RF engineers from both wireless carriers.

5. John Doe's home MSC will then send a handoff request to the neighboring system's MSC over the point-to-point (DS1) link between the two MSCs; most likely over a channel dedicated to signaling and call setups/teardowns.

6. The two MSCs will exchange handoff signaling parameters, and the call will be handed off to the neighboring system over the dedicated circuit connecting the two MSCs. In that instant, the call is transferred from the home system cell to a border cell in the neighboring carrier's system. The call in progress has been handed off from one system to the neighboring system transparently.

15.7   International Roaming

With the world shrinking at the rate it is in lieu of the globalization of the world economy and the blazing pace of evolution in the information age, international roaming for wireless carriers has started to become a major action item for forward-looking wireless carriers. GSM carriers have a much easier time of it than carriers using any other digital wireless technology, because GSM by default requires strict adherence to its standards. This, along with the fact that there are more GSM carriers worldwide than any other type of wireless carrier, makes international roaming in the GSM world almost effortless. It is very easy for the GSM carriers across the world to link their systems together.

International roaming agreements are not too difficult to reach per se. But once roaming agreements are signed between two carriers from different countries, the technical aspects of the partnership take longer to work out. Tests and trials must be run with each partner. This can take about 4 to 6 weeks for each partner. One of the most trying aspects of the testing process is running trials with partners that operate in vastly different time zones. During these trials, carriers can run about 20 call scenarios to determine whether various features will function properly while roaming with the partner. It is also very important for international roaming partners to deliver call detail record (CDR) files to each other in a timely manner to ensure accurate and rapid billing.

To maximize the opportunities to increase revenues through international roaming, wireless carriers need to find partners with a style and pace that suit their customers' international roaming needs.

15.8   Wireless Intelligent Network (WIN)

Wireless intelligent network (WIN) is a concept being developed by the Telecommunications Industry Association Standard Committee. The charter of this committee is to drive intelligent network capabilities, based on the ANSI-41D standard currently embraced by wireless providers because it facilitates roaming. Basing WIN standards on the ANSI-41 protocol enables a graceful evolution to an intelligent network without making current network infrastructure obsolete.

15.8.1   The Current Status of WIN Standards

The movement to develop a WIN strategy was originally triggered by wireless carriers under the auspices of the Cellular Telecommunications Industry Association (CTIA). They developed a set of requirements calling for industry standards that defined a new network architecture incorporating the service flexibility of intelligent networks with the mobility aspects of wireless networks.

The first phase of WIN standards was published in 1999 and established the fundamental call models and operations required to support this flexible service architecture. Many service providers currently implement WIN Phase 1 in their networks. Examples of WIN Phase 1 services are calling name presentation and restriction, call screening, and voice-control services.

WIN Phase-2 standards are nearing completion, which will provide both additional service capabilities for wireless carriers as well as a greater harmonization of network capabilities and operations with emerging 3G network requirements. WIN Phase 2 includes MSC triggers for an intelligent network prepaid solution.

WIN Phase 3 is currently in requirements review by the WIN standards group. This phase incorporates enhancements to support location-based services. These requirements are based on four service drivers: location-based charging, fleet and asset management service, enhanced call routing service, and location-based information service.

15.8.2   WIN Functions

The WIN mirrors the wireline intelligent network. The distinction between the wireless and the wireline network is that many of the wireless call activities are associated with movement, not just the actual phone call. In the WIN, more call-associated pieces of information are communicated between the MSC and the service control point (SCP), or the HLR.

> *Key:* The WIN moves service control away from the MSC and up to a higher element in the network, usually the SCP.

   In a WIN, the SCP provides a centralized element in the network that controls service delivery to subscribers. High-level services can be moved away from the MSC and controlled at this higher level in the network. It is cost effective because the MSC becomes more efficient, does not waste cycles processing new services, and simplifies service development.

   In a WIN, there is also a device known as an intelligent peripheral (IP). This device gets information directly from the subscriber, be it credit card information, a personal identification number (PIN), or voice-activated information. The IP gets information, translates it to data, and hands it off to another element in the network like the SCP for analysis and control. See Figure 15-6 for an illustration of a WIN architecture.

15.8.3   WIN Applications

All the technologies already discussed in this chapter represent types of WIN applications. An intelligent network capability is required for various (precall) validations and the billing reciprocation of wireless calls. Along with transparent roaming services, the following lists other features that will be fostered by a WIN architecture:

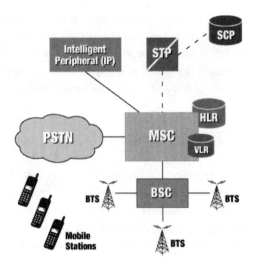

- Selective call screening and acceptance.

- Short message service.

- Speed-to-text conversion (hands-free operation), also known as voice-controlled dialing.

- Voice-based user identification (VUI), which permits a subscriber to place restrictions on access to services by using VUI.

- Incoming call-restriction or control, where incoming calls can be given certain treatments such as distinctive alerting, forwarded to voice mail or another number, routed to a subscriber-specific announcement, or even blocked.

- Calling name presentation (CNAP), which provides the name identification of the calling party to the called subscriber. This can be in the form of a personal name, company name, restricted, or not available.

- Password call acceptance (PCA), which is a call-screening feature that allows a subscriber to limit incoming calls to only those calling parties who can provide a valid password (i.e., a series of digits).

- Selective call acceptance (SCA), where incoming calls will only be accepted if they are on an approved screening list.

- Prepaid billing, and the ability to flag intelligent network calls where special billing records can be written right into the call record so that billing reflects the specific call handling.

The evolution of wireless networks to a WIN concept of service deployment delivers the following advantages, similar to the intelligent network benefits garnered by wireline providers:

- Multivendor product offerings that foster competition
- Uniform services to subscribers across service areas
- Efficient network utilization
- Rapid service creation and deployment

The following steps will need to occur before WIN will be a reality:

- Incorporation of SCPs and IPs into the wireless network architecture.
- Evolution of the MSC to an SSP. In the intelligent network, the SSP is the switching function portion of the network. The mobile switching center provides this function in the WIN.
- Separation of call control and transport from service control.
- Development of generic call models, events, and trigger points.

Test Questions

True or False?

1. _____ Automatic call delivery (ACD) defines the transparent routing of a call to a mobile customer when they are roaming.

2. _____ Use of the ANSI-41 standard for interswitch signaling is restricted to links between different cellular carriers.

3. _____ D links describe the links from a cellular carrier's MSCs to its own STP (or STPs).

4. _____ Automatic call delivery can occur using SS7 signaling between competing wireless carriers, when desired.

Multiple Choice

1. The links that attach the two members of one mated STP pair are known as:

   a.   C links

   b.   A links

   c.   B links

   d.   D links

   e.   F links

   f.   None of the above

2. Which standard describes interswitch networking in the wireless industry?

   a.   IS-54

   b.   IS-95

   c.   IS-45

   d.   ANSI-41

   e.   None of the above

3. The WIN architecture defines moving call processing away from the MSC to the:

   a.   VLR

   b.   SCP

   c.   TDM

   d.   ANSI-41

   e.   VoIP

   f.   None of the above

Wireless
Data
Technologies

The use of wireless data services is predicated on one main idea: We need to be as available and productive as possible, regardless of location. With the ability today to access information just about anywhere, at any time (i.e., Internet or corporate virtual private networks [VPNs]), mobile communication systems have been required to keep pace by offering the ability to access any network from just about anywhere. Today, these applications include the transmission of electronic mail (frequently with file attachments), faxes, web browsing, and access to real-time stock information, to name just a few. Tomorrow's applications will include video conferencing, location-specific advertising facilitated by new locator technologies that use global positioning system (GPS) or triangulation technologies, as well as personal entertainment services.

All of these applications can be broken down into three high-level categories:

- Query/response applications
- Batch file applications
- Streaming data, video, and/or voice applications

Examples of query/response applications include remote telemetry such as power meter readings, vending machine management, copier management, monitoring of burglar alarm systems, and data transfers from other sensing devices. Query/response applications also lie at the heart of most information services and wireless e-mail access.

Batch file applications are used to send large data files that are not time sensitive. Typical applications include the transmission of electronic documents (i.e., faxes), and shipping manifests. Batch file transport is not often used in personal, mobile communications.

Streaming applications contain time-sensitive information and are subject to latency concerns, but do not require guaranteed delivery due to their one-way transmission nature. This application area will be gaining serious traction in the marketplace by 2005, given the potential for new services such as the delivery of video programming (i.e., news clips), video conferencing, subscription-based streaming services, and streaming audio services.

The tools for accessing wireless data systems are also in a major state of change. Originally accessed by specialized cellular equipment slaved off of a laptop computer or personal digital assistant (PDA), wireless data systems can now be accessed using browser software or other proprietary, subscription-based software that is integrated or programmed right into wireless phones. New body local area network (LAN) personal area network (PAN) technologies, such as Bluetooth, now allow devices to automatically communicate without cables. Body LAN describes a LAN that operates within the immediate vicinity of a person's body. As a result of these new technologies, changes made to the calendar on a PDA can be automatically updated on a laptop computer located in a person's nearby briefcase and automatically sent to the corporate calendar service via a nationwide TCP/IP network.

16.1   Use Paradigms

Three paradigms exist for wireless data use:

- *Full user mobility* allows for session handoff while moving during automobile travel, for example.

- *Portable wireless data* allows access to a data service while in a given coverage area using a laptop, PDA-phone combination, or a Blackberry device. Wireless modems (transceivers) are now built into these devices so no extraneous equipment or parts are required. This paradigm does not offer full mobility or session handoff during online sessions. Access to the Internet while moving at a relatively high rate of speed will result in a dropped connection at the edge of the coverage area. Examples of portable wireless applications include browsing the Internet from a wireless fidelity (Wi-Fi) café, a Starbucks, a McDonalds, an airport, a hotel, or a convention center.

- *Fixed wireless data* is offered to an office or home through large customer-premise antennas. Sprint offered Sprint Broadband around the year 2000, which used multichannel, multipoint distribution system (MMDS) technology, also known as *wireless cable*, to support Internet access from homes or businesses at high-speed, broadband data rates. Sprint has since then grandfathered the service. However, a perfect example of fixed wireless data is the upcoming 802.16 WiMAX standard (see Chapter 21).

Within these paradigms, there are technically two media access methods in wireless data communications:

- Circuit-switched access
- Packet-switched access

Circuit-switched wireless access establishes a standard air link (air interface) connection to the nearest cell site, to a switch (mobile switching center [MSC] or peripheral), and then into the public switched telephone network (PTSN). This is accomplished by using modem contention and data transmission, usually employing Microcom Networking Protocol class 10 (MNP 10) or Enhanced Throughput Cellular (ETC) protocols. With the exception of the noise and distortion inherent in the air link, the circuit-switched data call is no different from a landline-based circuit-switched data call (i.e., dial-up Internet access). Of course, the protocols that are used to negotiate, transmit, and receive user data in a wireless context are designed specifically for this purpose as well.

Key: Circuit-switched access is all but dead today as users have grown dependent on, and used to, easy-to-use packet-based wireless data transport technologies, or streaming applications. A perfect example of this type of technology was Cellular Digital Packet Data (CDPD). CDPD was a circuit-based wireless access technology that is rapidly fading in the marketplace, if not already dead (as of late 2004). Packet-based access technologies are also better for wireless carriers as packet transport can be converged with voice transport, it uses less network resources than circuit-switched data, and it costs less due to the fact that dedicated circuits are not required to support every data transport session.

Circuit-switched wireless applications can run the spectrum of various remote access applications, including online services, e-mail, fax, LAN access, and real-time file transfers.

Packet-switched wireless access establishes a session, but does not use a dedicated bearer channel for the user data as circuit-switched data does. Instead, the wireless access device is able to request access to a network cloud (IP) to send data, as packets are generated by the source application. For mobile-terminated data traffic, the base station will send data addressed to the wireless access device on a time slot when the base station receives it. This allows the base station controller to use the bearer channel to service multiple wireless access devices with a limited number of bearer channels.

16.2 Signal Quality

The greatest difference between landline-based data transmission and mobile wireless data transmission is the quality of the connection to the user's device. In landline communications this is generally achieved by using a twisted pair of copper wires, which has both the advantage and disadvantage of being physically fixed between two points, an advantage because the communications path is of known, fixed quality. It is a disadvantage because in today's highly mobile work environment, it is very inflexible.

The wireless air link, however, is just the opposite. The air link is very flexible but has quality of service (QoS) problems that are inherent with all radio frequency (RF) systems. This is due to the inherently dynamic nature of radio itself. The characteristics of a wireless radio channel that affect quality include signal attenuation (free-space loss) and a very noisy environment that can cause a significantly higher bit error rate (BER) than a landline call. Also, because the data that is transmitted over this medium is digital, this causes major concerns for data integrity. As discussed in Chapter 4, the composition and density of objects between the source of the transmission (mobile data terminal) and the cell site are extremely variable by location, time of day, season, and many other factors.

Wireless signals also suffer a wide range of attenuation within buildings, ranging from 30 to 40 decibels. The air link is plagued by multiple sources of interference and noise, including man-made and white or Gaussian noise, as well as effects of multipath, or Rayleigh, fading.

16.3 Short Message Services

SMS is a globally accepted wireless service that enables the transmission of alphanumeric messages between mobile subscribers and external systems such as e-mail, paging, and voice-mail systems.

16.3.1 Overview

SMS first appeared on the wireless scene in 1991 in Europe. GSM included short messaging services as part of its standard from the outset.

In North America, SMS was initially made available on digital wireless networks built by early pioneers such as BellSouth Mobility, PrimeCo, and Nextel, among others. (Note: Bellsouth Mobility is now part of Cingular Wireless, PrimeCo was dissolved, and Sprint PCS has made a bid for Nextel). These digital wireless networks were based on every major digital wireless technology: GSM, CDMA, and TDMA standards (IS-54, IS-136).

Network consolidation from mergers and acquisitions has resulted in large wireless networks with footprints spanning the entire United States or even international coverage, and these networks sometimes support more than one wireless technology. This new category of service providers demands network-grade products that can easily provide a uniform solution, enable ease of operation and administration, and reliably accommodate existing subscriber capacity, message throughput, future growth, and services. Short messaging service center (SMS-SC) services based on an intelligent network approach are well suited to these requirements.

Figure 16-1 represents the basic network architecture for an ANSI-41 SMS-SC deployment supporting multiple input sources, including a voice-mail system, Web-based messaging, e-mail integration, and other external short message entities (ESMEs). Communication with other network elements such as the home location register (HLR) and MSC is achieved through the signal transfer point (STP).

SMS provides a mechanism for the transmission of short messages to and from wireless devices. The service makes use of an SMS-SC, which

Figure 16-1
Basic SMS network architecture

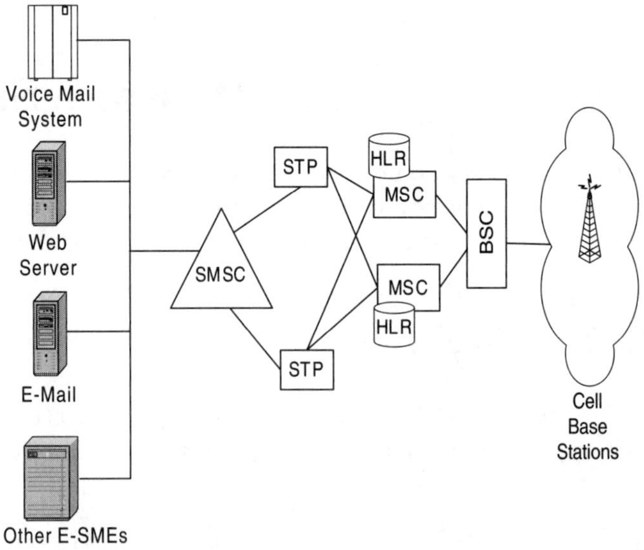

acts as a store-and-forward system for short messages. The wireless network provides the mechanisms required to locate the destination station(s) (i.e., paging) and transports short messages between the SMS-SCs and the wireless stations (mobile phones or similar devices). SMS service elements are designed to provide guaranteed delivery of text messages to the destination. SMS also supports several input mechanisms that allow interconnection with different message sources and destinations.

Key: A differentiating characteristic of SMS service is that an active mobile handset is able to receive or submit a short message at any time, regardless of whether a voice or data call is in progress (in some implementations, this may depend on the MSC or SMS-SC capabilities). Temporary failures due to unavailable receiving stations are identified, and the short message is stored in the SMS-SC until the destination device becomes available.

SMS is characterized by out-of-band packet delivery and low-bandwidth message transfer, which results in a highly efficient means for transmitting short bursts of data. Initial applications of SMS focused on eliminating alphanumeric pagers by permitting two-way general-purpose messaging and notification services, primarily for voice mail. As technology and networks have evolved, a plethora of services has been introduced (see Chapter 18).

Wireless data applications supported by SMS include the downloading of subscriber identity module (SIM) cards for activation, debit transactions, wireless points of sale (POS), and other field-service applications such as automatic meter reading, remote sensing, and location-based services. Integration with the Internet also spurred the development of Web-based messaging and other interactive applications such as instant messaging, gaming, and chatting.

16.3.2 Network Elements and Architecture

Like any network infrastructure, SMS systems have multiple (network) elements that are all required to make SMS systems operational.

16.3.2.1 External Short Messaging Entities An ESME is a device that may receive or send short messages, and will usually take the form of some type of server, likely housed at the MSC. The short message entity (SME) may be located in the fixed network, a mobile device, or another service center (i.e., where a BSC might also be located). Types of ESMEs are described in the following list.

- **VMS**—The VMS is responsible for receiving, storing, and playing voice messages intended for a subscriber that was busy or not available to take a voice call. It also sends voice-mail notifications for SMS subscribers.

- **Web**—The growth of the Internet has also affected the world of SMS. Therefore, it is almost mandatory to support interconnections to the World Wide Web for the submission of messages and notifications. The increasing number of Internet users has a positive impact on incremental SMS traffic. The growth of Internet applications will likely lead to the inevitable growth of SMS services.

- **E-Mail**—Probably the most demanded application of SMS is the ability to deliver e-mail notifications and to support two-way e-mail, using an SMS-compliant terminal. The SMS-SC must support an interconnection to e-mail servers acting as message transfer systems.

Key: Data services in general, and especially a service like SMS, have simply not taken off in the United States. They have not gained a market foothold, in spite of advertising and promotional efforts by some wireless carriers. These types of services are extremely popular in many other parts of the world such as Europe and Japan. But the United States always seems to lag the rest of the world when it comes to the widespread adoption of mobile data services. With good marketing and a sound technological base (i.e., data services based on General Packet Radio Service [GPRS], SMS, 3G-based applications), this trend may yet be reversed between 2005 and 2008.

16.3.2.2 The Short Message Service Center (SMS-SC) The SMS-SC is a combination of hardware and software responsible for the relaying, storing, and forwarding of a short message between an SME and a mobile device.

The SMS-SCC must have high reliability through inherent system redundancy and dual-homed connections to other critical network elements. It must also have abundant processing and storage capacity, the ability to support many thousands of subscribers, and message throughput. The system should also be easily scalable to accommodate the growing demand for SMS in the network.

Normally, an IN-based solution such as SMS will allow for a lower entry cost because it can support other applications on a single hardware platform and share resources, thereby distributing the deployment cost over several services and applications.

The SMS-SC is easy to operate and maintain, with the flexibility to activate new services and upgrade to new software releases.

16.3.2.3   Signal Transfer Point (STP)   The STP is a network element normally available on intelligent network deployments that allows IS-41 interconnections over SS7 links with multiple network elements. SMS makes use of the mobile application part (MAP) of the SS7 standard, which defines the methods and mechanisms of communication in wireless networks, and makes use of SS7 transaction capabilities application part (TCAP).

16.3.2.4   HLR   and   Visitor   Location   Register   (VLR)   As noted in Chapter 15, the HLR is a database used for permanent storage and management of home user subscriptions and service profiles. Upon inquiry by the SMSC, the HLR provides routing information for designated subscribers. Also, if the destination station is not available when the message delivery is attempted, the HLR informs the SMS-SC that the station will be recognized by the mobile network as accessible when it becomes available again, and the message can be delivered. The SMS-SC has a true store-and-forward capability. The VLR is a database that contains temporary information about subscribers homed in one market/HLR who are roaming in another market/HLR. This information is needed by the MSC to service visiting subscribers.

16.3.2.5   MSC, the Base Station Subsystem, and the Mobile Terminal   Like all wireless networks, SMS also requires the use of the fundamental components of any wireless system: The MSC performs the switching functions of the system and controls calls to and from other telephony and data systems. The MSC delivers the short message to the specific mobile subscriber through the proper base station. The base station subsystem consists of base station controllers (BSCs) and the base transceiver stations (BTSs) themselves, also known as cell sites or simply cells. The mobile device is the wireless terminal (phone or similar device) capable of receiving and originating short messages. These devices have commonly been digital wireless phones, but more recently the application of SMS has been extended to other terminals such as point of sale (POS), handheld computers, and PDA.

The capabilities of the mobile terminal may vary depending on the digital wireless technology being used by the subscriber. Some functionality, although defined in the SMS specification for a given wireless technology, may not be fully supported in the terminal, which may represent a limitation in the services that the carrier can provide. This trend, however, is disappearing as service providers' merger and acquisition activity demands uniform functionality across all the parent companies.

16.3.3 Service Elements

SMS is comprised of several service elements relevant to the reception and submission of short messages:

- *Message Expiration*—The SMS-SC will store and reattempt delivery of messages for unavailable recipients until either the delivery is successful or the expiration time is reached, which is set on a per-message basis or on a platform-wide basis.

- *Priority*—This is the information element provided by an SME to indicate the urgency of messages and differentiate them from the normal priority messages. Urgent messages usually take priority over normal messages, regardless of the time of arrival to the SMS-SC platform.

- *Message Escalation*—The SMS-SC stores the message for a period no longer than the expiration time (it is assumed that the escalation time is smaller than the expiration time associated with the message), and after said escalation time expires, the message will be sent to an alternate message system (such as an e-mail server) for delivery to the user.

SMS also provides a time stamp reporting the time of submission of the message to the SMS-SC and an indication to the handset of whether or not there are more messages to send (GSM) or the number of additional messages to send (ANSI-41).

16.3.4 Subscriber Services

SMS comprises two basic point-to-point services:

- Mobile-originated short message (MO-SM)
- Mobile-terminated short message (MT-SM)

MO-SM are transported from the MO-capable handset to the SMS-SC and can be destined to other mobile subscribers or for subscribers on fixed networks such as IP networks (including the Internet and private e-mail networks). MT-SM are transported from the SMS-SC to the handset and can be submitted to the SMS-SC by other mobile subscribers via MO-SM or by other sources such as voice-mail systems, paging networks, or operators.

Key: Depending on the access method and the encoding of the bearer data, the point-to-point short messaging service conveys up to 190 characters to an SME in GSM networks and from 120 to 205 in ANSI-41 networks. *Typically, the average length of an SMS message is quoted as 160 characters.*

Many service applications can be deployed by combining these service elements. Aside from the obvious notification services, SMS can be used in one-way or interactive services that provide wireless access to any type of information, anywhere. By using new emerging technologies that combine browsers, servers, and new markup languages designed for mobile terminals, SMS can enable wireless devices to securely access and send information from the Internet or intranets quickly and cost-effectively. One of these technologies where SMS can provide a cooperative method, versus a competitive approach, is the Wireless Application Protocol (WAP), which is a protocol based on Extensible Markup Language (XML) that supports the transport of data for mobile wireless users.

16.3.5 SMS Applications

SMS was initially designed to support limited-size messages, mostly notifications and numeric or alphanumeric pages. Although these applications are still important and widely used, there are now more recent service niches that can make use of SMS technology:

Information Services—A wide variety of information services can be provided by SMS, including weather reports, traffic information, entertainment information (e.g., cinema, theater, concerts), financial information (e.g., stock quotes, exchange rates, banking, brokerage services), and directory assistance. SMS can support both push (MT) and pull (MO) approaches to allow delivery under specific conditions and even delivery on demand, as a response to a request.

E-mail Interworking—Existing e-mail services can be easily integrated with SMS to provide e-mail to short messaging conversion and mobile e-mail and message escalation.

Notification Services—Notification services are currently the most widely deployed SMS services. Examples include the following:

- Voice/fax message notification, which indicates that voice or fax mail messages are present in a voice mailbox.
- E-mail notification, which indicates that e-mail messages are present in an e-mailbox.
- Reminder/calendar services, which issue reminders for meetings and scheduled appointments.

WAP Integration—SMS can deliver notifications for new WAP messages to wireless subscribers and can also be used as the transport mechanism for WAP messages. These messages can contain information from diverse sources that include databases, the World Wide Web, e-mail servers, and so forth.

Mobile Data Services—The SMS-SC can also be used to provide short message wireless data. The wireless data may be in interactive services where voice calls are involved. Some examples of this type of service include fleet dispatch, inventory management, itinerary confirmation, sales order processing, asset tracking, automatic vehicle location, and customer contact management. Other examples may be interactive gaming, instant messaging, mobile chat, query services, mobile banking, and so forth.

Customer Care—The SMS-SC can also be used to transfer data in binary form that can be interpreted by the mobile device without presentation to the customer. This resource allows the operators to administer their customers by providing a means for programming the mobile device. Such services could include mobile device programming, which allows customer profiles and subscription characteristics to be downloaded to the mobile device (customers can be activated/deactivated based on the data downloaded). This capability could also include *advice of charge*, which enables SMS to report charges incurred for a phone call (e.g., calls made when roaming).

Customer Reminders—A wireless service provider could also deliver short messages to subscribers to remind them of past-due payments, for example, instead of reminding them over traditional mail or courier delivery. This would reduce cost and ensure that the message is delivered to its destination in a timely manner.

Restaurant POS—Short bursts of data are at the heart of many applications that were restricted to the world of data networks with fixed terminals attached to a LAN or wide area network (WAN). But many of these applications are better served if the data communication capabilities could be added to the mobility of the terminals. A great example of this concept is a waiter who can charge a customer's credit card right at the table, at any time, instead of going to a fixed POS terminal located by the register. This helps both the waiter *and* the customer meet their goals in a more efficient way.

GPS-Based Tracking—SMS is ideal for sending GPS information such as longitude, latitude, bearing, and altitude. GPS coordinates are typically about 60 characters in length. The ability to track the location of a moving asset such as a truck or its load is very valuable for both providers and clients. Again, this type of application just needs to interchange small amounts of information (i.e., 60 character bursts), such as the polar coordinates of a vehicle (longitude and latitude) at a current time of the day, and perhaps other parameters like temperature or humidity. This application does not necessarily require the monitored entity to be in movement. The requirements are basically short, bursty data and a location that has digital network coverage. For example, in a neighborhood, it would be faster, easier, and cheaper to drive a truck from the local power company, which sends inquiries to intelligent meters to obtain its current readings and then forward them via short message to a central data processing center to generate the billing. (Note: This type of application could be competing with a technology like Cellemetry, which offers the same type of service). Similarly, delivery trucks could be alerted to the inventory of a customer running low, when the truck is close to the customer's facilities. The truck driver could place a quick phone call to the customer to offer a temporary replenishment at a low cost for the distributor.

Banking—Another suite of applications that can use SMS as a data transport mechanism is banking. It is no secret that automated teller machine (ATM) and Internet transactions are less costly than transactions completed at a branch bank location. Internet transactions are even cheaper than ATM transactions. Enabling wireless subscribers to check their balances, transfer funds between accounts, and pay their bills and credit cards is valuable not only for the subscriber, but also for financial institutions. This ultimately lowers everyone's costs (bank) and charges (customers).

Entertainment—Entertainment applications are also good drivers of SMS usage. Examples of these are simple short message exchanges between two parties (texting) or between multiple participants (chat).

Today, wireless Web surfing allows users to search for information without the physical constraints of a PC. College students would certainly appreciate not having to go to the computer lab or their dorm to check e-mail or find out what the required book is for the semester that is about to start.

E-mail continues to be by far the most used wireless data application. But handsets are evolving *very* quickly and are including more and more functionality that supports newer applications at the same time that the ergonomics factor improves. The next big success beyond wireless Web will probably be Internet shopping and other e-commerce applications such as electronic coupons, advertising, and so forth.

16.3.6   Benefits of SMS

In today's competitive world, differentiation is a significant factor in the success of any wireless service provider. Once basic service such as voice telephony is deployed, SMS can provide a powerful vehicle for service differentiation. If market forces allow for it, SMS can also represent an additional source of revenue for the service provider. The best approach for assessing fees for a service like SMS would be in the form of a small monthly service fee, versus charging per message. A wireless carrier would not want to put the customer in a position of counting their SMS transactions, much like many customers used to (and still do) count their minutes in their calling plans.

The benefits of SMS to subscribers include convenience, flexibility, and seamless integration of messaging services and data access. So a primary benefit is the ability to use the mobile handset as an extension of the computer. SMS also eliminates the need for separate devices when messaging because services can be integrated into a single wireless device: the mobile terminal. These benefits normally depend on the applications that the service provider offers.

SMS benefits to mobile subscribers include the following:

- Delivery of notifications and alerts
- Guaranteed message delivery
- Reliable, low-cost communication mechanism for concise information
- Ability to screen messages and return calls selectively
- Increased subscriber productivity
- Delivery of messages to multiple subscribers at a time
- Ability to receive diverse information
- E-mail generation
- Creation of user groups
- Integration with other data and Internet-based applications

The benefits of SMS to the service provider are as follows:

- Potential to increment average revenue per user due to increased number of calls on wireless and wireline networks by leveraging the notification capabilities of SMS
- Ability to enable wireless data access for corporate users
- New revenue streams resulting from the addition of value-added services such as Web-based application integration, reminder service, stock and currency quotes, or even airline schedules
- Provision of key administrative services such as advice of charge; over-the-air downloading of audio, video, or image files; and over-the-air service provisioning
- Protection of important network resources (i.e., voice channels), due to the careful use of the control and traffic channels by SMS

16.4 Cellular Digital Packet Data (CDPD)

This section offers a historical look at the earliest wireless data service offered by legacy 850 MHz cellular carriers in the 1990s. It should be noted that circuit-switched analog wireless services such as CDPD are rapidly dying and being replaced by other wireless data technologies such as SMS, GPRS, and 3G-based technologies. So this section is essentially a brief history lesson. However, it should be noted that many wireless carriers still offer some form of circuit-switched data, for the time being anyway.

The CDPD access protocol was developed jointly in the early 1990s by Ameritech Mobile Communications, Bell Atlantic Mobile Systems, GTE Mobilenet, Contel Cellular, McCaw Cellular Communications, NYNEX Mobile Communications, AirTouch Cellular, and Southwestern Bell Mobile Systems to provide data services on an existing analog cellular infrastructure. It should be noted that none of these carriers exist today due to mergers, buyouts, and industry consolidation.

The CDPD protocol was standardized in the 1995 to 1996 time frame; CDPD transmissions were commonly used to support query/response, small batch entry, and telemetry applications. The carriers listed in the preceding passage designed the network to support the following goals:

- Seamless service for users in all network systems
- Future growth and scalability based on standard open network protocols, including open systems interconnection (OSI) Connectionless Network Protocol (CLNP) and TCP/IP
- Maximum use of available network infrastructure
- Protection of data and user's identity
- Minimal configuration parameters

The CDPD architecture extended to the OSI Network Layer 3. Mobile end systems (M-ES) supported user application data recovery in the higher layers of the OSI model. This functionality is also common elsewhere in data communications.

CDPD end systems (ESs) were the logical endpoints of communications and were the entities addressed as source or destination (service access points). M-ES generally took the form of laptops or palmtops with CDPD modems. Fixed end systems (F-ESs) hosted data applications on server or mainframe systems and were completely isolated from the mobility issues of the M-ESs they communicated with during transmissions. Intermediate systems (ISs) were network entities providing network relay functions by receiving data from one correspondent network entity and forwarding it to another. Intermediate systems could be physically located anywhere in the wireless system: at the MSC or at the base station. However, it was critical that they were situated behind the mobile data intermediate system (MD-IS) in the network hierarchy. ISs were basically routers and were unaware of mobility. Mobile data intermediate systems were the intermediate systems, which provided mobility management for the M-ESs in a given service area by defining the home and service (location) functions, which can span different carrier's CDPD networks. Mobile data base stations (MD-BSs) were the cellular radios designed specifically for CDPD connections.

Channel-hopping technology utilizes the empty spaces (i.e., silence) in voice conversation to carry one or more datagrams. Packetized data can efficiently be placed in these unused portions of the voice channel to optimize the use of available bandwidth. To do this effectively the data transmission must never interfere with the voice call. The MD-BS (radio) had a contention method that was to allow packetized data to be applied to the unused portions of the voice channel in either a planned hop or forced hop mode.

A planned hop involves the auto-sensing of unused bandwidth, and data would be inserted in a continuous flow without buffering. A planned hop means that there is bandwidth available on a given RF channel, but another channel has more unused bandwidth, so the system (the CDPD protocol) plans to move, or hop, the data transmission to another channel to make use of even more unused bandwidth.

A forced hop means that CDPD transmission is occurring within the unused bandwidth of a channel that is carrying a voice call. Because voice traffic is considered higher in priority (quasi QoS), if no channels are available for an incoming voice call, the new voice call will come into the channel and kick the CDPD transmission to another RF channel.

Forced channel hops require that the data be buffered until unused bandwidth is available. This action could cause a loss of the connection and for this reason most wireless carriers over time chose to avoid channel-hopping technology altogether in their CDPD networks. They ended up assigning fixed, static radio channels to carry CDPD traffic.

Key: Wi-Fi, SMS, Blackberry, and Mobile IP technologies have taken the place of CDPD, which has gone the route of the horse and buggy from a wireless perspective.

16.5   Cellemetry

Cellemetry is a unique application of wireless carrier networks that uses the control channel as its foundation.

16.5.1   Overview

Cellemetry® data service provides nationwide, economical two-way communications between remote equipment and a central processing facility by using an underutilized portion of wireless networks: the (overhead) control channels. Cellemetry was developed by Bellsouth RF engineers when they ran tests and confirmed the underutilization of the control channels in AMPS networks.

Cellemetry uses wireless control channels to transport telemetry messages. The message-handling capacity of the control channels is far greater than what is required by the wireless system, even during the busiest times of the day. Empirical tests in wireless markets demonstrate that there is sufficient capacity to operate both the wireless system operations and Cellemetry data service.

A key operation of Cellemetry systems is that they perform an operation that the wireless system performs thousands of times every day: verification of wireless subscribers, even roamers.

Key: Cellemetry radios, which are a special type of transceiver, function exactly like standard (or roaming) mobiles except the MINs are specially assigned so that the MIN and ESN are routed to a Cellemetry gateway connected to the same intrawireless network. With Cellemetry, the MIN serves to identify the Cellemetry radio and the ESN is the data field. The Cellemetry gateway adds a timestamp and the SS7 network adds an approximate location of the message's point of origin, known as mobile switching center identification (MSCID).

The Cellemetry gateway processes, stores, and/or routes messages according to profile type and customer requirements. Some applications may require immediate processing, such as alarm monitoring. Others, such as utility meter reading, may only require the messages be stored and transferred in a batch file once a day.

As the name implies, Cellemetry data service reflects the coupling of wireless and telemetry technologies. From a remote location to the central processing location, Cellemetry provides reporting capability for short, telemetry messages. From the central location to the remote location, Cellemetry provides a remote control capability as well. Cellemetry operations are virtually transparent to the wireless carrier and are transparent to wireless customers. This service utilizes a centralized service bureau connected via the SS7 network and requires no equipment additions or modifications by the wireless carrier—only standard database translations, which are normally done for any roaming mobile subscriber. More significant is the fact that Cellemetry covers the entire wireless system's coverage area from the first day of operation. There is no need to concentrate users to make the system economically feasible, and there is no need to consider deploying a wireless overlay technology. Likewise, there are no delays while the system infrastructure is being installed at every wireless base station.

Cellemetry can be integrated with many services that require one-way or two-way short telemetry messages, such as SMS. Flexibility is further enhanced because the Cellemetry gateway can process messages according to the specific needs of each individual application. For instance, if the service is used to convey a message from an alarm panel, the Cellemetry gateway will process the message on a real-time, immediate basis and pass the message to the central alarm-monitoring service. On the other hand, if a soft drink vending company utilizes Cellemetry to poll its machines each night for its stock status, the Cellemetry gateway will accumulate all of the responses from the individual vending machines at night and provide them in batch form when requested from the vending company the next morning. Moreover, individual applications can have different responses from the same Cellemetry radio. While the vending machine uses batch processing for its stock status, it could have an alarm message sent to the vending company on an immediate basis.

A similar scenario applies to utility meter reading. Normal meter readings can be obtained on a batch basis during the night and delivered to the utility company the following morning. However, real-time meter readings can be made any time during the day for customers who desire to close out or open service and require an immediate, current meter reading. In fact, Cellemetry can even be used to turn on or turn off service remotely from the customer service center. With Cellemetry, waiting a day for the utility technician to travel to a residence and reactivate a meter to begin service can be a thing of the past.

16.5.2 System Description

Every wireless network has a Forward Control Channel (FOCC) and a Reverse Control Channel (RECC). The FOCC is used to send general information from the wireless network (through the base station) to the mobile user. The RECC is used to send information from the mobile phone to the base station. The control channels are used to initiate and manage wireless calls (or sessions when transmitting data). The wireless control channels are more robust than voice channels. First, each message is transmitted five times via the RECC. If the wireless base station receives the same message from any three of the five transmissions, it deems the message as correct. Second, the frequency reuse plan for control channels is different than the reuse plan for voice channels, and it reduces interference potential on the control channels. Third, most wireless operators use the control channels at a proportionally higher transmitted power than voice channels. So Cellemetry can be reliably used in environments in which a wireless call could never be made due to low signal levels.

The data flow with Cellemetry service is the same as the data flow for roaming wireless customers. The forward control channel broadcasts the system identification (SID) of the wireless system on a frequent basis. Every time a mobile is turned on, it compares the SID of its home system, which is stored in its nonvolatile memory, to the SID being sent over the forward control channel. If a match is obtained, indicating that the mobile is operating in its home system, nothing happens. If a match is not obtained, indicating that the mobile subscriber is a roamer, the mobile phone illuminates its roam light to alert the customer that roaming wireless rates will apply. Depending on the wireless carrier's preference, roamers may be required to register as often as each call or as infrequently as once a day.

Key: By connecting the Cellemetry gateway to the ANSI-41 intersystem signaling network, validation request messages can be captured in exactly the same manner as a roaming mobile phone validation request. The Cellemetry gateway is a fully redundant, fault-tolerant system, centrally interconnected to the SS7/ANSI-41 network. The gateway can operate reliably with just one single link, even with hardware or software failures.

Mobiles are replaced by Cellemetry radios at the required point of transmission (e.g., a utility meter or vending machine, for example). The signal path is the same for all Cellemetry applications.

The MIN now becomes the 10-digit equipment ID (i.e., utility meter ID) and the ESN becomes the data payload. Information is added to the database of the roaming MSC to direct the dedicated MINs assigned to the Cellemetry gateway. The Cellemetry gateway processes the equipment ID (MIN) and data payload (ESN) as determined by the particular Cellemetry application profile. The Gateway uses IP via frame relay, the Internet, or dial-up modem to communicate with the Cellemetry customer (e.g., the security company, electric utility company, vending company, etc.).

In order to make the Cellemetry system appear transparent to the wireless system, the gateway must send the proper validation response back to the (roaming) MSC. But in the case of Cellemetry service, the gateway sends back a validation response that reports the MIN and ESN as being valid, but that the customer *cannot make or receive any telephone calls*. This validation is an additional safeguard against fraud. Even if someone were to intercept a Cellemetry unit's MIN and ESN, they could not use this MIN/ESN combination to make wireless calls.

16.5.3 Modes of Operation

There are two modes of operation of the Cellemetry radio: the modem mode and the stand-alone mode. In the modem mode, the Cellemetry radio acts exactly like a modem, passing information in both directions without modification. A host controller will be required in the modem mode to direct the action of the Cellemetry radio to initiate the transmission of the Cellemetry messages via the RECC and to encode the information to be transmitted in the ESN field. The host controller also interprets the MINs transmitted via the forward control channel to the Cellemetry radio and takes appropriate action.

In the standalone mode, the Cellemetry radio uses internal software to provide a specific functionality without the need of an external host controller. This mode is designed for applications where device size, power consumption, and cost must be minimized. The Cellemetry radio can be used to respond with a pulse count, such as a utility meter reading or copier counter reading, either on an immediate basis or on a delayed basis (e.g., zero to three hours) depending on the MIN transmitted via the FOCC. The Cellemetry radio can also be used to turn on or turn off a remote device in the standalone mode, as well as report a contact closure (i.e., security or system alarms).

In either mode, the Cellemetry radio can respond to a broadcast page. This design feature permits one page via the FOCC to initiate thousands of registrations via the RECC. Many other current data systems require one forward channel message for the initiation of one reverse channel message (a one-to-one ratio). Forward channel traffic can be greatly reduced using Cellemetry service.

16.5.4 Applications

Cellemetry service can provide communication for many different applications. The list of potential applications is extensive. For example, Cellemtry can report security alarms, copy counts for photocopiers, utility meter readings, pipeline corrosion monitoring, or railroad crossing-gate monitoring, to name just a few applications.

Cellemetry data service applications fall into four general categories:

- Monitoring—copiers, meters, vending machines
- Alarm/threshold—home alarm, temperature, pollution
- Location/geo-positioning—container tracking, vehicle tracking using automatic vehicle locator (AVL)
- Remote control (outbound) —turn electricity on/off, flow control

Let us look at an electric meter-reading application. A Cellemetry radio is installed in the electric meter. Operating in the stand-alone mode, circuitry within the electric meter sends pulses to the Cellemetry radio, reflecting the power usage at the residence or business. During the night at a desired interval (typically once a month for residential customers), the Cellemetry gateway transmits a page over the wireless system using a special MIN that is assigned to a number of Cellemetry radios in the system. This special MIN, called the group MIN, causes the Cellemetry radios to read the electric meter immediately but reply randomly over the next three hours. In this manner, all of the Cellemetry radios with the same group MIN will not try to respond at once, which reduces transmit collisions. The Cellemetry gateway will accumulate all of the meter readings and report them to the electric utility company when requested. If an immediate meter reading is required, such as with closing out or opening an account, a MIN page corresponding to the Equipment ID of the meter is transmitted over the wireless network. The Cellemetry radio responds almost immediately with its current meter reading. There are other special MINs, which, when transmitted via the FOCC, cause a switching signal to be delivered to the electric meter, providing a means of controlling a device such as a contactor who wants to disconnect or connect electric service. See Figure 16-2.

Another Cellemetry application is security alarm panel reporting. In this scenario, the Cellemetry radio operates in the modem mode connected to the host controller in the security panel. Following industry-established protocol, the Cellemetry radio sends a message when directed by the host controller. The MIN identifies the account or customer while the ESN identifies the type of alarm, such as intrusion, panic, fire, and so forth. Once the Cellemetry message is received at the Cellemetry gateway, the gateway places a call to the alarm central monitoring facility. The alarm central monitoring facility handles the Cellemetry-reported alarm exactly as it would an alarm reported over a landline phone line. The alarm central monitoring facility returns a call to the Cellemetry gateway, indicating that the alarm has been received and is being handled by the alarm company. The Cellemetry gateway in turn requests the wireless system to page the Cellemetry radio, which originally sent the alarm via the forward control channel. Once the page

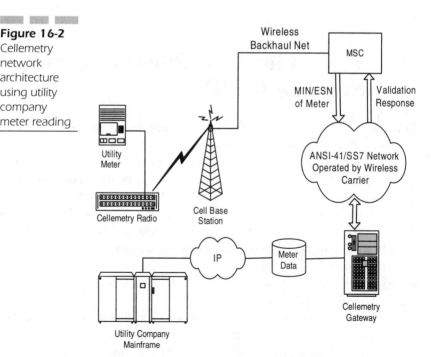

Figure 16-2
Cellemetry network architecture using utility company meter reading

is received by the Cellemetry radio, the page is reported to the security panel, thus completing the round-trip transmission, indicating that the alarm has been properly received and reported.

Measurements have shown that the capacity of the control channels far exceeds the needs of the wireless system. To further eliminate the possibility of having any impact on the wireless system, the Cellemetry radio utilizes the busy-idle bit, which is sent over the forward control channel. The busy-idle bit is one bit multiplexed in the FOCC data stream, which indicates that the wireless base station is communicating with a wireless user. A mobile phone will not attempt to register with the wireless base station if the busy-idle bit is set high. Only if it is set low will the mobile attempt a registration. The Cellemetry radio looks at the busy-idle bit over a multisecond window. If the busy-idle bit is set high for greater than a certain percentage of the time, the Cellemetry radio will defer its registration until the busy-idle bit activity is reduced. In this manner, regular wireless customers always will obtain the control channel and service first.

Cellemetry may compete with other wireless technologies over the long haul (e.g., SMS, WiMAX), but for now its use and availability make it a good option for companies seeking remote telemetry using a wireless network.

16.6 General Packet Radio Service (GPRS)

GPRS is a nonvoice, value-added service that allows information to be sent and received across a mobile telephone network.

GPRS is a service that provides *packet* radio access over GSM and time-division multiple access (TDMA/IS-136) networks. The radio and network resources are only accessed when data actually needs to be transmitted between the mobile user and the network. This data is divided into packets and is then transferred via the radio and core network to its destination. Between alternating transmissions, no network resources need to be allocated. GPRS facilitates instant connections where information can be sent or received immediately as the need arises, subject to radio coverage.

Theoretical maximum speeds of up to 171.2 kilobits per second are achievable with GPRS using all eight GSM timeslots at the same time, depending on network availability, channel coding scheme, and terminal capability. This is 10 times as fast as legacy circuit-switched data services on GSM networks. GPRS is an important migration step toward third-generation wireless networks. It will allow network operators to implement an IP-based core architecture for data applications, which will continue to be used and expanded upon for 3G services to support integrated voice and data applications. GPRS will also prove to be a testing and development area for new services and applications, which will also be used in the development of 3G services. Some wireless carriers will provide automatic access to GPRS; others will require a specific knowledge of how to use their specific model of mobile phone to send or receive information through GPRS.

16.6.1   GPRS Architecture

The network architecture of a GPRS system is described in the following section.

16.6.1.1   GPRS Network   In the core network, existing MSCs are designed to support circuit-switched traffic using central-office (CO) technology and cannot process packetized traffic. Enabling GPRS on a GSM network requires the addition of two additional core modules (network nodes). The two new components, called GPRS support nodes, are required in a GPRS-enabled wireless network:

- The serving GPRS support node (SGSN)
- The gateway GPRS support node (GGSN)

The SGSN is essentially a packet-switched MSC—it delivers packets to mobile stations within its service area. SGSNs send queries to HLRs to obtain profile data of GPRS subscribers. SGSNs detect new GPRS mobile stations in a given service area, process the registration of new mobile subscribers, and keep a record of their location inside a given area. So the SGSN performs mobility management functions such as mobile subscriber attach/detach and location management. The SGSN is connected to the base-station subsystem via a frame relay connection to the packet control unit (PCU) in the BSC.

The GGSN is used as an interface to external IP networks such as the public Internet, other mobile service providers' GPRS services, or business intranets. As the word gateway in its name suggests, the GGSN acts as a gateway between the GPRS network and IP-based PDSN (the Internet is the ultimate PDSN). GGSNs also connect to other GPRS networks to facilitate GPRS roaming. The SGSN provides packet routing to and from the SGSN service area for all users in that service area.

GGSNs maintain routing information that is necessary to tunnel the protocol data units (PDUs) to the SGSNs that serve particular mobile stations. Other GGSN functions include network and subscriber screening and address mapping. One or more GGSNs may be provided to support multiple SGSNs.

In addition to adding several GPRS nodes (SGSN, GGSN) and a GPRS packet backbone, other network nodes need to be added to a GSM network to implement a GPRS service. These include

- The addition of PCUs often hosted in the GSM base station subsystems
- Mobility management to locate the GPRS mobile station
- A new air interface for packet traffic
- New security features such as ciphering
- New GPRS-specific signaling systems

Key: From a high level, GPRS can be thought of as an overlay network onto an existing GSM network. This data overlay network provides packet data transport at rates from 9.6 to 171 Kbps. Multiple users can share the same air-interface resources.

GPRS attempts to reuse existing GSM network elements as much as possible, but in order to effectively build a packet-based mobile wireless network, some new network elements, interfaces, and protocols that handle packet traffic are required. Therefore, GPRS requires modifications to numerous network elements, as described in Table 16-1. See Figure 16-3 for an illustration of a typical GPRS network topology.

16.6.1.2 GPRS Base Station Subsystem Each BSC will require the installation of one or more PCUs and a software upgrade. The PCU provides a physical and logical data interface out of the base station subsystem (BSS) for packet data traffic. The BTS may also require a software upgrade but typically will not require hardware enhancements.

When either voice or data traffic is originated by a wireless GPRS/GSM subscriber, it is transported over the air interface to the BTS

Table 16-1

Modifications
Required in
GSM
Networks to
Support GPRS

GSM Network Element	Modification or Upgrade Required for GPRS
Subscriber Terminal	A totally new subscriber terminal is required to access (Mobile Phone) GPRS services. These new terminals will be backward compatible with GSM for voice calls.
Base Station	A software upgrade is required in the existing BTS.
BSC	The BSC will also require a software upgrade, the installation of a new piece of hardware called a PCU. The PCU directs the data traffic to the GPRS network and can be a separate hardware element associated with the SBC.
Core Network	The deployment of GPRS requires the installation of new core network elements called the SGSN and the GGSN.
Databases (i.e. HLR, VLR)	All the databases involved in the network will require software upgrades to handle the new call models and functions introduced by GPRS.

and from the BTS to the BSC in the same way as a standard GSM call. However, at the output ports of the BSC, the traffic is separated: Voice is sent to the mobile switching center per standard GSM voice call processing, and data is sent to the SGSN via the PCU over a frame relay interface. Refer to Figure 16-3.

16.6.2   GPRS Data Communication

Synergies exist between network elements of GSM services and GPRS. On the physical layer, resources can be reused and some common signaling parameters exist. In the same radio carrier, there can be time slots reserved simultaneously for circuit-switched use (voice) and for GPRS use. Ideal resource utilization is obtained through the dynamic sharing of radio resources between circuit-switched channels and GPRS

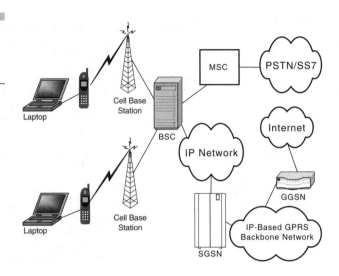

Figure 16-3
Basic GPRS
network
architecture

channels. During the setup of a circuit-switched GSM call, there is enough time to preempt the GPRS resources, effectively putting them on hold and allowing circuit-switched calls, which have higher priority, to go through.

GPRS provides bearer service from the edge of a data network to a GPRS mobile subscriber. The physical radio interface consists of a flexible number of time slots (from one to eight, conforming to the GSM standard) and thus provides a potential raw data rate of 171 kilobits per second (remember that GSM is a TDMA technology). A Media Access Control (MAC) utilizes the resources of the physical radio interface and provides a service to the GPRS Logical Link Control (LLC) protocol, between the mobile station and the SGSN. The GPRS LLC is a modification of a High-Level Data Link Control (HDLC)-based Radio Link Protocol (RLP) that supports variable frame sizes. The two most important features offered by the LLC are the support of point-to-multipoint addressing and the control of data-frame retransmission. From the standpoint of the application, GPRS provides a standard interface for the Network Layer—layer three of the OSI model.

16.6.2.1 Data Routing One of the key activities in a GPRS network is the routing of data packets to and from a mobile subscriber. GPRS routing can be divided into two areas: data packet routing and mobility management.

The main function of the GGSN involves interaction with external data networks (e.g., the public Internet). The GGSN updates the network location directory using routing information supplied by the SGSNs about the location of a mobile station, and it routes the external data network protocol packets encapsulated over the GPRS backbone to the SGSN currently serving the mobile station. It also deencapsulates and forwards external data network packets to the appropriate data network and collects billing data that is forwarded to a billing gateway.

GPRS operators will allow roaming through an interoperator backbone network. The GPRS operators connect to the interoperator network via a boarder gateway (BG), which can provide the necessary interworking and routing protocols (for example, Border Gateway Protocol, or BGP). It is also possible that GPRS operators will implement QoS mechanisms over the intercarrier connections network to support service-level agreements (SLAs).

The GPRS network encapsulates all data network protocols into its own encapsulation protocol, called the GPRS Tunneling Protocol (GTP). This is done to ensure security in the backbone network and to simplify the routing mechanism and the delivery of data over the GPRS network. Like all other encapsulation activity in data networks, GRPS headers are stripped off at distant end termination points.

16.6.2.2 GPRS Mobility Management

The operation of GPRS is partially independent of the GSM network. Some GPRS procedures share network elements with existing GSM systems to increase efficiency and make optimum use of unused GSM resources, such as unallocated time slots.

A mobile station has three states in the GPRS world: idle, standby, and active. The three-state model represents the nature of packet radio relative to the GSM two-state model (idle or active).

Data is transmitted between a mobile terminal and the GPRS network only when the mobile is in the active state. In the active state, the SGSN knows the cell location of the mobile station. But in the standby state, the location of the mobile is known only as to which routing area it is in. (The routing area consists of multiple cells within a GSM location area.)

When the SGSN sends a packet to a mobile station that is in the standby state, the mobile must be paged (over the control channel). Because the SGSN knows the routing area in which the mobile station is located, a packet paging message is sent to that routing area. After receiving the packet paging message, the mobile station gives its cell location to the SGSN to establish the active state. Packet transmission to an active mobile is initiated by packet paging to notify the mobile of an incoming data packet. The data transmission proceeds right away after packet paging, over the channel indicated by the paging message. The purpose of the packet paging message is to shorten the process of receiving packets. The mobile has to listen to only the packet paging messages, instead of all the data packets in the downlink channels. This reduces battery use significantly.

Key: The main reason for having the standby state is to reduce the load in the GPRS network caused by cell-based routing update messages and to conserve the battery in the mobile. When a mobile is in the standby state, there is no need to inform the SGSN of every cell change—only of every routing area change. The operator can define the size of the routing area and, in this way, adjust the number of routing update messages. In the idle state, the mobile can receive only those multicast messages that can be received by any GPRS mobile station.

A cell-based routing update procedure is called upon when an active mobile station enters a new cell. In this case, the mobile sends a short message containing information about its move. Through GPRS channels to its current (serving) SGSN, the message contains the identity of the mobile and its new location. This procedure is used only when a mobile is in the active state.

When a mobile in an active or a standby state moves from one routing area to another in the service area of one SGSN, it must again perform a routing update. The routing area information in the SGSN is updated and the success of the procedure is indicated in an acknowledgement (ACK) message.

The inter-SGSN routing update is the most complicated of the three routing updates. In this case, the mobile station changes from one SGSN area to another, and it must establish a new connection to a new SGSN. This involves creating a new logical link between the mobile station and the new SGSN, as well as informing the GGSN about the mobile's new location.

Mobility management within GPRS builds on the existing structures used in GSM networks. As a mobile station moves from one area to another, mobility management functions are used to track its location within each mobile network, which could include roaming scenarios. The SGSNs communicate with each other and update the user location. The mobile station profiles are preserved in the visitor location registers that are accessible by the SGSNs via the local (roamer) GSM MSC. A logical link is established and maintained between the mobile station and the SGSN in each mobile network. At the end of a transmission or when a mobile station moves out of the area managed by a specific SGSN, the logical link is released and the resources associated with it can be reallocated to other users.

16.6.3 GPRS Terminals

A complete understanding of GPRS requires understanding of its terminal types and availability. *Terminal equipment* is generally used to refer to the variety of mobile phones and mobile stations that can be used in a GPRS environment. The equipment is defined by terminal classes and types.

Certain GSM phones today are capable of supporting GPRS, unlike earlier GSM phones that could not handle the enhanced air interface, or the ability to packetize traffic directly. A variety of wireless GPRS terminals exist, including a high-speed version of current phones to support high-speed data access, a new kind of PDA device with an embedded GSM phone, and PC cards for laptop computers. All these terminals will be backward compatible with GSM for making voice calls (using GSM).

16.6.3.1 GPRS Terminal Classes   A GPRS terminal can be one of three classes: A, B, or C.

- *Class A* terminals support GPRS and other GSM services (such as SMS and voice) simultaneously. This support includes simultaneous connection, activation, monitoring, and traffic. Therefore, a Class A terminal can make or receive calls on two services simultaneously. In the presence of circuit-switched (voice) services, GPRS virtual circuits will be held or placed on busy rather than being cleared.

- A *Class B* terminal can monitor GSM and GPRS channels simultaneously but can support *only one* of these services at a time. As such, a Class B terminal can support simultaneous connection, activation, and monitoring, but not simultaneous traffic. As with Class A, the GPRS virtual circuits will not be closed down when circuit-switched traffic is present. Instead, they will be switched to busy mode. Therefore, users can make or receive calls on either a packet or a switched call sequentially, but not simultaneously.

- A *Class C* terminal supports only nonsimultaneous connections. The user must select which service to connect to. Therefore, a Class C terminal can make or receive calls from only the manually (or default) selected service (e.g., *either* voice or GPRS data). The service that is not selected is not reachable. The GPRS specifications state that support of SMS is optional for Class C terminals.

16.6.3.2 Device Types In addition to the three variables defined in the preceding section, each GPRS-capable handset will have a unique form factor. Some of the form factors will be similar to current wireless devices, while others will evolve to use the enhanced data capabilities of GPRS (EDGE).

The first types that were available were similar to a standard mobile phone, available in the standard form factor with a numeric keypad and a relatively small display.

The next device type is the PC card. PC cards are credit card-sized hardware devices that connect via a serial cable to the bottom of a mobile phone. Data cards for GPRS phones enable laptops and other devices with PC card slots to be connected to mobile GPRS-capable phones. Card phones provide functionality similar to that offered by PC cards, without needing a separate phone. These devices may need an earpiece and microphone to support voice services.

Smart phones are mobile phones with built-in voice, nonvoice, and Web-browsing services. Smart phones integrate mobile computing and mobile communications into a single wireless terminal. They come in various form factors, which may include a keyboard or an icon drive screen. The Nokia 9000 series is a popular example of this form factor.

PDAs such as the Palm Pilot series or Handspring Visor are data-centric devices that are adding mobile wireless access. These devices can either connect with a GPRS-capable mobile phone via a serial cable or have GPRS capability built directly into the device.

16.6.3.3 Service Access To use GPRS, users specifically need

- A mobile phone or terminal that supports GPRS
- A subscription to a GSM wireless service that supports GPRS (e.g., Cingular Wireless) must be enabled for the user. Automatic access to GPRS may be allowed by some wireless carriers; other carriers may require customers to subscribe to the service for a small, additional monthly fee.

- Knowledge of how to send and/or receive information via GPRS, using a specific model of mobile phone, including software and hardware configuration. This creates a customer service requirement that wireless carriers should support.

- A destination to send to or receive information from through GPRS, which is likely to be an Internet address, because GPRS is designed to make the Internet fully available to mobile users. GPRS users can access any Web page or other Internet applications.

16.6.4 Network Attributes of GPRS

This section describes the attributes of GPRS network technology.

16.6.4.1 Immediacy GPRS offers instant connections where information can be sent or received immediately as the need arises, subject to network coverage. Dial-up wireless modem connections are not necessary. Immediacy is an obvious advantage of GPRS when compared to circuit-switched data systems of yesteryear. A high level of immediacy is a very important feature for time-critical applications such as remote credit card authorization, where it would be unacceptable to keep the customer waiting for even 30 extra seconds. Immediacy can also be described as always on, similar to broadband Internet connections such as DSL or cable modems.

16.6.4.2 Packet Switching Instead of Circuit Switching GPRS is characterized by the overlay of a packet-based air interface over an existing circuit-switched GSM (voice) network. This gives the user an option to use a packet-based data service. To supplement a circuit-switched network architecture with packet switching in the past involved a major upgrade to GSM networks.

With GPRS, information is split into separate but related packets before being transmitted and reassembled at the receiving end. The GPRS standard is offered in a very elegant manner—with network operators needing only to add a couple of new infrastructure nodes and making a software upgrade to some existing network elements.

16.6.4.3 Spectral Efficiency Packet switching means that GPRS radio resources are used only when users are actually sending or receiving data.

Key: Rather than dedicating a radio channel to a mobile data user for a fixed period of time like CDPD (a legacy circuit-switched wireless data standard), available radio resources can be shared simultaneously between multiple users. This efficient use of scarce radio resources means that large numbers of GPRS users can theoretically potentially share the same bandwidth and be served from a single cell. The actual number of users supported depends on the application being used by each subscriber and how much data is being transmitted.

GPRS therefore allows GSM wireless carriers to maximize the use of their network resources in a dynamic and flexible way.

GPRS should improve the peak capacity of a GSM network because it simultaneously

- Allocates precious radio resources more efficiently by supporting virtual connectivity.

- Migrates traffic that was previously sent using circuit-switched data to GPRS instead.

- Reduces SMS-SC and control channel loading by transferring some traffic that previously was sent using SMS to GPRS instead, using the GPRS/SMS interconnect that is supported by the GPRS standard.

16.6.4.4 Internet Aware GPRS fully enables mobile Internet functionality by allowing interworking between the Internet and the GPRS network. Any service that is used over the Internet today—File Transfer Protocol (FTP), Web browsing, chat, e-mail, telnet—will be available over a GSM wireless network because of GPRS.

Because it uses the same network protocols, GPRS can be viewed as a subnetwork of the Internet, with GPRS-capable mobile phones being viewed as mobile hosts. This means that each GPRS terminal can essentially have its own IP address and will be addressable as such.

16.6.4.5 Supports Both GSM and IS-136/TDMA It should be noted that GPRS is not only a service designed to be deployed on mobile networks that are based on the GSM digital wireless standard, but the IS-136/TDMA standard also supports GPRS. This reflects an agreement to follow the same evolution path toward third-generation mobile phone networks concluded in early 1999 by the industry associations that support these two network types. But as noted in Chapter 6, this point is becoming moot as IS-136 has one foot firmly planted in the graveyard of wireless standards whose time has come and gone. Any existing IS-136 network still in existence today will be supplanted by GSM or even 3G technology soon, if this has not already occurred.

16.6.5 Limitations of GPRS

It should be clear that GPRS is an important mobile data service that offers a major improvement in spectrum efficiency, capability, and functionality. But it is also important to note that there are some limitations with GPRS, as summarized in the following section.

16.6.5.1 Limited Cell Capacity

GPRS does impact a network's existing cell capacity. There are limited radio resources that can be deployed for different uses—use for one purpose precludes simultaneous use for another. In other words, voice and GPRS calls both use the same network resources. The extent of the impact depends upon the number of timeslots, if any, that are reserved for exclusive use of GPRS in a GSM network. But GPRS dynamically manages channel (time slot) allocation. It also allows for a reduction in busy-hour control channel loading by sending short administrative/overhead messages over actual GPRS bearer channels instead of control channels. This improves network efficiency.

16.6.5.2 Actual Transmission Speeds Much Lower Than Specs

Achieving the theoretical maximum GPRS data transmission speed of 172 Kbps would require a single user consuming all eight GSM timeslots without any error protection. It is highly unlikely that a wireless carrier would allow all timeslots to be used by a single GPRS user. Also, early GPRS mobile terminals (e.g., phones or other devices) have limited functionality, supporting the ability to use only one, two, or three timeslots. The bandwidth available to a user with an older GPRS-capable phone will therefore be severely limited.

Key: As such, the theoretical maximum GPRS speeds should be checked against the reality of constraints in the networks and terminals. The reality is that wireless networks are always likely to have lower data transmission speeds than fixed networks.

16.6.5.3   Latency   Like TCP/IP transmissions, GPRS packets are sent in all different directions over different network routes to reach the same destination. This exposes the potential for one or more of those packets to be lost or corrupted during data transmission over the radio link. GPRS standards recognize this inherent feature of wireless packet technologies and incorporate data integrity and retransmission strategies into the standards. But the result is that potential transit delays can occur when using GPRS.

16.6.5.4   Modulation   GPRS is based on a modulation technique known as Gaussian minimum-shift keying (GMSK). The successor technology of GPRS, EDGE is based on a new modulation scheme that allows a much higher bit rate across the air interface, known as eight-phase-shift keying (8 PSK) modulation. Because 8 PSK also is used for 3G (see Chapter 7), GSM network operators will need to incorporate it at some point to make the transition to 3G mobile phone systems.

16.6.5.6   Lack of Store and Forward Capability   Where the store and forward engine in SMS is the heart of the SMS center and a key feature of SMS service, there is no storage mechanism incorporated into GPRS, apart from the incorporation of interconnection links between SMS and GPRS.

16.6.6   Applications Supported by GPRS

Although Chapter 18 delves into the many possible applications that can be supported by today's wireless technology, many applications can and are driven solely by a standardized wireless data technology such as GPRS. As such, it is not to say that other technologies that exist now or in the future could not also support the applications listed in the following passages. It is simply a matter of many GPRS documents listing the following applications as being GPRS-enabled.

GPRS facilitates several applications that have not previously been available over GSM networks, due to the limitations in the speed of circuit-switched data running at a measly 9.6 Kbps and message length limitations of SMS (160 characters). GPRS will fully enable Internet applications people are accustomed to, from Web browsing to instant messaging.

> *Key*: It should be noted that all the applications listed in the following section, although capable of being used with GPRS technology, would certainly operate more efficiently and faster with 3G-based technologies such as UMTS or CDMA2000. Using Internet access as an analogy, GPRS can be compared to a dial-up modem, whereas 3G technologies would be comparable to dedicated broadband connections such as DSL or cable modems, in terms of transmission speeds.

Instant Messaging—Instant messaging, also known as chat, can be distinguished from general information services because the source of the information is a person with chat, whereas for information services the source tends to be from an Internet site. The information intensity — the amount of information transferred per message tends to be much lower with chat. In the same way that Internet chat groups have proven a very popular application of the Internet, groups of like-minded people — so called communities of interest — have begun to use nonvoice mobile services as a means to chat, communicate, and discuss issues.

Because of its synergy with the Internet, GPRS would allow mobile users to participate fully in existing Internet chat groups rather than needing to set up their own groups that are dedicated to mobile users. Because the number of participants is an important factor determining the value of participation in the group, the use of GPRS here would be advantageous.

Content and Visual Information—A wide range of content can be delivered to mobile phone users, ranging from share prices, sports scores, weather, flight information, news headlines, prayer reminders, lottery results, jokes, horoscopes, traffic, location-sensitive services, and so on. This information need not necessarily be textual; it may be maps, graphs, or other types of visual information.

The length of a short message of 160 characters—the SMS limitation—suffices when it comes to delivering information that is quantitative. When the information is of a qualitative nature, however, 160 characters are too short. As such, GPRS will likely be used for qualitative information services when end users have GPRS-capable devices, but SMS will continue to be used for delivering most quantitative information services. Interestingly, chat applications are a form of qualitative information that may still be delivered using SMS to limit people to succinctness and reduce the incidence of bogus and irrelevant posts to the mailing list that are a common occurrence on Internet chat groups.

Still Image Transmission—Still images such as photographs, pictures, postcards, greeting cards, presentations, and static Web pages can be sent and received over the mobile network just like they are over land-line telephone networks. It will be possible with GPRS to post images from a digital camera connected to a GPRS device directly to an Internet site, allowing near real-time desktop publishing.

Web Browsing—Using circuit-switched data for Web browsing was never an appealing application for mobile users. Because of the slow speed of circuit-switched data technology (i.e., CDPD), it took a long time for data to arrive from the Internet server to the browser. One option was for users to switch off the images and just access the text on the Web. As a result, they would end up with difficult-to-read text layouts on screens that are difficult to read from. Therefore, mobile Internet browsing is better suited to a technology such as GPRS.

Collaborative Working—Mobile data facilitates document sharing and remote collaborative working. This lets different people in different places work on the same document at the same time. Multimedia applications combining voice, text, pictures, and images are also envisioned for the future. Anytime someone can benefit from having and being able to comment on a visual depiction of a situation or matter, such collaborative working can be useful. By providing sufficient bandwidth, GPRS facilitates multimedia applications such as document sharing.

Job Dispatch—Nonvoice mobile services can be used to communicate information about new jobs from office-based staff to mobile field staff. Customers typically telephone a call center whose staff take the call and categorize it. Those calls requiring a visit by field sales or service representatives can then be escalated to those mobile workers. Job dispatch applications can optionally be combined with vehicle positioning applications—such that the nearest available suitable personnel can be deployed to serve a customer (e.g., plumber or electrician or Sears appliance repair). Nonvoice mobile transmissions services can be used not only to send the job out, but also as a means for the service engineer or sales person to keep the office informed of progress toward meeting the customer's needs.

The 160 characters of a short message are sufficient for communicating most delivery addresses, such as those needed for sales, service, or some other job dispatch application such as mobile pizza delivery and courier package delivery. But 160 characters requires manipulation of the customer data like the use of abbreviations such as "St" instead of "Street." 160-character messages also do not leave much space for giving a field representative any information about the problem that has been reported or the customer profile. The field representative is able to arrive at the customer premises but is not very well briefed beyond that. This is where GPRS will come into play, to allow more information to be sent and received more easily. For example, with GPRS, a photograph of the customer and their premises could be sent to the field rep to assist in finding and identifying the customer. As such, it is expected that job dispatch applications may be an early adopter of GPRS-based communications.

Remote LAN Access—Remote LAN applications for mobile workers encompass access to any applications that employees would use when sitting at their desks, such as access to the company intranet, corporate e-mail services such as Microsoft Outlook, and database applications. The mobile terminal (e.g., handheld or laptop computer) has the same software programs loaded onto it as the desktop, or *thin* client versions of the applications accessible through the corporate LAN. This application area is therefore likely to be a conglomeration of remote access to several different information types—e-mail, intranet, and databases. The ideal bearer for remote LAN access depends on the amount of data being transmitted, but the speed and latency of GPRS make it ideal.

File Transfer—File transfer applications encompass any form of downloading sizeable data across the mobile network. Regardless of source and file type, this kind of application tends to be bandwidth intensive. The source of this information could be one of the Internet communication methods such as FTP, telnet, HTTP, or Java—or from a proprietary database or legacy platform. It therefore requires a high-speed mobile data service such as GPRS (EDGE or UMTS) to run satisfactorily across a mobile network.

Home Automation—Home automation applications combine remote security with remote control. Basically, a home can be monitored from

anywhere—on the road, on vacation, or at the office. If your burglar
alarm goes off, not only can you get alerted, but you get to see who the
perpetrators are via real-time photos or video streaming and possibly
even lock them in. Not only can you *see* things at home, but you can *do*
things to your home from a remote location. You can program your DVD
recorder, turn your oven on so that the preheating is complete by the
time you arrive home (traffic jams permitting), and so on. The mobile
phone really can become similar to remote control devices we use today
for our television, video, stereo, and so on. IP will soon be everywhere
—not just in mobile phones because of GPRS, but in all manner of
household appliances and in every machine. Eventually, all devices will
be addressable via IP. A key enabler for home automation applications
will be Bluetooth, which allows disparate devices to work together.

    Key: This all assumes that a person's home is already
preconfigured and set up to support home automation.
GPRS—any wireless technology—is a value-add in this
context and an enabler, as it allows homeowners to manage
their homes remotely. For more information on home networking, see Chapter 22.

16.6.7   Benefits of GPRS

The main benefits of the GPRS architecture are its flexibility, scalablil-
ity, interoperability, and roaming capability. GPRS allocates radio
resources only when there is data to send and it reduces reliance on traditional circuit-switched network elements. The increased functionality
of GPRS will decrease the incremental cost to provide data services, an
occurrence that will, in turn, increase the use of data services among consumer and business wireless users. GPRS will also improve the quality
of data services as measured in terms of reliability, response time (lower
latency), and features supported. The unique applications that will be
supported for GPRS will appeal to a broad base of wireless subscribers
and allow carriers to differentiate their services; these new services will
increase capacity requirements on the radio and base-station subsystems. One method GPRS uses to address capacity issues is to share the
same radio resource among all mobile stations in a cell, optimizing the
scarce resources. New core network elements will also be deployed to
more efficiently support the bursty nature of data communications.

    Relatively high mobile data speeds may not be available to individual
mobile users until Enhanced Data Rates for Global Evolution (EDGE) or

Universal Mobile Telephone System (3GSM) is introduced. The increased functionality of GPRS will decrease the incremental cost to provide data services.

16.7   EDGE: Enhanced Data Rates for Global Evolution

EDGE is a third-generation (3G) wireless technology that is capable of high-speed data transmission. EDGE occasionally is called E-GPRS because it is an enhancement of the GPRS network—*Enhanced* GPRS.

Like GPRS, EDGE divides the spectrum into time slots, but EDGE squeezes more data into each time slot. Each GPRS time slot can handle a maximum of 20 Kbps of user data, for a theoretical peak rate of 172 Kbps when all eight time slots are used simultaneously. By comparison, a single EDGE time slot can handle up to 59.2 Kbps, for a total of 473.6 Kbps with all eight time slots in use.

GPRS is based on the modulation technique, GMSK. EDGE is based on a modulation scheme that allows a much higher bit rate across the air interface, eight-phase shift keying (8PSK) modulation. Because 8PSK will also be used for 3G, network operators will need to incorporate it at some stage in their networks to make the transition to 3G mobile phone systems.

The objective of EDGE is to increase data transmission rates and spectrum efficiency and to facilitate new applications and increased capacity for wireless systems. With the introduction of EDGE in GSM networks, existing services such as GPRS and high-speed circuit switched data (HSCSD) are enhanced by offering a new physical layer. The services themselves are not modified.

Key: EDGE is introduced within existing specifications and descriptions rather than by creating new ones. GPRS allows data rates of 115 Kbps and, theoretically, of up to 160 Kbps on the physical layer. EDGE is capable of offering data rates of 384 Kbps and, theoretically, of up to 473.6 Kbps. A new modulation technique and error-tolerant transmission method, combined with improved link adaptation mechanisms, make these higher rates possible. This is the key to increased spectrum efficiency and enhanced applications, such as wireless Internet access, e-mail, and file transfers.

As the Third-Generation Partnership Project continues standardization toward the GSM/EDGE radio access network (also known as GERAN), GERAN will be able to offer the same services as wideband CDMA (WCDMA) by connecting to the same core network.

16.7.1 GPRS and EDGE: The Technical Differences

Regarded as a subsystem within the GSM standard, GPRS introduced packet-switched data into GSM networks. Many new protocols and new nodes have been introduced to make this possible.

Key: EDGE defines an approach to increase the data rates on the radio link for GSM. Basically, EDGE only introduces a new modulation technique and new channel coding that can be used to transmit both packet-switched and circuit-switched voice and data services. EDGE is therefore an add-on to GPRS and cannot work alone. In other words, a GSM-based wireless carrier cannot deploy EDGE unless they have first deployed GPRS. GPRS is a required stepping-stone to implement EDGE technology. So, for example, an operator could deploy GSM/GPRS/ EDGE but not GSM/EDGE.

GPRS has a greater impact on the GSM system than EDGE. By adding the new modulation, coding to GPRS, and by making adjustments to the radio link protocols, EDGE offers significantly higher throughput and capacity.

GPRS and EDGE have different protocols and different behavior on the base station system side. But on the core network side, GPRS and EDGE share the same packet-handling protocols. Therefore, they behave in the same way.

Key: Reuse of the existing GPRS core infrastructure (serving GRPS support node/gateway GPRS support node) underscores the fact that EDGE is only an add-on to the base station system and is therefore much easier to introduce than GPRS.

In addition to enhancing the throughput for each data user, EDGE also increases capacity. The same GSM time slot can support more users with EDGE. This decreases the amount of radio resources required to support the same traffic, which frees up capacity for more data or voice services. EDGE makes it easier for circuit-switched and packet-switched traffic to coexist while making more efficient use of the same radio resources. So in tightly planned networks with limited spectrum, EDGE may also be seen as a capacity booster for the data traffic.

16.7.2   EDGE Technology

EDGE uses the knowledge gained through the use of the existing GPRS standard to deliver significant technical improvements. See Table 16-2 for a description of how EDGE improves on GPRS technology.

 Key: EDGE can transmit three times as many bits as GPRS during the same period of time. This is the main reason for the higher data rates offered by EDGE.

The differences between the radio and user data rates with EDGE versus GPRS are the result of whether or not the packet headers are taken into consideration when making these calculations. These different ways of calculating throughput often cause misunderstandings within the industry about actual throughput numbers for GPRS and EDGE.

The data rate of 384 Kbps is often used in reference to EDGE. The International Telecommunications Union has defined 384 Kbps as the data rate required for a service to fulfill the International Mobile

Table 16-2

GPRS and EDGE, A Comparison of Technical Data

	GPRS	EDGE
Modulation	GMSK	8-PSK/GMSK
Modulation bit rate	270 Kbps	810 Kbps
Radio data rate, per time slot	22.8 Kbps	69.2 Kbps
User data rate per time slot	20 Kbps	59.2 Kbps
User data rate (all 8 time slots)	160 Kbps	473.6 Kbps

Telecommunications-2000 standard in a pedestrian environment. This 384 Kbps data rate corresponds to 48 Kbps per time slot, assuming an eight-time slot terminal.

The modulation type that is used in GSM is GMSK, which is a kind of phase modulation. To achieve higher bit rates per time slot than those available in GSM/GPRS, the modulation method requires change in an EDGE environment. EDGE specifies reuse of the channel structure, channel width, channel coding, and the existing mechanisms and functionality of GPRS. The modulation standard selected for EDGE, 8PSK, fulfills all of those requirements. EDGE channels can be integrated into an existing frequency plan, and new EDGE channels can be assigned in the same way as standard GSM channels. Only under very poor radio environments is GMSK more efficient. Therefore, the EDGE coding schemes are a mixture of both GMSK and 8PSK.

Another improvement that has been made with the EDGE standard is the ability to retransmit a packet that has not been decoded properly with a more robust coding scheme. For GPRS, resegmentation is not possible. Once packets have been sent, they must be retransmitted using the original coding scheme even if the radio environment has changed. This has a major impact on throughput. Additionally, with EDGE, the addressing numbers have been increased to 2048 and the window has been increased to 1024 in order to minimize the risk for stalling. This, in turn, minimizes the risk for retransmitting low-layer frames and prevents decreased throughput.

16.7.2.1 Measurement Accuracy

Like the GSM environment, GPRS measures the radio environment by analyzing the channel for (RF) carrier strength, bit error rate, and so forth. Performing these measurements takes time for a mobile station, which is of no concern in the speech world as the same coding is used all the time. But in a packet-switched environment, it is essential to analyze the radio link quickly in order to adapt the coding toward the new environment. The channel-analysis procedure that is used for GPRS makes the selection of the right coding scheme difficult because measurements for interference are performed only during idle bursts. As a result, measurements can only be performed twice during a 240-millisecond period.

For EDGE, the standard does not rely on the same slow measurement mechanism. Measurements are taken on each and every burst within the equalizer of the terminal, resulting in an estimate of the bit error probability (BEP).

Estimated for every burst, the BEP is a reflection of the current carrier-to-interference (C/I) ratio, the time dispersion of the signal, and the velocity of the mobile subscriber if they are in motion. The variation of the BEP value over several bursts will also provide additional information regarding velocity and frequency hopping. A very accurate estimation of the BEP is then possible to achieve.

To achieve the highest possible throughput over the radio link, EGPRS uses a combination of two functionalities: link adaptation and incremental redundancy. Compared to a pure link adaptation solution, this combination of mechanisms significantly improves performance.

16.7.2.2 Link Adaptation Link adaptation uses the radio link quality, measured either by the mobile station in a downlink transfer or by the base station in an uplink transfer, to select the most appropriate modulation coding scheme for transmission of the next sequence of packets using EDGE technology. For uplink packet transfers, the network informs the mobile station which coding scheme to use for transmission of the next sequence of packets.

16.7.2.3 Incremental Redundancy Incremental redundancy initially uses a coding scheme, with very little error protection and without consideration for the actual radio link quality. When information is received incorrectly, additional coding is transmitted and then soft combined in the receiver with the previously received information.

Key: Soft combining is just like hard combining, but it is accomplished entirely through software. Soft combining increases the probability of decoding the information correctly. This procedure will be repeated until the information is successfully decoded.

16.7.3 EDGE Impact on GSM/GPRS Networks

Due to the minor differences between GPRS and EDGE, the impact of EDGE on the existing GSM/GPRS network is limited to the base station subsystem. The base station is affected by the new transceiver unit capable of handling EDGE modulation as well as new software that enables the new protocol for packets over the radio interface in both the base station and BSC. The core GSM network does not require any modifications. Due to this simple upgrade, a network capable of EDGE can be deployed with limited investments and within a short time frame.

16.7.3.1 EDGE Requirements From the beginning, the standardization of EDGE was restricted to the physical layer and to the introduction of a new modulation scheme. Because EDGE was intended as an evolution of the existing GSM radio access technology, the requirements were set accordingly:

- EDGE- and non-EDGE-capable mobile stations should be able to share one and the same time slot.
- EDGE- and non-EDGE-capable transceivers should be deployable in the same spectrum.
- A partial introduction of EDGE should be possible.

To ease the implementation of new terminals while taking into account the irregular characteristic of most services currently available, it was also decided that two classes of terminals should be supported by the EDGE standard: a terminal that provides 8PSK capability in the downlink only, and a terminal that provides 8PSK in both the uplink and downlink.

16.7.3.2 EDGE Architecture EDGE does not bring about any direct architecture impacts. The PCU may still be placed either in the base station, the base station controller, or the GPRS support node, and the central control unit is always placed in the base station. However, the radio link control retransmission request function on the network side is located in the PCU. Because of this, any delay introduced between the PCU and the radio interface will directly affect the radio link control acknowledged/unacknowledged round-trip times. The maximum radio link control automatic repeat request window size has been extended for EDGE in an attempt to mitigate this risk and to allow the operator to optimize network behavior.

16.7.3.3 Alignment with WCDMA The next step in the evolution of GSM/EDGE includes enhancements of service provisioning for the packet-switched domain and increased alignment with the service provisioning in UMTS/UTRAN (UMTS terrestrial radio access network).

These enhancements are currently being specified for the coming releases of the 3GPP standard (see Chapter 7).

> *Key*: Based on EDGE high-speed transmission techniques and enhancements to the GPRS radio-link interface, GERAN will provide improved support for all QoS classes defined for UMTS: interactive applications, background operations, streaming audio and video, and voice. This will allow a new range of applications to be adequately supported, including IP multimedia applications.

This part of the GSM/EDGE evolution focuses on support for the conversational and streaming services, because adequate support for interactive and background services already exists. Multimedia applications will also be supported via parallel, simultaneous bearer paths to the same mobile station, which could have different QoS characteristics.

A driver for such evolution on the packet-switched side is the paradigm shift within the telecommunications world from circuit- to packet-switched communications.

Both the core network defined for GPRS and the current GSM/EDGE radio access network require modifications to support enhanced packet services. The GPRS/EDGE networks can quickly and cost-effectively evolve with market needs and align with services provided by WCDMA networks. The current evolution of GSM/EDGE, which covers all of the preceding aspects, is being standardized in 3GPP.

16.7.3.4 Short-Term Benefits: Capacity and Performance

EDGE introduces a new modulation technique, along with improvements to the radio protocol, that will allow wireless carriers to use existing frequency spectrums more effectively (e.g. 800, 900, 1800, and 1900 MHz).

The simple improvements of the existing GSM/GPRS protocols make EDGE a cost-effective, easy-to-implement upgrade to a GSM network. Software upgrades in the base station system enable use of the new protocol. New transceiver units in the base station enable use of the new modulation technique. It should be noted that the software upgrades could be implemented even easier if the BTS radios were software-defined radios.

Key: EDGE technology triples the capacity of GPRS. The introduction of EDGE enables bit rates that are approximately three times higher than standard GPRS bit rates. This was simply handled by reusing the GPRS QoS profiles and extending the parameter range to reflect the higher bit rates. In other words, introducing higher throughput values. This capacity boost improves the performance of existing applications and enables new services to be offered such as multimedia services. It also enables each transceiver to carry more voice and/or data traffic. EDGE enables new applications at higher data rates. Providing the best and most attractive services will increase customer loyalty.

16.7.4   Benefits of EDGE

EDGE and WCDMA are complementary technologies that together will sustain a wireless carrier's need for 3G network coverage and capacity nationwide. Enhancing a GPRS network with EDGE is accomplished within the existing spectrum and by deploying WCDMA in a new frequency band. Rolling out the two technologies in parallel enables faster time to market for new high-speed data services as well as lower capital expenditures. The installed base evolves; it is not replaced or built from scratch, which makes implementation seamless. With EDGE, operators can offer more wireless data applications, including wireless multimedia, e-mail, Web infotainment, and locator services, for both consumer and business users.

Long term, EDGE can be seen as a foundation toward one seamless GSM and WCDMA network with a combined core network and different access methods that are transparent to the end user.

Cingular Wireless launched the world's first commercial EDGE network in June 2003 in Indianapolis, IN. In September 2003, CSL deployed EDGE in Hong Kong. Both operators introduced their services with a single handset model—the Nokia 6200 and Nokia 6220, respectively—although they say more models will be available.

Cingular's launch is noteworthy, if only because EDGE has been repeatedly promised and then postponed so many times. For example, in 1998, Ericsson forecast EDGE deployments by 2000. Three years later, AT&T Wireless and Nokia forecast commercial launches by 2002, but Cingular stepped up and made it happen.

16.7.5 Skepticism About EDGE

With the first commercial launch of EDGE on June 30, 2003, the technology's advocates finally had a success story. But there are skeptics. Despite claims by proponents that EDGE provides a smooth, cost-effective means to provide 3G services, the technology's business case and market potential have some major flaws. EDGE is a data-only technology, but it does affect voice capacity in the GSM network.

> *Key*: One of the reasons that today's EDGE networks deliver barely one-quarter of their peak rate is that higher throughput comes at the expense of voice capacity. A wireless carrier could give each EDGE user all eight GSM time slots rather than the current two or four time slots, but that would reduce the amount of overall network capacity that can be devoted to voice calls.

This limitation creates difficult choices for the wireless carrier. They could charge EDGE users a significantly higher rate than GPRS because they are using more than their share of network capacity, but that would limit customer adoption and revenue streams. It also would result in more blocked and dropped voice calls, which is not an option because voice is still the killer app in wireless today and will continue to drive the majority of revenue for the foreseeable future—at least in the United States. Wireless carriers could also charge just a small premium for EDGE subscribership and limit the number of timeslots per users. The downside to this approach is that to potential customers EDGE will not look like much of an improvement over GPRS.

By being late to market, EDGE may have missed its window of opportunity in several respects:

▪ First, EDGE has to catch up with other 3G technologies such as CDMA2000 and WCDMA, which have been commercially deployed since 2000.

▪ Many GSM operators have decided to go directly from GPRS to WCDMA because WCDMA offers greater benefits, better infrastructure, and devices are already available. In the case of most European operators, the tight timetables for their 3G licenses force them to devote all of their resources to building WCDMA networks. Most European operators hold 3G licenses that have stringent timetables for launching commercial WCDMA service, so they do not have time for an EDGE detour. And WCDMA is a markedly different technology than GSM/GPRS, so European carriers already have their hands full learning the nuances of a new technology. Some of those networks are already in commercial service, so it is difficult for a wireless carrier to make a business case for going back and adding EDGE to its network when WCDMA provides the same (or better) functionality.

Another mark against EDGE: As of September 2003, only 50 wireless carriers throughout the world have expressed interest in EDGE and the majority of them were in the Americas. Therefore, unlike GSM, GPRS, or WCDMA, EDGE does not have a global cost structure. This means that devices and infrastructure will cost more, hampering its ability to compete with more widely used and available technologies. This makes EDGE more like TDMA than GSM. Without the support of European carriers, EDGE will likely become a niche technology.

The fact that the availability of EDGE devices is still so limited, even down to the scarcity of accessories such as modem cables, suggests that EDGE still was not ready when the first network launched in June 2003. Speculation exists that EDGE was launched prematurely in an attempt to appease its critics. Its underwhelming performance has just given them more ammunition.

16.7.5.1 A Weak Business Case? One of EDGE's key selling points has been that it is part of the GSM technology family, which has the largest worldwide market share in terms of users and networks. Therefore, the argument goes, EDGE will be able to leverage GSM's cost structure and selection of devices and infrastructure.

But that argument fades when seeing the amount of current and proposed EDGE deployments. First, although 50 operators worldwide have committed to launching EDGE, there is a big difference between a commitment and a commercial launch. Time will tell whether the operators that have committed to EDGE actually launch it or go straight to WCDMA, which has a much stronger business case. In other words, the capital expense of rolling out WCDMA will bear revenue fruit much more quickly and at higher margins. Meanwhile, rival 3G technologies such as CDMA2000 and WCDMA are already in commercial service, thus driving equipment volumes (which drives down wholesale and retail prices) and user adoption. Also, even if all 50 operators do launch EDGE, that is only a fraction of the 400 wireless carriers around the world that already use GSM. So it is difficult to understand how EDGE's cost structure could approach GSM's cost structure.

EDGE already carries a premium simply because it is a brand-new technology. For example, the wholesale price of hardware necessary to add EDGE to handsets currently is about 15 percent more than GPRS, according to vendors such as Broadcom. EDGE does not appear to be in a position to achieve the volumes necessary to reduce that premium. Deutsche Bank Securities expects worldwide shipments of EDGE devices to hit 19.2 million units by the end of 2004 and 61.3 million by the end of 2005. By comparison, its forecast for CDMA2000 is 134 million in 2005.

Many GSM (and IS-136) carriers in North and South America have saturated networks, so deploying GPRS/EDGE or GSM/GPRS/EDGE overlays, respectively, may not be an option because there is not enough spectrum to accommodate high-bandwidth data services. This constraint suggests one reason why the initial EDGE networks and devices deliver only a fraction of the technology's theoretical peak data rate: The operators may have only enough spectrum to launch a bare-bones version of EDGE, let alone support a version that runs over all of the available time slots. If device vendors believed that EDGE networks will soon support the technology's maximum throughput, they would have already announced devices that support all of the time slots.

EDGE's real-world data rates are far lower than its theoretical peaks of 473 Kbps. For example, Cingular acknowledges that its EDGE network can support peaks of only 170 Kbps and average rates of 75 Kbps to 135 Kbps.

Key: So although its average rates are faster than the average rates for GPRS, EDGE is not well positioned to compete with CDMA2000 1xEV-DO, which provides average rates of 500 Kbps. In fact, at 75 Kbps, EDGE is barely competitive with CDMA2000 1X, which provides average rates of 60 to 100 Kbps.

16.7.5.2 The Coverage Problem

Part of EDGE's lackluster data performance stems from the fact that it is different than GSM and GPRS. For example, with EDGE, signal strength is not an accurate indicator of performance.

The task and cost of adding EDGE to an existing cell site varies and helps determine whether a wireless carrier can make a business case for deploying the technology. For example, although EDGE's backers say that the cost of adding it to a GSM/GPRS base station is only $1 to $2 per POP, that claim assumes that the operator has infrastructure that is no older than 1999, depending on the vendor. This would allow for a software upgrade rather than a forklift upgrade.

Wireless carriers also cannot simply add EDGE to a GSM/GPRS cell site and assume that EDGE coverage will be the same as the GSM/GPRS coverage. An obvious solution for plugging coverage holes is to add cell sites to cover large gaps and repeaters for smaller gaps. That may be a viable option in small geographic areas, such as a business district, but if the holes are dotted throughout an entire market or multiple markets, the costs of additional infrastructure and creating a separate RF engineering plan for EDGE quickly add up, undercutting the technology's business case.

Spotty coverage reduces the data rate because EDGE sends data only at speeds that channel conditions can bear. So if the user is in an area where the signal is weak, the network will throttle back the speeds so that it does not have to retransmit lost packets. If a carrier's EDGE coverage is inconsistent, users will notice the change in throughput.

CDMA2000 is a better alternative than EDGE. Unlike EDGE, CDMA2000 can be easily and cost-effectively deployed throughout an entire market, and it supports advanced, high-speed data applications.

CDMA2000 has been commercially deployed since 2000 and serves more than 60 million users on 71 commercial networks across the world. One of the key reasons for its commercial success is that the technology can be deployed rapidly throughout the coverage area with small capital outlays.

CDMA2000 also offers far higher data rates than EDGE, and even WCDMA. With a typical data throughput of 60 to 100 Kbps on CDMA2000 1X and 300 to 600 Kbps on CDMA2000 1xEV-DO, operators can deliver a wide variety of high-bandwidth services, such as video on demand (VOD), music on demand (MOD), videoconferencing, multimedia services (MMS), and even TV *broadcasts*. Verizon Wireless deployed CDMA2000 1xEV-DO in San Diego and several other cities in 2004.

CDMA2000 also does not require a large chunk of new spectrum, so it is an attractive option for carriers that need to launch 3G quickly but have no new spectrum. CDMA2000 already has a wider variety of devices to appeal to a broad range of demographics and user needs. For example, as of October 2003, more than 425 models of CDMA2000 phones and PC card modems were commercially available, including 40 for CDMA2000 1XEV-DO. By comparison, less than a half-dozen EDGE devices were available at the same time. For an operator that needs to launch 3G today and cannot wait for device supplies to catch up, CDMA2000 is the best option.

 Key: The intent of this section is not to denigrate EDGE as a technology but more to present a realistic, practical view of where the technology stands commercially, especially compared to other 3G technologies that seem to have more traction in the wireless marketplace.

16.8   The Mobile IP Standard

Mobile IP is an open standard, defined by the Internet Engineering Task Force (IETF), that allows users to keep the same IP address, stay connected, and maintain ongoing applications while roaming between IP (wireless) networks. Mobile IP is scalable because it is based on IP—any media that can support IP can support Mobile IP.

The number of wireless devices for voice or data is projected to surpass the number of fixed devices. Mobile data communication will likely emerge as the technology supporting most communications including voice and video. In other words, Mobile IP may eventually work with Voice over IP (VoIP) and streaming media applications. Mobile data communication will be pervasive in 3G wireless networks and in wireless LANs using 802.11. It is easy to envision Mobile IP also becoming intertwined with 802.16 WiMAX. Though mobility may be enabled by link-layer technologies, data transmitted across networks or different link layers is still a problem. The solution to this problem is a standards-based protocol, Mobile IP.

In IP networks, routing is based on fixed IP addresses, similar to how a postal letter is delivered to the fixed address on the envelope. A device on a network is reachable through normal IP routing by the IP address it is assigned on the network.

It becomes problematic when a mobile user roams into a visiting network and is no longer reachable using normal IP routing. This results in the active sessions of the device being terminated. Mobile IP was created to enable users to keep the same IP address while traveling to a different network (which may even be on a different wireless operator), thus ensuring that a roaming individual could continue communication without sessions or connections being dropped.

Because the mobility functions of Mobile IP are performed at the network layer rather than the physical layer, the mobile device can span different types of wireless and wireline networks while maintaining connections and ongoing applications. Remote login, remote printing, and file transfers are some examples of applications where it is undesirable to interrupt communications while an individual roams across network boundaries. Also, certain network services, such as software licenses and access privileges, are based on IP addresses. Changing these IP addresses could compromise the network services.

Mobile IP has the following three components:

- Mobile node
- Home agent
- Foreign agent

The Mobile IP process has three main phases:

- *Agent discovery:* A mobile node discovers its foreign and home agents during agent discovery.

- *Registration:* The mobile node registers its current location with the foreign agent and home agent during registration.

- *Tunneling:* A reciprocal tunnel is set up by the home agent to the care-of address (current location of the mobile node on the foreign network) to route packets to the mobile node as it roams.

Together these components and operations phases work through a standardized process to allow roaming mobile subscribers to access and use IP-based wireless applications anywhere they are located, whether it is across town or cross-country.

16.9 Wireless Application Protocol (WAP)

Although all the protocols described heretofore allow for greater mobility, they all provide relatively low data rates. New application protocols that take the data rate into account are therefore desirable. The first such protocol is the WAP. The WAP Forum was developed jointly by Ericsson, Nokia, Motorola, and Phone.com in June 1997. The objective of the founders of the WAP Forum was to create standards allowing small, consumer-class wireless devices to access the Internet anywhere, anytime.

WAP devices are limited in the sense that they do not have the luxury of accommodating large code space to run applications, unwieldy kernels, and expansive displays. Developing applications *in the small* has given new meaning to a WAP device.

In the WAP environment, wireless devices do not need to have a complete Web browser implementation to provide Web access. Instead, the wireless service provider will provide WAP protocol gateways in order to reduce the amount of data that needs to be sent to the wireless device to the bare essence of a Web page (in terms of the total number of bytes of data required). This is accomplished by looking for specific wireless markup language (WML or Wireless XML) tags embedded in the Web page being accessed. As a result, the transmission of high-resolution graphics, wordy text descriptions, and objects with complex formatting can be skipped in the transmission through the wireless carrier's infrastructure to the wireless handset.

Once reduced, the Web page is then sent to the wireless device using a lightweight transport stack based on the User Datagram Protocol (UDP) and compressed IP. The use of compression is desirable as it allows the bearer channel to be used as efficiently as possible. Instead of a typical 64-byte TCP packet containing 15 bytes of payload, the compressed packet will require approximately 20 bytes of total data to be transported to the mobile handset.

16.9.1 WAP Subprotocols

Bandwidth conservation is crucial in the wireless world. WAP provides a protocol stack designed to minimize bandwidth requirements while offering secure connections over a variety of transport protocols. The WAP architecture is straightforward and looks similar to the pervasive ISO network models. These subprotocols and their functions are as follows.

The WAP application environment is the topmost layer and the one that application developers are most interested in. This layer holds general device specifications, wireless markup language (WML) and WML script programming languages, and the telephony application programming interfaces, and it defines content formats, including graphics and personal information management (PIM) information.

The Wireless Session Protocol in the session layer is a tokenized version of HTTP 1.1, designed specifically for limited bandwidth and longer latency applications while still allowing guaranteed delivery and push content.

The transaction layer encompasses the Wireless Transaction Protocol (WTP), which limits the overhead of packet sequencing. It supports a variety of message types, including various types of guaranteed and nonguaranteed one- and two-way requests.

The security layer supports wireless transport layer security, a protocol based on standard transport layer security, formerly known as secure sockets.

WAP's lowest level contains the Wireless Datagram Protocol, which provides a consistent interface between various over-the-air protocols.

WAP's future is in question, because other technologies are being developed that may supplant it. One criticism of WAP is that it is too difficult to view scaled-down Web pages on a tiny mobile phone (or PDA-type) device. But at the same time, many services and technologies are being developed with an eye toward being WAP-compatible.

Test Questions

True or False?

1. _____In a typical wireless system, a significant amount of the control channel bandwidth is unused.

2. _____ IS-95 and IS-136 are the protocols that are used to carry data traffic on the control channel in telemetry systems.

3. _____ In the realm of mobile wireless data, the two most common forms of data transmission are circuit-switched (portable) communications and packet data (fully mobile) communications.

4. _____ GPRS technology is enabled mainly through the addition of two new core wireless network elements: the SGSN and the GGSN.

5. _____ EDGE technology will only require changes to BTS transceivers in order to make EDGE fully operational. No changes to the core network are required.

Multiple Choice

1. The predominant types or applications of wireless data include:
   a. Bursty communications (e.g., file transfers)
   b. Terminal server sessions
   c. Query/response applications
   d. Batch file applications
   e. Web server access
   f. a, c, and d only
   g. a, b, and d only
   h. e only
2. Which wireless data technology has a message length of 160 characters?
   a. GPRS
   b. WAP
   d. EDGE
   e. Mobile IP
   f. None of the above

The New Age of Cell Phones

17.1 Introduction

Cell phones hold the promise of making life easier by ultimately giving mobile subscribers a wide range of possibilities.

In 2004, 650 million cell phones were sold across the world. Americans racked up 500 billion minutes on their cell phones in the first half of 2004 —97 percent of their total cell phone bill charges were for time spent talking. In the not-so-distant future, though, talking may be the last priority of a cell phone user.

A small but growing number of U.S. wireless subscribers use their cell phones to play games, e-mail photographs, and send text messages. But industry analysts claim that within the next few years, an array of new services could radically transform the way we use cell phones.

We may be able to tune in to as many as 100 channels of TV, find the nearest coffee shop using locator technology, get directions, and even pay for purchases using Bluetooth *e-wallet* technology.

This is already happening in Asia and Europe, home to some of the fastest wireless networks in the world. In Asia, wireless subscribers can find everything from English–Japanese translation dictionaries, to dating services on their cell phones. Belk believes that a good rule of thumb is that anything you see commercially available in Japan or Korea usually takes around 18 months before it finds its way to the United States and subscribers begin embracing it. By the same token, the Europeans have been vigorously using SMS and e-wallet applications for years, yet SMS is still struggling to be embraced in the United States. But the blame does not lie completely with mobile subscribers: Wireless carriers have made only half-hearted efforts to advertise and promote SMS service.

17.2 The All-in-One Cell Phone

Today's cell phones can do just about anything except make coffee. Just about any popular electronic application or toy has been integrated into today's cell phone. They are not just cell phones anymore. They are also, all-in-one devices:

- MP3 players, TVs (soon), digital cameras
- PDAs, GPS receivers, answering machines (voice mail)
- Game consoles, mapping devices, pagers
- FM radios, speakerphones

17.2.1 Cutting-Edge Mobile Innovations— The Future of the Cell Phone

At a truly stunning rate, wireless innovations move from the cutting edge to the routine. Some analysts suggest that the current 1.5 billion phones that exist in the world as of 2005 are paving the way to actually replace the personal computer (PC). It is plain to see that the mobile phone is already on its way to becoming an all-in-one microcomputer.

Mobile subscribers in Tokyo, who have access to the fastest mobile networks in the world, use their cell phones to read magazines, watch TV, read books, and play games.

Today's cell phones are used in Germany to support a service called *Symbian Dater*, where mobile customers program their dating profile into their cell phone. When a male mobile subscriber gets within distance of a woman in a nightclub (for instance) whose dating profile matches his (Symbian compares the profiles and decides if they are a match), his phone vibrates, showing a video of the woman whose profile matches his profile. He then scans the nightclub for the woman and introduces himself.

Sales of cell phones far outweigh the sales of TVs, stereos, and even the revered PC. Cell phone sales are now three times that of personal computers. Today's most sophisticated phones have the processing power of a 1995 PC, while using 100 times *less* electricity. However the cell phone— especially the keyboard—has to evolve a bit more in order to reach the point where people are willing to abandon their personal computers. The Palm Treo® is becoming hugely popular. The Treo 600 has a tiny keyboard, slots for added memory, and the following other features:

- Palm OS® Organizer
- Built-in camera and speakerphone
- Color display
- Quad-band world phone (850/900/1,800/1,900 MHz operation)

Figure 17-1
Palm Treo 600

- POP3 e-mail access
- Web browser
- 24 Mb available memory

See Figure 17-1 for a picture of the Palm Treo.

Nokia's N-Gage, launched in Autumn 2003, has a newer version that plays video games. Motorola's MPx has a dual-hinge design, where the handset opens in one direction and looks like a regular phone, but it also flips open along another axis and looks like an e-mail device (e.g., Blackberry). The Motorola RAZR phone was launched in late 2004 and is also the thinnest flip phone ever developed, at just over a ½-inch thick. It is similar to the Treo600 in many respects, such as MPEG4 video capability, long-range Bluetooth capability, a speakerphone, a color screen, and of course a digital camera.

Key: One key fact needs to be remembered when discussing the potential of the mobile phone to replace the PC: Mobile phones keep getting smaller, but people are not. Millions of people are not going to replace the full screen, mouse, and keyboard experience with staring at a tiny little screen. It is simply not as practical.

One way around this issue is to develop and deploy speech-recognition technology for cell phones, which currently exists for personal digital assistants (PDAs). Phones do not have the processing power yet to adopt speech recognition, but Moore's Law says that this day is not far off (Moore's Law states that computer processing power doubles approximately every 12 months, with no appreciable increase in cost). Another hindrance to the PC replacement concept is the battery of the cell phone

—this technology also needs to catch up. Projection keyboards are under development, where a laser inside the phone emits the pattern of a large keyboard onto a flat surface, and the phone's camera reads and understands the user's finger movements.

Innovation in the mobile industry is full of twists and turns, so all these enhancements may (or may not) be far off. This is because the cell phone industry is not a monopoly. That is a good thing, because no one company could conceive of and develop all these leading-edge capabilities.

Digital mapping is another application that is being used with cell phones today. The real world is documented and then integrated into the mobile phone. Driven by space photography, global satellite positioning, locator technology, search engines, and new ways of marking information for the Internet, cartography is now on the leading edge. One application making use of this concept is a mobile application called Dodgeball, which allows a mobile subscriber to know if any friends are within 10 blocks of you (cool, but scary?). Microsoft has a new division called Map-Point, with 150 engineers (including many cartographers) researching simple ways for developers to put mapping information into their software applications, which means they will undoubtedly find their way into the mobile world eventually.

There are also smart phones under development that have video cameras (yes—video cameras!), GPS antennas, and automatic access to wireless fidelity (Wi-Fi) hot spots. Smart phone sales currently constitute only 5 percent of the cell phone market, but this share doubles every year. The mobility and always-on aspect of today's smart phones are what make them more appealing to the populace.

Samsung announced in May 2004 that it is developing a cell phone that will receive 40 satellite TV stations. By 2007, mobile phones will pack a gigabyte of flash memory, making the phone essentially a huge photo album or music player and give the vaunted iPod a major marketplace challenge. Cell phones will sport a new look, more memory, and more power.

One new design concept is called *metaphoring*. For example, a phone that is also an MP3 player will look more like an MP3 player. A phone that is also a camera will look more like a camera than a cell phone. Last fall, Kyocera introduced such a phone, the Koi, a 1.2-megapixel camera phone that sports a sleek swivel design intended to be held like a camera. Organic shapes may become popular, where the phones are more rounded instead of rectangular. And, after decades of shrinking cell phones, experts believe phones are going to begin growing, just a bit, to accommodate larger screens to support new applications.

Removable storage devices and hard drives for cell phones are under development. Hard drives as large as 80 Gb are expected in future cell phones, no different than what we have at home on a DVD recorder. Stereo audio, faster processors, higher-resolution screens and, of course, the longer-life batteries needed to power these hungry digital devices are also on their way, according to industry experts.

Wireless carriers naturally hope all these features and functions will generate more revenue. They are counting on the video generation of 18- to 34-year-olds who grew up using cell phones and playing video games to keep the revenue from games, e-mail, video, photos, and other data flowing.

Here are some other really unique, even odd, developments related to today's cell phone:

- NTT DoComo offers what they call a skull phone, which allows users to place the phone directly up against their head in order to hear the caller. This phone is supposed to be good for noisy environments such as train stations, city streets, or shopping malls, for example.

- Researchers in the chemistry department at St. Louis University have developed a new type of biofuel cell phone battery that runs on ethanol (alcohol) and enzymes. It is renewable and enviro-friendly. Ethanol is a good fuel source because it is fairly easy to break down, and it is readily available. This concept has been around since the 1960s, but the key challenge has been stabilizing the enzyme so that it can produce power for more than a few hours before *denaturing* and falling apart. The research group has built a test fuel cell that has run for 250 days with direct refills of vodka, gin, wine, or beer every two weeks. The goal is to develop a system that lasts 2 years and requires refills only once a month. Whether this translates into a new cell phone battery type will depend on its weight.

- At the request of Motorola, a polymer has been developed that looks like any other plastic, but which degrades into soil when discarded (environmentalists will *love* this). University of Warwick (Britain) researchers helped develop a phone cover that contains a real sunflower seed, which will feed on the nitrates that are formed when the polyvinylalcohol polymer cover turns to waste. It is biodegradable, nontoxic plastic. The new plastic can be rigid or flexible in shape.

17.2.2 The Camera Phone Revolution

In 2004, the year many consider to be the breakout year for the camera phone, 84 *million* cell phones with digital cameras were shipped.

There are now video-camera phones that can play back video on a television, a phone with surround sound, and a phone that can go into Disney mode, complete with Mickey Mouse's laugh and a Goofy icon on the screen.

The camera phone is truly revolutionizing the whole concept of photography. It is providing the ability to instantly, digitally capture everyday events that in yesteryear might have otherwise elicited thoughts and remarks such as, "Sure wish I had my camera with me." This is facilitated by the fact that the vast majority of mobile customers carry their phones with them everywhere. Everywhere.

Key: Cam phones now provide a window into moments once ignored, customs once overlooked. In the process, they're changing our laws (they are outlawed in the U.S. Supreme Court), our relationships, and the way we experience (and document) our worlds. Many public restrooms have "No Camera Phone" signs now—so do many courtrooms, gyms, and now some schools.

Camera phones are potent inventions, the first affordable device that merges imagery with the ability to transmit images to mass audiences instantaneously. In 2004, it was not a camera company but a cell phone manufacturer (Nokia) that sold the most cameras in the world. In 2005, the cell phone industry expects to sell 420 million camera phones, a huge leap from the 260 million phones sold in 2004. Cam phones have reached critical mass and are reminding many of the giddiness they felt when the Internet came of age in the late 1990s.

Camera phones increasingly have more megapixels, which means they are capable of taking higher-quality pictures for printing too. A megapixel (a million pixels, or picture elements) is a unit of image sensing capacity in a digital camera. A pixel is the basic unit of programmable color on a computer display or in a computer image. In general, the more megapixels in a camera, the better the resolution when printing an image in a given size. But they also are beginning to boast features found on stand-alone cameras, such as red-eye reduction, auto-focus, and automatic flashes.

17.3 Ringtones

Love 'em, hate 'em. Those often obnoxious but never boring ringtones we hear all around us every day are here to stay. They play everything from the "I Dream of Jeannie" theme song to sounds of the rainforest, to well-known audio clips from movies to actual song snippets. Full songs and noises are both available. Some of them even sound like a real, old-fashioned telephone ring (no!).

Ringtones were first popularized by tech-savvy teenagers in the late 1990s. The subsequent success of ringtones is a rare bit of good news for a music industry that has been stung by Internet piracy and fussy fans who would rather spend their hard-earned money on videos versus CDs. Billboard Music now has a new category on its Web site that is just for downloading ring tones. People, especially teens, view the ringtone as an identity statement.

Ringtone revenues are divided between the music labels, their artists, and wireless operators. The average price of a ringtone is anywhere from 60 cents to $2.00, but prices vary widely by country and region. In 2004, ring tone sales were projected to hit $3.5 billion (yes, billion) around the globe—becoming 10 percent of the $32 billion global music market. In the United States, $375 million was generated by ringtone downloads in 2004. Just 1 year before, the number was only $18 million. Ringtone sales are projected to reach $1 billion by 2007, according to research firm IDC.

 Key: Mobile subscribers can even download many ringtones and program their phones to ring with certain tones (and photos) depending on who is calling.

Ringtones are so popular in India that 6 months after their introduction in 2004, 1 in 10 mobile users have a custom ringtone. The fastest-growing part of the ringtone trend is called "Master Tones," or "Real Music Ringtones," which have MP3 clarity.

17.4 Cell Phone Viruses

Because mobile phones have become so computerized, they are a new target for the pathetic, misguided, bored hackers and virus writers who have plagued the Internet (and end user PCs) for years now.

In June 2004, a group of virus writers somewhere in eastern Europe unveiled the first worm that attacks mobile phones. The Cabir worm hit a few hundred people with phones that run the Symbian operating system and are equipped with Bluetooth technology.

There are now mobile viruses, known as *mobile malware*, that can infect phones that are within 10 meters of each other that are Bluetooth capable. This virus is known as Lasco.A, which was announced in January 2005. Analysis of the code shows that this virus is based on an earlier cell phone worm called Cabir.H. But like many desktop computer viruses, the program not only spreads automatically, but also infects other .sis files on a host mobile handset. This type of mobile phone virus can render the phone completely unusable. An antivirus company in Russia recommends that a (smart) phone using the Symbian operating system and the series 60 platform should have Bluetooth's discovery mode disabled and accept no files from unnamed devices. Sharing games between handsets can infect phones too. The bottom line: Just as you would be wary of accepting or opening an unknown e-mail or file sent to you, the same cautious approach should be taken when dealing with anything sent to your mobile phone.

17.5   Is All This Really Necessary?

Like any major innovation, there are benefits and disadvantages too. The good side of this new age of cell phones is the new and cutting-edge innovations and the multipurpose functionality of mobile terminals today. The downside is that the new capabilities create a new support challenge. Sprint PCS has gotten to the point where they seek out new hires in technical and customer support who have an IT background, because the phones have become so sophisticated and full of computing capabilities. They are very software dependent. This raises huge economic challenges for today's wireless carriers. Consider the following impact that smart phones have on service:

- User service calls last four times longer.
- Customer care costs could rise by $1 billion across the industry.
- Four thousand additional tech support reps are needed across the industry to support the current explosion of smart phones in the marketplace.

But wireless carriers are optimistic. In contrast to the issues listed in the preceding passage, carriers believe they will meet the smart phone challenge by

- Developing more reliable phones and applications
- Hoping customers become more proficient in their use of the phones
- Urging customers to use self-diagnostic tools
- Emphasizing self-service

One question remains, which will be answered over time: Is packing all this stuff into one device too much? Do people really want that much functionality in their cell phones? It is likely the 18- to 34-year-old demographic may say yes, but what about the rest of us?

No matter how many tasks cell phones are able to do, there will always be consumers who simply want them for their original purpose—to talk.

Test Questions

True/False

1. _____ A new type of cell phone launched in Japan allows users to listen to the caller by pressing the phone up against their skull.

2. _____ Future cell phones are being developed with the intent of having an 80GB hard drive.

3. _____ Ringtones generated $375 million in revenue in the United States in 2004.

Multiple Choice

1. Motorola commissioned a research group to develop a phone cover that contains a real _____, which will feed on the nitrates that are formed when the polyvinylalcohol polymer cover turns to waste.

   a. Worm
   b. Sunflower seed
   c. Bacteria
   d. Virus
   e. None of the above

The Business Side and Wireless Applications

18.1 The CTIA

The Cellular Telecommunications and Internet Association (CTIA) is the main advocacy group for the wireless industry in the United States. The CTIA supports traditional 850 MHz cellular carriers nationwide, as well as 1,900 MHz PCS carriers nationwide. There used to be a companion trade group known as the Personal Communications Industry Association that supported the upstart PCS (1,900 MHz) carriers, but it folded its activities into the CTIA to present a more unified approach to supporting the wireless industry in the United States. The CTIA acronym used to stand for Cellular Telecommunications Industry Association. It changed its name—but not the acronym—because it better reflects the rapid merging of applications from the wireless industry and the Internet.

Wireless carriers must become member companies of the CTIA in order to reap the benefits offered by this trade group. Membership benefits include regular access to industry reports regarding business and technology issues, industry statistics, and contact information for other carriers that lists persons by title and department. The CTIA is involved in representing the wireless industry to the FCC and the Congress. It funds lobbying efforts related to legislation that will impact the industry by having its representatives attend hearings and testifying before Congress regarding issues and/or laws that will impact the wireless industry. The CTIA is headquartered in Washington, DC, because that is where it makes sense for it to be located, based on its mission.

18.2 Cultural Impact and Hot Buttons

Commercial wireless service, available to both consumers and businesses, is the fastest growing consumer technology in history. This is proven by the fact that as of late 2004, over 175 million people in the United States subscribe to wireless service.

Who among us can say that somehow wireless phone use has not changed our lives (for better or worse?). One thing that is for certain . . . cell phones have made our lives much more productive. How often have you been out at the store, and your spouse calls you on the cell phone to ask you to add an item to the grocery list? How many times have you checked your voice mail *on the way in* to work to save time? How many parents buy their kids a cell phone, so they can have instant access to their kids for peace of mind's sake? Consider the following facts that bear out the popularity of wireless in society today:

- Twice in one year, the magazine *Consumer Reports* has run *cover stories* that advertise an article inside that analyzes wireless service as a whole, by handset, then carrier by carrier (region by region). These are in-depth reports designed to help consumers make the best choice.

- At least three times since 2003, the *Chicago Tribune* has run editorials that relate directly to the user of wireless service and its impact on society as a whole. Some of these editorials address the perceiving addictive effect that wireless phone use has on many people (can this really be denied?). Editorials have also coined the term "cell yell" to describe people's indiscriminate use of cell phones in all manner of public places—"loud, boorish conversations occurring everywhere."

- More and more people are now using their cell phones—*and expecting coverage*—in U.S. national parks. Many people call ranger stations asking for directions when they get to forks in paths and are unsure which path is the best route. One goof even scaled Mt. Whitney in California in April 2004, a 14,000-foot peak, and could not wait to start making calls on her cell phone, instead of taking in the breathtaking view.

- More and more people who attend sporting events actually know where the television camera angles will be and obtain tickets to sporting events with the sole purpose of doing nothing but trying to get on TV so they can call all their friends and tell them to look for them on TV. (See how goofy I look?)

- Movie theatres now specifically request patrons to silence cell phones prior to the movie's beginning.

- A recent announcement by the FCC that they were even *considering* allowing the use of cell phones while in-flight caused a minor uproar among the general public. Many people wrote letters to their newspapers stating the only peace they got while traveling was on the plane, and they asked that the FCC not eliminate this last bastion of a cell phone-less environment.

- Because it has been determined that talking on a cell phone while driving increases the chances of a traffic accident by 400 percent, many local governments have outlawed handheld wireless phone use while driving in a vehicle. For instance, Hong Kong instituted a law in 2000 that fines anyone caught using a wireless phone while driving $256. New York City has a similar law banning the use of handheld cell phone talking while driving in a vehicle. This is one of the reasons that the earpiece option and in-car speakerphones have become very popular with wireless phones. Users can talk on the phone while having their hands free to drive.

- The first question that many policemen now ask when reporting to the scene of an auto accident is "Were you using your cell phone while driving?"

- Many restaurants now offer cell-free zones.

- Some clothiers (mostly catalog) now offer cell phones that are built into jackets and other clothing items.

- Due to the immense popularity of picture phones, many health clubs and gyms have now banned cell phones altogether in their establishments, because too many creative people used their picture phones for questionable uses.

- Wireless 411 is on the radar screen. A U.S. Senate Commerce Committee approved a proposed national wireless 411 directory assistance service, but wireless subscribers cannot be added to the database without their consent. In this age of ever-increasing invasions of privacy, this hot button could be dead in the water before it even gets a chance to be resuscitated. Verizon Wireless is adamantly opposed to the idea. The wireless industry is strongly in favor of the idea, as they contend that an estimated eight million wireless customers do not have a wired phone and are not listed anywhere. Many wireless carriers argue that customers should be able to keep their numbers private. By the way, the federal do-not-call list *can* apply to wireless phone numbers as well as landline numbers.

- Legislation is being considered to outlaw the transmission of spam to cell phones. Currently, our cell phones are one place where we can quietly go where we control exactly what we do. No unwanted ads, yet. No spam. It will be interesting to see how this issue plays out, especially in an era where more wireless subscribers are expected to subscribe to push video streaming services that will contain news, weather, stock quotes, and so forth.

The problem that is apparent when examining all the preceding facts is that too many of us have become inconsiderate when using our cell phones in public. It is as if we have forgotten our manners and that we believe the world is our phone booth. Or in many cases we are just trying to multitask too much (driving while talking). Why else would all the preceding activities be happening? How often have any of us been behind someone on the highway who was going too slow or casually weaving in and out of his or her lane? How many of us get frustrated listening to loud, one-sided conversations while riding public transportation? In any event, somehow, someway over time we can hope that the wireless craze will eventually taper to the point where people moderate, right?

18.3 The Wireless Explosion: The Growth of the Most Popular Consumer Technology in History

Consider the following facts and statistics that bear out the growth surge wireless has experienced:

- The wireless industry is now a $10 billion per year industry in the United States alone.

- It is projected that by 2006 the amount of wireless phones in operation will exceed the number of landline phones in operation for the first time ever.

- In 2003, the Telecommunications Industry Association (TIA) reported that wireless spending exceeded wireline-based spending in the United States for the first time ever.

- Twenty-six percent of business telecommunication budgets will be spent on wireless services and infrastructure in 2006; this includes both voice and data. It is projected that by 2007, there will be over 10 million wireless business users. This data reflects wide area wireless Internet/intranet access from PCs and data devices for the business segment.

- Growth in wireless data for the small-medium business segment is expected to be $135 million through 2006. That is *just* the small-medium segment!!!

- Total spending in the United States in 2004 on wireless communications was expected to grow by 7.6 percent, achieving a total of $144.7 *billion*!

- The United States wireless market is projected to become a $191 billion industry by 2007.

- The number of wireless subscribers in the United States is expected to reach 195.5 million by 2007.

- The handset market is expected to grow from $10.2 billion in 2004 to $13.2 billion in 2007, growing at an 8.3 percent compound annual rate.

- Overall wireless equipment spending is project to grow from $12.7 billion in 2004 to $26.3 billion in 2007, growing at a 4.4 percent compound annual rate.

- In late 2004, Siemens set a mobile communication speed record by transmitting one gigabit per second over the air to a mobile device. By comparison, wireless LAN (WLAN) networks presently offer the fastest wireless links to mobile devices at speeds of around 50 Mbps. To accomplish this feat, Siemens used a smart antenna system. See Figures 18-1 through 18-5 for graphs that reflect the truly explosive growth the wireless industry has enjoyed in its 22 years of existence in the United States.

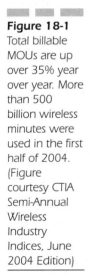

Figure 18-1
Total billable MOUs are up over 35% year over year. More than 500 billion wireless minutes were used in the first half of 2004. (Figure courtesy CTIA Semi-Annual Wireless Industry Indices, June 2004 Edition)

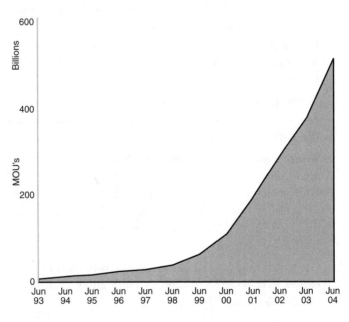

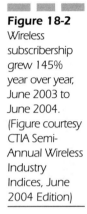

Figure 18-2
Wireless subscribership grew 145% year over year, June 2003 to June 2004. (Figure courtesy CTIA Semi-Annual Wireless Industry Indices, June 2004 Edition)

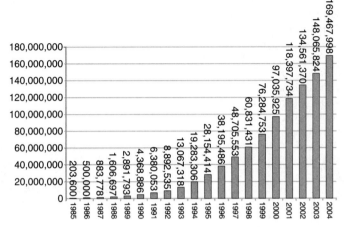

Figure 18-3
Direct carrier employment is up 13.4% year over year from June 2003 to June 2004. (Figure courtesy CTIA Semi-Annual Wireless Industry Indices, June 2004 Edition)

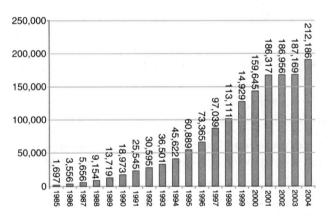

Figure 18-4
Cell sites in service are up 18% year over year, 2003–2004. (Figure courtesy CTIA Semi-Annual Wireless Industry Indices, June 2004 Edition)

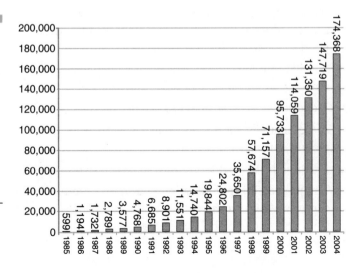

Figure 18-5
Digital sub-
scribership is up
12.3% year
over year June
2003 to June
2004. (Figure
courtesy CTIA
Semi-Annual
Wireless
Industry
Indices, June
2004 Edition)

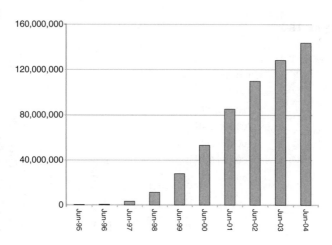

18.4 Key Industry Drivers

The following list denotes key drivers that are making wireless an increasingly ubiquitous part of everyday life:

- An expectation of pervasive broadband for anyone who desires connectivity, especially broadband based on Wireless Fidelity (Wi-Fi)
- The spreading of an instant, always-on, on-demand, multitasking culture shift
- Increasing adoption of data-capable devices
- Increasingly rapid integration of voice, data, and image functions—multimedia—into single devices (e.g., combination PDA/cell phone/Blackberry/digital camera)
- A shift from wireline to wireless usage plans and patterns, especially among younger people (see Section 18.6)
- A desire for instant inventory updates
- A requirement for presence-aware services (e.g., fleet management, technician tracking)
- Multimedia applications that serve the real-estate industry
- The projected rapid adoption of unified communication (UC) services, where voice, data, and fax can be accessed and managed from any phone or PC anywhere

■ The adoption of disruptive technologies such as Wi-Fi and Voice over IP (VoIP)

18.5   The Commoditization of Wireless

As wireless service has become so ubiquitous in the United States and around the world since the mid-1990s, this rapid growth has served as a catalyst to making wireless a commodity service. Another factor that has led to this phenomenon is the fact that in the United States, anyway, there are so many wireless carriers competing for the consumer (and business dollar) that they have been forced to compete in the only area where they can really compete—price.

Key: A service becomes commoditized when the main factor that differentiates it from its competition is price. This applies to any industry, and the wireless industry is no exception.

Because all wireless carriers offer roughly the same service features over roughly the same types of phones, the only competitive leg left for wireless carriers to stand on is the price of the service.

The downside to this is that low prices have severely cut into wireless carrier margins. This results in less capital available for investments to upgrade the networks, which would enable carriers to offer newer, more advanced services. Sound like a catch-22? It is. The way around it: The FCC lifting the spectrum cap in January 2003. Up to that point in time, each wireless carrier in every market could own no more than 45 MHz of spectrum in each market area. The FCC lifted this cap, many speculate, in order to encourage merger and acquisition activity. It has worked (see Section 18.8). Another thing that has—and will—contribute to the commoditization of wireless service is the current and impending emergence of disruptive technologies such as Voice over IP (VoIP). VoIP service offers very slick features such as Web-based management of vertical calling features like find me/follow me, click to call, softphones, and so forth. And the prices offered with this type of service are right on par with wireless service.

Also, when 3G networks are rolled out in a more widespread manner, that should result in the offering of more advanced services, which should drive revenues up and pay for the 3G networks.

18.6 Landline Replacement

In Spring 2001, the *Chicago Sun Times* ran a cover story that stated the incumbent local exchange carrier (ILEC) operating in Illinois, claimed it was losing 30,000 access lines per month to competition. Make no mistake, this competition was largely wireless carriers who were snagging a large part of the coveted 18- to 34-year-old demographic. This was due to the confluence of several factors:

- 2001 saw huge growth in broadband Internet access, mainly driven by cable companies and their high-speed modems. Many young people (and some middle-aged people) saw no need for their home phones because they had broadband Internet and a cell phone.

- Why spend the extra money every month on a home phone, especially if they are on the go much of the time? Plus, with their cell phone, they only need to give out one phone number—not two. When they get home, they put the cell phone in the charger.

A huge part of this mobile-only market segment is college students, military members, or people starting their careers. These people may fine it is just easier to maintain a mobile phone and avoid dealing with roommates, reestablishing service after apartment moves, and so on. This is the market segment that has been driving landline loss to a large extent.

The key challenge with wireless phones replacing landline phones is quality assurance. It is a fait accompli that wireless service will never offer true toll quality that is incumbent with the landline phone network. This is not simply due to the fact that the landline telephone network has had over 100 years to evolve. This is about the air interface—the dynamic physics of radio (waves) that even the best engineering cannot overcome. If (or when) consumers accept the fact that their cell phones will never be as perfect as their landline phones, true widespread landline replacement by wireless could possibly occur. People will need to accept that the potential for a dropped call will always be lingering, depending on their operating environment (e.g., Are they on a train? Are they moving through a tunnel? Are they hiking on a mountain?). They may have static on their calls or choppy speech. Is this day on the horizon? Probably. Why? Because people most likely value the mobility aspect of wireless more than they are bothered by the potential of a dropped call or choppy speech on a wireless call.

18.7 The Death of Long-Distance and Roaming Charges

In order to gain competitive advantage, many carriers have adopted a "no roaming fees, no long-distance charges" marketing strategy. If this marketing strategy continues to work—as it is now—roaming fees as they are known today will become a thing of the past—even if people roam in rural areas, where roaming fees are prevalent.

The key drivers behind the elimination of roaming and long-distance charges are as follows:

- Carriers simply have to cut these fees to attract new customers. It is purely a competitive thing in many respects.
- More carriers now have nationwide network footprints. This allows them to more economically serve their customers when they are roaming.
- Wholesale long-distance rates are now so low, wireless carriers have the ability to absorb this additional cost of doing business when they assess their customers the monthly service fee.

The beauty of this evolution for mobile subscribers is this: The industry has taken this step; now there is no going back. In other words, if carriers determine down the road that they would like to begin charging for roaming and long distance again, their customers will not stand for it. Customers would simply change carriers, which is even easier and more tempting today given the ability to port their mobile telephone numbers with them to a new carrier. Carriers may be tempted into this situation as more and more mergers are occurring, driving the industry to a smaller core of competitors on the playing field. In other words, they may believe they do not need to absorb the roaming and long-distance charges anymore because there are less competitors in the industry. The bottom line is that all carriers are now obtaining huge synergies and economies of scale as they build nationwide networks supported by SS7 networks. They can now afford to subsidize these charges.

18.8 Mergers, Acquisitions, and the Rule of Three

The speculation in the industry when the spectrum cap was lifted was that it would lead to a flurry of mergers and acquisitions. The speculation was correct. There are fewer and fewer players left in this industry from a competitive viewpoint, and it is quite possible that the industry will end up with four major competitors, possibly fewer. Industry consolidation status is listed as follows:

- Cingular purchased AT&T Wireless and the deal was formally closed in October 2004. SBC's CEO Ed Whitacre has even publicly said he would like to buy out Bellsouth's share of Cingular if he had the opportunity.
- Sprint announced its intention to purchase Nextel in December 2004.
- Alltel Mobile announced its intention to purchase Western Wireless in January 2005.
- Verizon Wireless is standing pat for now.
- US Cellular is rumored to be acquisition bait by Alltel Mobile, which is understandable because both carriers have large rural footprints that likely complement each other.
- T-Mobile, the German-owned US GSM carrier, is also idle in terms of acquisition activity.

But all in all, the activity summed up in the preceding passage paints a picture of only four major wireless carriers with nationwide footprints and deep pockets (which enables them to be bold in any number of ways): Verizon Wireless, Sprint PCS, Cingular, and T-Mobile. But there is more —rumors abound in the industry that T-Mobile may counter Sprint's bid for Nextel. Bear in mind that if T-Mobile were to merge with Nextel that would open the way for Verizon to bid on Sprint PCS. Only the passage of time will determine how this all plays out. But the consolidation also allows bigger companies to operate with fewer (human) resources—this is an economy of scale unto itself. Though not a very people-friendly one, it improves overall efficiency and improves economics. Also, the consolidations feed on themselves: When the first big merger occurs, the other players have to merge just to be big enough to compete with the new big kid on the block. It is a numbers and size game.

This consolidation was necessary, per industry insiders. As mentioned in Section 18.5, wireless carriers had razor-thin margins due to the intense competition in the marketplace. No company or industry can operate endlessly with little or no margin to show its stockholders. Like other major industries (automakers, long distance in its heyday, computer makers), the rule of three may very well likely come into play in the wireless industry as well. In late 2003 there were six major carriers. In early 2005, there are four major carriers. Is three on the horizon?

18.9 Cost per Gross Add and ARPU

One of the financial indicators of a healthy wireless company is a low cost per gross add, known as cost of acquisition in other industries. The cost per gross add is the gross cost to the company to obtain one new customer, in terms of the following costs and overhead: executive management, marketing programs, engineering, network operations, customer service, information technology (IT) systems maintenance and development, corporate administration, and capital outlay for network buildouts and improvements. The average cost per gross add in the wireless industry is approximately $350 to $400 per subscriber. It is important to note that in any service industry such as wireless, it costs a company approximately three times normal cost per gross add to win back old customers. Cost per gross add is a statistic that is usually listed in wireless carrier annual reports.

Another financial indicator that is an offset to cost per gross add is average revenue per unit (ARPU). This statistic indicates how much revenue per subscriber a wireless carrier obtains on a monthly basis. The higher the ARPU, the better. ARPU can also be analyzed in conjunction with cost per gross add, so a wireless carrier can determine its net earnings per subscriber. Like cost per gross add, the ARPU number shows up frequently in wireless carrier internal financial reports. Average wireless industry ARPU is about $50 per month at the end of 2004. The good news for wireless carriers is that the numbers are trending upwards. Widespread rollout of 3G services and/or new service features should foster continued growth in these numbers.

18.10   Price per Minute Rates

The industry average cost to build an entire base station ranges anywhere from $250,000 to $400,000. This cost is dependent on the type of digital wireless technology being used, whether it is CDMA, GSM, or another technology. It includes land for the site (usually a lease), taxes, shelter, base station equipment, tower, antennas, and construction labor. The huge cost of wireless infrastructure is why customers used to pay such high rates per minute for wireless phone use, compared to landline rates. This is because the cost per base station used to be approximately three times what it is today. The cost to build, maintain, operate, and continuously expand and improve a wireless network is enormous. This is especially true in urban areas with large populations, a denser concentration of cells, and higher expectations from customers.

Key: As wireless coverage has saturated the United States (and the world), the cost per minute charged to wireless subscribers has trended downward. This is because as coverage increases, the manner in which new coverage is deployed has become cheaper. For example, carriers can collocate on other carrier's towers or they can deploy microcells or enhancers instead of full-fledged cell base stations. This is also because there is so much demand for base station components; the cost to wireless carriers for base transceiver stations (BTS) components has also dropped over the years due to economies of scale achieved in the manufacturing and procurement process. Ultimately, the overall decrease in costs is passed on to the consumer in the form of lower rates.

Wireless rates per minute that are charged to subscribers in 2004 average around $0.04 to $0.05 per minute. This cost reflects both local and long-distance calls. In 1995, the average cost per minute charged to wireless subscribers was anywhere from $0.22 per minute to $0.35 per minute. As wireless coverage approached the saturation point in the United States, the rate—but not necessarily the overall quantity—at which new base stations are required has slowed, driving down costs. This evolutionary development, along with increased competition, has resulted in wireless service becoming more of a commodity—a more common, inexpensive product that can be easily obtained. The more any product or service becomes a commodity, the lower the prices charged to customers. The lower rates are one catalyst that is causing many people to permanently abandon their landline phones for wireless phones.

Competition in the wireless industry has become so fierce that many industry observers believe there is a need for mergers or consolidation. The main reason is because the extremely low rates charged to wireless customers today reflect intense competition within the industry. The problem with this commoditization—where price is the key differentiator —is that profit margins for all carriers are extremely low. When profit margins are low, capital investment that is required for network expansion and improvements is at a very low level. Because of this, an eventual migration to a playing field of only three to four competitors is likely to occur by 2006 or 2007. The industry simply cannot sustain the miniscule profit margins they have been living with in the early twenty-first century. So it is likely that the economic rule of three may eventually apply to the wireless industry. Most large industries with huge markets maintain a playing field of three competitors. Think about long distance in the United States—for a long time there were three major players: AT&T, Sprint, and MCI. Think about the domestic auto industry in the United States: same thing. For the longest time there was Chrysler, GM, and Ford. All major industries usually sustain three major competitors. The wireless industry is likely on that same course as well.

18.11   Churn

"Churn" is a marketing term that refers to the number of customers who cancel their wireless service. In other words, it describes the number of customers who churn off the carrier's network. Churn is a term used in any service industry that serves vast numbers of customers where customers are usually bound by some form of term agreement or contract for the service. There are two types of churn:

- Intercarrier churn occurs when customers, for whatever reason, cancel service with their existing service provider and obtain service from a competing service provider in the marketplace. In other words, I cancel my Verizon Wireless service and sign up with Cingular Wireless.

■ System churn occurs when customers, for whatever reason, cancel service with their existing service provider and do not obtain service from a competing wireless carrier in their market. These are customers who no longer desire any wireless service whatsoever. It is very rare for this type of churn to occur, because wireless is the type of service where once customers experience the many benefits of the service, they usually find it hard to do without these benefits. The one exception to this theory would be a customer who has extremely unrealistic expectations for the service itself (e.g., insisting they should never experience a dropped call or static, or that they should never experience a loss of signal). All wireless subscribers experience network inefficiencies at some point and no wireless carrier can perfectly engineer the network to avoid this problems—radio is a very dynamic, sometimes finicky technology.

Key: The key to making a competitive difference in the wireless marketplace is to provide value or features—or the perception of value or features—that your competition does not provide.

▩ 18.12   Agents and Resellers

Many businesses have entered the wireless marketplace since 1983 to act as sales agents for wireless carriers. These companies purchase huge blocks of air time from wireless carriers at a discounted, wholesale rate. They then resell the service—the airtime—to their customers at a markup, acting as their own independent wireless company from the customer's perspective. When marking up the prices, the reseller will always ensure that its rates are equal to and consistent with its carrier sponsor (their wholesale provider) to avoid a scenario where they are competing with each other in the marketplace. This alignment would also likely prevent possible legal action involving one or both parties—the carrier or the reseller.

Reselling wireless service is a win-win-win scenario for all involved parties:

- The wireless carrier wins because it has an expanded business base —a larger retail presence.

- The wireless carrier also wins because it is able to generate additional business—more minutes of use (MOU) on its network.

- Customers win because they have more service locations to choose from and patronize. This could mean added convenience for the customers. For example, the nearest Cingular Wireless store might be 6 miles from the customer's house but a Cingular Wireless dealer is 2 miles away from his or her house.

- The reseller and the economy also win because resellers represent business growth, jobs, an additional revenue-generating engine for the wireless carrier, and ultimately a contribution to the economy at large.

Key: It is easy to tell who the resellers are in any given market, because they have names that are distinct and separate from the major carriers themselves. In other words, they'll have names like Windy City Wireless or Areawide Cellular. But underneath their names, they'll have phrasing such as "A Cingular Wireless Dealer."

▬▬ 18.13   Prepayment Offerings

Around 1995, industry studies revealed that up to 25 percent of potential wireless customers fell into the category of credit challenged. Because of this information, many wireless carriers have sought to sign on credit-challenged customers in a manner that minimizes risk to the carriers. With prepayment plans, customers buy a mobile phone and purchase blocks of minutes of use or blocks of use based on dollar bundles (e.g., $40 of minutes). No credit check is necessary, and that administrative task is avoided by the wireless carrier.

At periodic intervals, the wireless switch and the carrier's billing system will notify the customer how much airtime they have left, in dollars or minutes of use. Customers can also pull that information up on demand on their phones.

Prepayment systems offer important benefits. First, wireless carriers can harness a large portion of their potential customer base essentially risk-free. Second, this system becomes an option even for the potential customer who has good credit but who wants to maintain strict control over his or her wireless expenditures.

18.14 Wireless Market Ownership

Wireless markets are bought, sold, and traded on a regular basis among wireless carriers nationwide. This includes entire market clusters, or portions of market clusters, as well. When clusters are traded or purchased, the motive is for the wireless carrier to create or expand a *seamless*, contiguous regional or national footprint. When markets trading occurs, naturally the trade is for markets of equal or comparable value. The FCC must approve all market sales or trades. There are advantages and disadvantages to trading markets with other wireless carriers.

The main advantage to trading wireless markets is that it allows wireless carriers to build out their existing market clusters. Remember, a market cluster is an area of geographically adjacent markets, all owned by the same wireless carrier. The key benefit of enlarging market clusters is that it allows carriers to attain a competitive edge and to attain economies of scale in terms of network design and configuration, marketing, and system operations.

There is one main disadvantage to buying or trading for other wireless markets: The incoming wireless carrier is inheriting a legacy system. The previous owner of a market and the employees hired to design and operate the system may not have been very knowledgeable or competent about wireless system design and operations. There may be an extensive amount of rework involved in redesigning the system where deficiencies existed under the previous owner. This could cost a lot of time, expend a large amount of labor, and possibly a large amount of capital to invest in new equipment. However, those carriers who purchase or trade for other markets usually go into these situations with their eyes wide open. Ideally, they do their homework on the system(s) they intend to purchase and factor in any rework or any new equipment requirements prior to signing any purchase agreements. Retraining network operations and surveillance personnel may be required if the market being purchased contains different network infrastructure than all the other markets the carrier owns. Purchasing all-new equipment in the market so that the equipment matches what is in all of the carrier's other markets would be cost prohibitive.

Another disadvantage is that the previous owner of the market might have signed bad contracts and may deal with any number of other businesses:

- There may be unfavorable interconnection arrangements with area LECs.
- Unfavorable leases may have been signed for property, buildings, or tower occupancy.

18.15 Wireless Applications

Let us be clear about one thing: Voice is still the killer app in the wireless industry, at least in the United States. The United States has always been slow to appreciate and adopt data-based wireless services such as short message service (SMS), for example. Where people in Finland use their mobile phones (via e-wallet applications) to pay for items at a vending machine, this type of vending machine did not even exist in the United States as of early 2005.

The nature of today's wireless service requires several basic attributes in order to spur widespread adoption of new applications, especially those that are 3G enabled. These are all prerequisites to being able to offer advanced and value-added services:

- First among them is mobility, pure and simple: the ability to maintain error-free, constant voice and data communications while on the move. In other words, no dropped calls.
- Immediacy, which allows subscribers to obtain connectivity when they need it, regardless of location and without a lengthy login session. In other words, a service similar to the always-on nature of DSL or cable modems.
- Finally, localization will allow subscribers to obtain information relevant to their current location (think locator technology).

The combination of these characteristics provides a wide choice of possible applications that can be offered to mobile subscribers.

In general, applications can be separated into two high-level categories: business segment and consumer segment. These include the following:[1]

- Communications: E-mail, fax, unified messaging, intranet/Internet access (business)

- Value-added services (VAS): Information services, games (consumer)

- E-commerce: Retail, ticket purchasing, banking, financial trading (business and consumer)

- Location-based applications: Navigation, traffic conditions, airline/rail schedules, location finder (consumer)

- Vertical applications: Freight delivery, fleet management, sales-force automation

- Advertising (consumer)

Applications can be broken down into two high-level categories:

- Those where it appears to end users that they are using the mobile communications network purely as a *pipe* to access messages or information.

- Those applications where users believe that they are accessing a service provided or *forwarded by (through)* the network operator, such as video streaming.

18.15.1   E-Mail and Internet Access

E-mail on mobile networks may take one of two forms. It is possible for e-mail to be sent to a mobile user directly, or users can have an e-mail account maintained by their network operator or their Internet service provider (ISP). In the latter case, a notification will be forwarded to their mobile terminal; the notification will include the first few lines of the e-mail as well as details of the sender, the date/time, and the subject. This is how the Blackberry service works. Fax attachments can also accompany e-mails.

Corporate E-Mail   With up to half of employees typically away from their desks at any one time, it is important for them to keep in touch with

---

[1]Source: ARC Group

the office by extending the use of corporate e-mail systems beyond an employee's desktop. General Packet Radio Service (GPRS)-capable devices will be more widespread in corporations than amongst the general mobile subscribers, so there are likely to be many corporate e-mail applications using GPRS versus Internet e-mail applications whose target market is more generalized.

Internet E-Mail   Internet e-mail services will be available through a gateway service where the messages are not stored or a mailbox service in which messages are stored. In the case of gateway services, the wireless e-mail platform simply translates the message from Simple Mail Transfer Protocol (SMTP), the Internet e-mail protocol, into SMS form and sends the message to the SMS Center for processing. In the case of mailbox e-mail services, the e-mails are actually stored, and the user gets a notification on his or her mobile phone and can then retrieve the full e-mail by dialing in to collect it, forward it, and so on. And by linking Internet e-mail with an alerting mechanism such as SMS or GPRS, users can be notified when a new e-mail is received.

Internet Access   As a critical mass of users is approached, applications aimed at general consumers are increasingly being placed on the Internet. The Internet is becoming an invaluable tool for accessing corporate data as well as for the provision of product and service information.

18.15.2   Intranet Access

The first stage of enabling users to maintain contact with their office is through access to e-mail, fax, and voice mail using unified messaging systems. Increasingly, files and data on corporate networks are becoming accessible through corporate intranets that can be protected through firewalls by enabling secure tunnels via virtual private networks (VPNs).

18.15.3   Unified Messaging

Unified messaging (UM) is an application that is getting its toehold on today's VoIP networks. UM uses a single mailbox for all messages, including voice mail, faxes, e-mail, and SMS messages. With the various mailboxes accessible from one place—one portal—UM systems then allow for a variety of access methods to recover messages of different types. Some will use text-to-voice systems to read e-mail and, less commonly, faxes over a normal phone line. Most will allow for inquiries of the contents of the various mailboxes through data access, such as the Internet. Others may be configured to alert the user on the terminal type of his or her choice when messages are received.

18.15.4   Value-Added Services

Value-added services refer strictly to content provided by network operators to increase the value of their service to their subscribers. Two terms that are frequently used with respect to the delivery of value-added data applications are "push" and "pull" as defined in the following:

- *Push* refers to the transmission of data at a predetermined time or under predetermined conditions. It could also apply to the unsolicited supply of advertising such as the delivery of news as it occurs or stock values when they fall below a preset value.

- *Pull* refers to the demanding of data in real time by the user (for example, requesting stock quotes or daily news headlines). Video streaming could fall into either of these two categories.

To be valuable to subscribers, push/pull content must posses several characteristics:

- Personalized information tailored to user-specific needs with relevant information. A stock ticker, focusing on key quotes and news, or an e-commerce application that knows a user's profile are two examples of personalized information.

- Content based on a user's current location, which can include maps, hotel finders, or restaurant reviews. (This content is locator-based.)

- Convenience means that the user interface and menu screens are intuitive and easy to navigate.

- Trust pertains primarily to e-commerce sites where the exchange of financial or other personal information is required.

Some value-added services are outlined in the following sections.

18.15.4.1   E-Commerce   E-commerce is defined as conducting business transactions over the Internet, or through a type of data service. This would only include those applications where a contract is established over the data connection, such as for the purchase of goods or services. This also includes online banking applications because of the similar requirements of user authentication and secure transmission of sensitive data such as account numbers.

18.15.4.2 Banking The incentive among banks to encourage electronic banking stems from the comparable costs of making transactions in person in a bank to making them electronically. Specific banking functions that can be accomplished over a wireless connection include balance checking, moving money between accounts, bill payment, and overdraft alert. The Federal Reserve reported that in 2004, for the first time in U.S. history, electronic payment by check exceeded standard check payments to the tune of $7.8 billion. Wireless banking transactions grew 4 to 7 percent from 2002 to 2005. Mobility enhancements will only take this evolution further.

18.15.4.3 Financial Trading The immediacy with which transactions can be made using the Internet and the requirement for up-to-the-minute information has made the purchasing of stocks a popular application. By providing push services and coupling these with the ability to make secure transactions from a mobile terminal, a valuable service unique to the mobile environment can be provided.

18.15.5 Video Streaming

Video streaming is now entering the mobile realm. Some view the ultimate application of the unwired age as live TV delivered to millions of handsets via streaming media. Customers are still mulling over their commitments to live TV over mobile phones. They are still evaluating the need for audio and video streaming. In other words, this type of application *could* be a technology in search of a market, instead of the other way around. The results of a survey in *Wireless Review* magazine in the December 2004 issue stated the following regarding "Consumer Interest in watching TV or video on portable devices:"

26 percent—Not at all interested

27 percent—Not very interested

32 percent—Somewhat interested

9 percent—Very interested

5 percent—Extremely interested

Could it be that we have simply had enough TV in our lives, and we do not really need or want it available to us 24 hours a day, *even if the capability exists to deliver it to us*? When looking at the survey results, maybe TV-to-the-handset falls under the same category of all the people who are buying that gadget that can remotely turn TVs off in public places (covertly, of course).

Nevertheless, analysts are still pumped up about the future of mobile video, even if they are still uncertain how the business model will play out. Strategy Analytics predicts that the global market for mobile video content will reach $4.6 billion by 2008, with mobile music sales reading $2.2 billion. But these numbers include both streamed and downloaded content, with streaming representing a small fraction of the total. Wireless carriers' thinking about mobile multimedia of all types should be similarly cautious. On a worldwide basis, one of the biggest multimedia applications on phones is music. But this is not to say that streaming has no place in music. Using streaming or progressive downloading (pseudo-streaming) to the handset can create a radio-like service. This approach is becoming very popular in Europe. But streaming songs to the mobile handset remains the exception today, not the rule.

Two basic types of services send video data to wireless handsets. The first is one-way service, delivering commercial video content from service providers to users. The second type is two-way service, sending video from user to user. Either type may or may not use streaming.

Mobile television, such as the appropriately named MobiTV service available to customers of Sprint and AT&T Wireless (now Cingular), is the highest-profile example of one-way video. MobiTV uses streaming technology to deliver commercial television programming, such as news, sports, and weather—live to users. MobiTV essentially just repackages broadcast and cable programming. In-Stat/MDR expects the number of mobile video subscribers to reach 273,000 (out of 175 million wireless subscribers) by January 2005, exceeding 1 million by the end of 2005 and leveling off at around 22 million by 2009.

As with music, there is a nonstreaming counterpart to mobile TV. Cinema Electric creates original video and other content that users can download the way they do ringtones and wallpaper. It distributes eight channels with names like Electric Catwalk, Movie Messages, Pocket-Girls, and Sports Action. The video clips and animation usually amount to no more than 90 Kb. RealNetworks offers news, sports, movie trailers, and weather, which it streams to handsets on demand via Sprint PCS and AT&T mode using short clips (rather than live).

Two-way video between users may or may not require streaming. Video messaging can run over existing nonstreaming technology such as multimedia messaging service (MMS). This lets users create, send, and receive messages that include text, audio, graphics, and video. But other services will happen in real time, such as mobile videoconferencing. This is referred to as a *look where I am* application, where users will wave the camera around to show friends what the user is seeing at any given moment (great, more goofs at the ballpark).

Carriers need to clearly separate the technology from the service in the eyes of the consumer. Some issues have less to do with the technology than its commercial implications, such as pricing. For example, there might be a small price or fee to stream content to the handset only once. But if you want to keep it and view it more, you may have to pay four times the cost (e.g., $2 versus $0.50).

Content aggregators may not find the mobile business model very appealing in that people who receive streaming content will not be able to keep it, so they will not be prepared to pay a premium for it. The people who own the content are asking for a significant share of the revenue; and carriers want their share (30 percent retail margin). So that leaves very little for the guy in the middle that's providing the solution—the aggregator.

Many industry analysts expect two-way video to be slow in coming. As with other mobile technologies, the United States lags the rest of the world because what Americans really like to do, still, is call people and talk. That is about it. As with everything else in wireless, only time—and the fickle consumer—will determine how streaming media plays out.

18.15.6 Location-Based Services and Telematics

Location-based services provide the ability to link, push, or pull information services with a user's location. Examples include hotel and restaurant finders, roadside assistance, and city-specific news and information. This technology also supports vertical applications such as workforce management and vehicle tracking.

18.15.7   Vertical Applications for Business

In the mobile environment, vertical applications apply to systems utilizing mobile network architectures to support the implementation of specific tasks within the value chain of a company. Examples of vertical applications include the following:

- *Sales support*—The provision of stock and product information for sales staff, as well as the integration of their use of appointment details and the remote placement of orders. Appointment details could include information on product availability based on customer meeting agendas.

- *Dispatching*—The communication of job details such as location and scheduling or allowing the inquiry of information to support job function. For example, this type of application could be used by a landscaping firm or a construction firm.

- *Fleet management*—The control of a fleet of delivery or service staff, monitoring their locations and scheduling work based on location.

- *Parcel delivery*—Tracking the locations of packages for feedback to customers and performance monitoring.

18.15.8   Advertising

Advertising services will be offered as a push-type information service. Advertising may be offered to customers to subsidize the cost of voice or other information services. For example, it could be location sensitive. A user entering a mall, for example, would receive advertising specific to the stores in that mall.

▰▰▰ Test Questions

True or False?

1. _____ Video streaming in the mobile arena is an iffy proposition at this time, due to multiple factors that may impact its success.

2. _____ System churn describes when wireless customers cancel service with their current provider and obtain service from the competing wireless carrier in their market.

3. _____ Around 1995, industry studies revealed the fact that up to 75 percent of potential cellular customers fell into the category of credit challenged.

4. _____ The main advantage of buying/trading for new wireless markets is that the incoming wireless carrier is inheriting a legacy system.

5. _____ Wireless service is becoming commoditized partially because many young people find no need for a landline phone when they have cable modems and mobile phones.

6. _____ One of the financial indicators of a healthy cellular company is a high cost per gross add.

Multiple Choice

1. Which of the following acronyms denotes the statistic that indicates how much revenue per customer a wireless carrier obtains on a monthly (regular) basis?

   a. CPU

   b. RAM

   c. ARPU

   d. CFS

   e. None of the above

2. A survey conducted by *Wireless Review* magazine discovered that the general public has very little interest in what application?

   a.   Locator-based services

   b.   Streaming audio

   c.   Push advertising

   d.   Mobile TV

   e.   None of the above

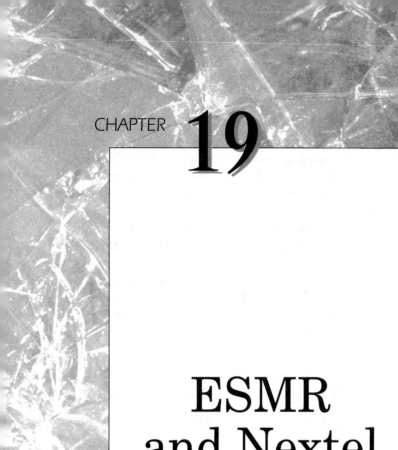

CHAPTER 19

ESMR
and Nextel

■■■ 19.1 SMR Overview

Specialized mobile radio (SMR) service was created by the FCC in 1974 for carriers to provide two-way radio dispatching to the public safety, construction, and transportation industries. SMR systems provide dispatch services with push-to-talk technology for companies with multiple vehicles. These services included voice dispatch, data broadcast, and mobile telephone service, but SMR had limited roaming capabilities. RAM and ARDIS are wireless data services that are licensed in the SMR frequency range. An SMR subscriber could interconnect with the PSTN much like a cellular subscriber.

SMR systems traditionally used one large high-power transmitter to cover a wide geographic area, like mobile telephone service (MTS)/ improved MTS (IMTS) systems. This limited the number of subscribers because only one subscriber could talk on one frequency (channel) at any given moment.

Key: SMR systems operate in the following frequency ranges: 806 to 821 and 851 to 866 MHz.

The number of frequencies allocated to SMR is smaller than for cellular and personal communication services (PCS), and multiple operators have been in each market in the past. Because dispatch messages are short by nature, an SMR system can handle many more dispatch-only customers than interconnect subscribers.

■■■ 19.2 Migration to Enhanced Specialized Mobile Radio (ESMR): Nextel Corporation

In 1987, a newly founded company, Nextel, began to revolutionize the SMR market. Formerly known as Fleet Call, Nextel acquired radio spectrum in six of the largest SMR markets in the United States: Los Angeles, New York, Chicago, San Francisco, Dallas, and Houston. In the early 1990s, Nextel accumulated more SMR licenses, allowing them to piece together a nationwide footprint.

In April 1990, Nextel asked the FCC for permission to build enhanced SMR (ESMR) systems in those six markets. The ESMR systems would consist of multiple low-power transmitters, which would allow the same frequencies to be reused some distance away very similar to cellular and PCS. The resulting cellular-like network would open new consumer and business communications markets to the SMR industry. Nextel also planned on implementing its new systems as all-digital (TDMA) systems to further expand calling capacity. In February 1991, the FCC approved Nextel's request, and its first digital mobile network came online in Los Angeles in August 1993.

Nextel has undergone respectable growth since the late 1990s, presenting a major competitive challenge to the traditional 850 MHz cellular and 1,900 MHz PCS carriers. At various times, Nextel has made significant leaps ahead of its wireless competitors in service development, marketing strategy, and, most recently, wireless data. Nextel has evolved from being a niche competitor that a lot of carriers dismissed to a company that a lot of carriers fear.

Nextel was among the first to pursue the buildout of a national network and bundle long-distance charges into its service package. It was the first to eliminate roaming charges and implement per-second billing. It remains the only carrier with a national footprint to offer both a *dispatch function* and a conventional digital wireless voice.

Nextel has clearly made great strides since its beginnings as a wireless dispatch operator saddled with technological problems and facing capable and competitive companies in the crowded cellular and PCS marketplace.

Services offered by Nextel are on par—or better than—many of its competitors. This includes Web services, mobile data service packages, and downloadable ringtones to name a few. There are multiple data packages, where two options are offered: pay a monthly fee for 100 messages or a pay-as-you-go package. Both options offer ability to transmit multimedia messages: text, image, and/or audio messages.

19.3   iDEN Technology

Another thing that makes Nextel different than its traditional wireless competitors, besides its uniquely assigned and used spectrum, is its use of a proprietary Motorola TDMA digital wireless technology known as Integrated Digital Enhanced Network, or iDEN™.

First introduced in 1994, Motorola's iDEN technology brought a digital wireless solution designed for a variety of vertical market mobile business applications to the marketplace. iDEN combines the functionalities of digital two-way radio, digital cellular, message service with acknowledgment, and wireless data into a single system.

 Key: iDEN is a TDMA technology that allows for up to six communication paths over a single RF channel.

iDEN uses Vector Summed Excited Linear Prediction (VSELP) vocoders. VSELP coding uses compression software that converts large segments of voice into smaller segments (packets). This increases channel efficiency by decreasing the amount of time it takes to transmit one conversation. Using a six-to-one compression ratio, 90 milliseconds of voice is compressed into a 15-millisecond voice packet. During calls, the audio quality in iDEN systems can be improved further by using a three-to-one compression ratio. In this mode, two time slots are used instead of one. This allows 45 milliseconds of voice to be compressed into a 15-millisecond time slot, improving the overall richness and tone of the call. To ensure VSELP packets do not get corrupted, iDEN uses forward error correction (FEC). FEC is added to the VSELP signal to allow the radio gear to correct for errors that occur over the air. This significantly improved audio quality even in weak signal and interference areas. The use of VSELP coding and FEC contributes to high-quality calls over six audio paths on a single RF channel.

Nextel (and iDEN's) unique service attribute is the use of trunked radio, otherwise known as the dispatch or push-to-talk service provided only by Nextel. Nextel also affectionately refers to this functionality as its walkie-talkie function. This service is formally known as Nextel's Direct Connect® feature and allows the carrier to charge a premium for its service due to this feature that only Nextel provides. In 2003, Nextel introduced the capability to employ Direct Connect nationwide with no roaming fees. When Nextel subscribers use this feature, the Motorola Nextel phones make a unique chirping sound after each transmission. This is known, strangely enough, as *chirping*. Nextel users can apply the Direct Connect feature either one on one or within groups. When one user chirps to other users who are not available, the system will let the user known with a display message and an audible tone. To use the dispatch feature described here, subscribers are organized into fleets and assigned to talkgroups. A fleet may contain hundreds of subscriber units. Fleets may then be segmented into talkgroups. Talkgroups are communication groups within a fleet. Each subscriber has an ID called *fleet member identifiers*. This ID is used to target a specific subscriber within that fleet.

Dispatch is further defined by location areas and service areas. There are three types of talkgroup service areas: wide service area, local service area, and selected service area.

iDEN/Nextel also offers all the standard features offered by other wireless carriers such as call forwarding, caller ID, SMS, circuit-switched data, and packet-switched data.

Nextel's iDEN network contains all the same network elements of other wireless systems: base stations, base station controllers (BSCs), a mobile switching center (MSC), a home location register (HLR), and visitor location register (VLR). iDEN also offers circuit-switched data service to send and receive faxes and communicate with remote computers. This is all accomplished through a process known as the Interworking Function (IWF), which is similar to a GSM network function. Three key elements are used to make dispatch calls (i.e., walkie-talkie calls). They are the Dispatch Application Processor (DAP), Metro Packet Switch (MPS), and the packet duplicator (PD). The DAP has its own HLR and VLR.

Nextel is open about its market strategy of pursuing the business segment versus the consumer segment. Because of the dispatch and wireless data capabilities, which include *conference call capability* with the talkgroup, blue collar industries such as landscaping and construction companies are especially good fits for Nextel service and its unique features. iDEN wireless handsets are also used in work environments ranging from manufacturing floors to executive conference rooms as well as mobile sales forces. Four-in-one iDEN technology allows business users to take advantage of multiple features with one pocket-sized digital handset that combines two-way digital radio, digital wireless phone, alphanumeric messaging, and data/fax capabilities that leverage Internet access. iDEN handsets and service combine speakerphone, voice command, phone book, voice mail, digital two-way radio, mobile Internet and e-mail, wireless modems, voice activation, and voice recordings so that subscribers can create a mobile virtual office if desired.

19.3.1   Assessment of iDEN Technology

iDEN technology gives the end user the benefit of having a wireless phone built on GSM technology with a walkie-talkie device, all in one handset. The walkie-talkie provides a robust push-to-talk product with extremely low latency in push-to-talk connect times, making this a phone that acts exactly like a two-way radio—but with a much greater range and footprint. Users can direct-connect to any other Nextel subscriber in the United States, Canada, Mexico, and in several Latin and South American countries with the same latency they incur when connecting to a user down the street.

The disadvantage of iDEN technology is that it is not compatible with any other wireless technology. Handsets are made by Motorola alone, so Nextel is beholden to Motorola and cannot get the handset discounts that other carriers get by leveraging their buying power among multiple handset vendors. Worst of all, there is no real 3G progression plan currently in place for iDEN, making future high-speed data plans difficult to achieve.

19.4   Sprint Buys Nextel

In December 2004, Sprint PCS announced its intention to purchase Nextel. Although this may seem like an odd pairing because Sprint's network is CDMA and Nextel's network uses proprietary iDEN technology, this purchase could be all about spectrum. Since 2003, Nextel has quietly been buying up as many multichannel, multipoint distribution system (MMDS) licenses (and spectrum) as they could. So Sprint may have an eye on that spectrum for future uses.

The new carrier will be the nation's third-largest carrier (behind Cingular and Verizon) and will be known as Sprint Nextel. The deal is a $35 billion merger, reflecting another step toward industry consolidation that has been a long time coming and is much needed. The merger will reduce costs for the new company, enabling it to compete with the two largest carriers. The combined company will have nearly 40 million subscribers, putting it in a close third behind Cingular's 47 million subs and Verizon's 42 million. Motorola will be developing a new phone specifically for the Sprint-Nextel combined company, softening any potential blow from the merger.

▰▰▰ Test Questions

True or False?

1. _____ A key disadvantage of Nextel's use of iDEN technology is that the carrier is beholden to Motorola, its single supplier for handsets.

2. _____ The most unique, attractive feature of Nextel service and iDEN technology is the satellite access feature.

Multiple Choice

1. Specialized mobile radio (SMR) service was created by the FCC in 1974 for carriers to provide what type of service?

   a. Wireless LAN systems

   b. Wireless PBX services

   c. Two-way radio dispatch

   d. CB radio

2. Which company asked the FCC for permission to develop and build enhanced SMR?

   a. AT&T Wireless

   b. Federal dispatch service

   c. Nextel

   d. Sprint PCS

   e. US Cellular

Wi-Fi (802.11 Wireless Fidelity)

20.1   Overview

So many technologies in the telecom world fall victim to hype that never materializes. But Wi-Fi is a perfect example of a technology that received (and continues to receive) a tremendous amount of hype while the hype simultaneously becomes reality. In other words, Wi-Fi went from hype to reality very, very quickly. Why? Two key reasons:

- It delivered a service people wanted—Internet access on the fly, available in many convenient public (and private places).
- It is very inexpensive to deploy Wi-Fi technology.

Wi-Fi is a form of wireless LAN. A wireless LAN (WLAN) is a data transmission system designed to provide location-independent network access between computing devices using radio waves rather than a cabled infrastructure.

In the business environment, wireless LANs are usually implemented as the final link between the existing wired network and a designated group of client computers, giving these users wireless access to the full resources and services of the corporate network across a building, between conference rooms, or in a campus setting. Publicly available Wi-Fi is essentially the same thing, but the wired network end point is the public Internet instead of a corporate network.

WLANs are on the cusp of becoming a mainstream connectivity solution for a broad range of business customers. The wireless market is expanding rapidly as businesses discover the productivity benefits of going wire free. According to the research firm of Frost and Sullivan, the wireless LAN industry will grow to $1.6 billion in 2005. To date, wireless LANs have been primarily implemented in vertical applications such as manufacturing facilities, warehouses, and retail stores. The majority of future wireless LAN growth is expected in healthcare facilities, educational institutions, and corporate enterprise office spaces. In the corporation, conference rooms, public areas, and branch offices are likely venues for WLANs.

802.11b is a WLAN standard known by the nickname Wi-Fi, for wireless fidelity. After being just a plaything for home hobbyists and technophiles since 1999, Wi-Fi has exploded in popularity.

Key: Put simply, Wi-Fi technology delivers a radio signal that beams Internet connections out 300 feet from a wireless base station to an 802.11-enabled terminal, which could be a laptop computer, a Blackberry device, a PDA, or a mobile phone.

The wireless base station in Wi-Fi environments is known as an access point, or hot spot, depending on the context. The access point is connected to a landline-based high-speed Internet access link (i.e., a DSL or DS1 circuit). In the future, this Internet link could be accessed via another complementary wireless technology known as WiMAX (see Section 20.4.2).

In 1999, 802.11b became the standard wireless Ethernet networking technology for both business and home. The Wi-Fi Organization was created to ensure interoperability between 802.11b products. Even though the number 11 in 802.11 stands for 11 Mbps, realistic throughput with 802.11(b) is 2.5 to 4 Mbps, which still makes Wi-Fi fast enough for most network applications and tolerable for file transfers. The data rates supported by the original 802.11 standard were too slow to support most general business requirements and slowed the adoption of WLANs. With 802.11b, WLANs are now able to achieve performance and throughput comparable to wired Ethernet LAN transmissions but in a wireless environment.

802.11b is a half-duplex protocol. Users can send *or* receive but not both at the same time. This makes Wi-Fi transmissions similar to basic e-mail transmissions. Wi-Fi also uses the same 2.4 GHz radio spectrum as many cordless phones and baby monitors, so the potential for cochannel interference exists in certain home environments.

Key: Because Wi-Fi transmissions operate in the 2.4 GHz range, 900 MHz cordless phones—not 2.4 GHz cordless phones—should be used when operating an 802.11b device in the same area.

20.2 Wi-Fi Operations and Hot Spots

Two basic pieces of equipment comprise 802.11 systems:

- A wireless terminal, which can be a laptop PC equipped with an 802.11b interface card (NIC), an 802.11-enabled PDA, or an 802.11-enabled Blackberry device.

- An access point (AP), sometimes called a *hot spot*, which acts as the bridge between the wireless end user and the Internet. The two terms—"access point" and "hot spot"—are interchangeable. For example, technically a home network using an 802.11b wireless router is both an access point *and* a hot spot. The same could be said for a small office. But a major airport with dozens of access points is itself called a hot spot because the entire airport is covered with Wi-Fi access points.

Key: An access point consists of a Wi-Fi radio transceiver; a wired network interface (e.g., 802.3 and/or direct PSTN link, depending on the venue) that functions as the backbone distribution system; and bridging software conforming to the 802.1d bridging standard. The access point is the Wi-Fi base station, which is essentially a picocell due to its small coverage footprint of around 300 feet. The wired network connection to the Internet can be implemented via DSL service, a DS1 circuit, multiples of DS1s, or even metro Ethernet interconnection circuits. It aggregates access for multiple wireless terminals (end users) onto the wired (LAN) backbone network and, ultimately, connects them to the Internet. In the future, this backbone interconnection may actually be to a 3G macrocell network or to a WiMAX base station. The physical footprint of an access point is extremely small, and the Wi-Fi transceivers are getting smaller all the time. Today a standard Wi-Fi access point transceiver is about the size of a laptop PC.

The wireless network adapters (wireless NICs) come in two major forms: PC cards for laptops and universal serial bus (USB) cards for desktop computers. There are also PCI adapters that let users plug a PC card into a desktop computer PCI slot. PCI stands for peripheral component interconnect. It's a standard designed by Intel that describes an interconnection system between a microprocessor and attached devices in which expansion slots are spaced closely for high-speed operation. Using PCI, a computer can support new PCI cards while continuing to support Industry Standard Association (ISA) expansion cards, an older standard. PCI slot adapters are often finicky about working correctly in anything but Windows 98/SE/ME. If using Windows XP or Windows 2000, the USB version is best in order to play it safe.

An access point is not needed for two wireless-enabled computers to communicate, but it is vital for free communication between wired and wireless networks. As more walls and distance are inserted between the user and the access point, the data rate (connection speed) will drop. Users should not expect to place more than a few walls between themselves and a hot spot in order to use Wi-Fi service effectively. See Figure 20-1 for a typical hot spot configuration.

20.2.1 Operation Modes

The 802.11 standard defines two operating modes: infrastructure mode and ad hoc mode. In infrastructure mode, the wireless network consists of *at least* one AP connected to a wired network infrastructure distribution system.

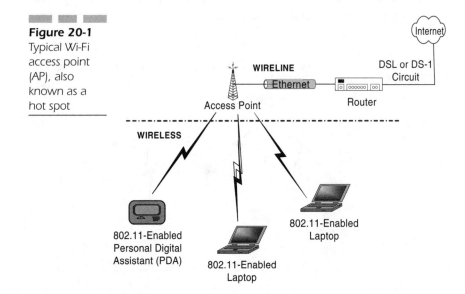

Figure 20-1
Typical Wi-Fi access point (AP), also known as a hot spot

Key: A set of wireless terminals connect to the access point via 802.11 radio signals. This configuration is called a *Basic Service Set* (BSS). An *Extended Service Set* (ESS) is a set of two or more BSSs forming a single subnetwork. Because most corporate WLANs require access to the wired LAN for services (file servers, printers, Internet access), they will operate in infrastructure mode. If service areas overlap, handoffs can occur similarly to how this is accomplished in a macrocellular network. So in-progress Wi-Fi sessions are moved from one AP coverage area to another, seamlessly.

See Figure 20-2 for an illustration of Wi-Fi infrastructure mode, Basic Service Sets, and Extended Service Sets.

Ad hoc mode, also known as an *Independent Basic Service Set* (IBSS), is a set of 802.11 wireless user terminals that communicate *directly* with one another without using an access point or any connection to a wired network. See Figure 20-3 for an illustration of ad hoc Wi-Fi mode.

This mode is useful for quickly and easily setting up a data network anywhere that a wireless infrastructure does not exist or is not required for services. Ad hoc mode would be useful in a conference room, hotel room, convention center, airport, or where access to the wired network is barred (such as for consultants at a client site).

Ad hoc mode is effectively wireless peer-to-peer networking. This would be similar to music file sharing on the Internet (i.e., Napster),

Figure 20-2
Wi-Fi infrastructure mode: basic service sets and extended service sets

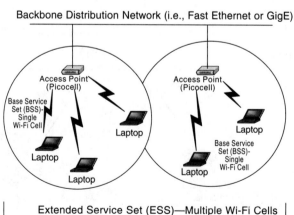

Figure 20-3
Wi-Fi ad-hoc
mode

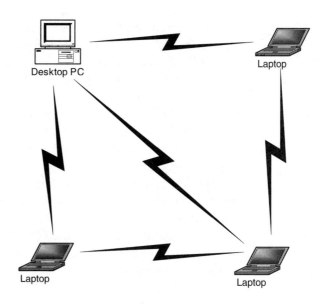

Wi-Fi Independent Basic Service Set (IBSS)
(Ad Hoc Mode)

except in much closer quarters versus through the expanse of the Internet. In the ad hoc network, computers are brought together to form a network on the fly. There is no structure to the network; there are no fixed points, and usually every node (computer) is able to communicate with every other node. An example of an ad hoc 802.11b network would be a meeting where employees bring laptop computers together to communicate and share design or financial information. To maintain order in these types of networks, one computer is usually elected as the base station master of the network, with the others being slaves. This is accomplished by using the *Spokesman Election Algorithm* (SEA).

20.2.2 Association, Cellular Architectures, and Roaming

The 802.11 *media access control* (MAC) layer—OSI layer 2—is responsible for how a user terminal associates with an access point.

Key: When an 802.11 end user enters the range of one or more APs (e.g., an airport), and the end user powers up his or her Wi-Fi-enabled laptop (or other terminal), a screen splash will occur showing the user if Wi-Fi access is available. If more than one Wi-Fi operator has access points in the area, all of the operators available at that venue will be shown on the screen splash, and the end user will have his or her choice of operators. The beauty of this setup is that the user may already have an existing account with one of the operators, where he or she has already paid his or her monthly fee. Then he or she can simply select that one operator and move forward. If not, once he or she selects an operator, he or she will be queried on what form of usage he or she would like to pay for, and he or she will have to input a credit card number to pay, then proceed.

All the preceding activity assumes that there is sufficient signal strength by one or all Wi-Fi operators to allow the end user to make a service selection. Once an operator is selected, the Wi-Fi users will be joined to a BSS. Once accepted by the access point, the client device (laptop, PDA, etc.) tunes to the radio channel of its designated access point. Periodically the user's mobile terminal will survey all 802.11 channels in order to assess whether a different access point would provide better performance characteristics. If it determines this is the case, it reassociates with the new access point, tuning to the radio channel to which that access point is set. See Figure 20-4.

Key: Reassociation usually occurs because the wireless station has physically moved away from the original access point, causing the signal to weaken. In other cases, reassociation occurs due to a change in radio characteristics in the building, or due simply to a high amount of network traffic on the original access point. In the latter case, this function is known as *load balancing,* because its primary function is to distribute the total WLAN load most efficiently across the available wireless infrastructure.

This process of dynamically associating and reassociating with access points requires Wi-Fi network managers to set up WLANs with very

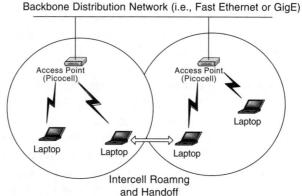

Figure 20-4
Wi-Fi access
point roaming

Wi-Fi Roaming Benefits:

- Easily Expanded Coverage
- Load Balancing
- Scalability and incremental Growth
- Transparent to End User

broad coverage by creating a series of overlapping 802.11b cells through-out a building or across a campus. This is similar to macrocellular design in carrier-grade wireless networks (i.e., Cingular or Verizon). To be successful, the network manager ideally will employ channel reuse, taking care to set up each access point on an 802.11 direct-sequence spread-spectrum (DSSS) channel that does not overlap with a channel used by a neighboring access point.

20.3 Standards and Tech Specs

When the IEEE ratified the 802.11a and 802.11b wireless networking standards in 1999, the goal was to create a standards-based technology that could span multiple physical encoding types, frequencies, and applications. The idea was to model the way the 802.3 Ethernet standard has been successfully applied to 10, 100, 1,000, and now 10,000 Mbps technology over fiber and various kinds of copper. Just a few years later, we had at our disposal a wide selection of 802.11b products from a multitude of vendors.

The 802.11b standard was designed to operate in the 2.4 GHz industrial, scientific and medical (ISM) band using direct-sequence spread-spectrum technology (see Section 6.4.1 for an explanation of direct-sequence spread-spectrum technology).

The 802.11a standard, on the other hand, was designed to operate in the more recently allocated 5 GHz UNII (or Unlicensed National Information Infrastructure) band. Specifically, the FCC has allocated 300 MHz of spectrum for unlicensed operation in the 5 GHz block, 200 MHz of which is at 5.15 MHz to 5.35 MHz, with the other 100 MHz at 5.725 MHz to 5.825 MHz. And unlike 802.11b, the 802.11a standard departs from the traditional spread-spectrum technology, instead using a frequency division multiplexing scheme that is intended to be friendlier to office environments. The 802.11a standard, which supports data rates of up to 54 Mbps (whoa!), is the Fast Ethernet equivalent to 802.11b, which supports theoretical data rates of up to 11 Mbps. Like Ethernet and Fast Ethernet, 802.11b and 802.11a use an identical MAC.

Along with standards body's activities, wireless industry leaders have united to form the Wireless Ethernet Compatibility Alliance (WECA). WECA's mission is to certify cross-vendor interoperability and compatibility of IEEE 802.11 wireless networking products and to promote the Wi-Fi standard for the enterprise, the small business, and the home. WECA members include WLAN semiconductor manufacturers, WLAN providers, computer systems vendors, and software makers. Membership includes companies such as 3Com, Aironet, Apple, Breezecom, Cabletron, Hewlett-Packard, Dell, Fujitsu, IBM, Intersil, Lucent Technologies, Nokia, Samsung, Wayport, and Zoom.

The three physical layers originally defined in the 802.11 standard include two spread-spectrum radio techniques and a dispersed infrared specification. The radio-based standards operate within the 2.4 GHz ISM frequency band, which is recognized internationally by regulatory agencies such as the FCC (USA), European Telecom Standards Institute (ETSI), and the MKK (Japan) for unlicensed radio operations. The 2.4 GHz ISM frequency band is unlicensed because the services and devices that use unlicensed spectrum are mandated to have very low RF power levels, which equates to low-range transmissions (i.e., no more than 300 feet max). At low range, the likelihood of interference is minimized or eliminated. 802.11-based products do not require user licensing or special training.

Spread-spectrum techniques satisfy regulatory requirements, increase reliability, and boost throughput. They also allow many unrelated products to share the spectrum without explicit cooperation and with minimal interference.

Conversely, the direct-sequence signaling technique divides the 2.4 GHz band into fourteen 22 MHz channels. Data is sent across the entire swatch of one of these 22 MHz channels without hopping to other channels.

20.3.1 Wi-Fi Security

For years, 802.11 historically provided for MAC layer (OSI layer 2) access control and encryption mechanisms, known as Wired Equivalent Privacy (WEP), with the objective of providing wireless LANs with security equivalent to its wired counterparts. For the access control, the ESSID (also known as a WLAN Service Area ID) was programmed into each access point and was required knowledge in order for a wireless client to associate with an access point. In addition, there is provision for a table of MAC addresses called an *Access Control List* to be included in the access point, restricting access to clients whose MAC addresses are on the list.

 Key: Any Wi-Fi network adapter (wireless NIC) coming within range of another 802.11b network adapter or access point could instantly connect and join the network unless WEP is enabled. WEP is secure enough for most homes and businesses, but it can still be hacked. There are several flaws in WEP making it unusable for high-security applications. Hackers can fairly easily decode WEP-encrypted information after monitoring an active network for less than one day.

WEP also slows down a Wi-Fi network. A *20 to 50 percent reduction in speed will occur*, depending on the products being used. The speed reduction problem is often the result of an access point (hot spot) that does not have enough processing power. With WEP enabled, expect total throughput of approximately 2.5 to 3.5 Mbps over 802.11b.

Encryption comes in 64-bit and 128-bit key varieties. All nodes must be at the same encryption level with the same key in order to operate. An encryption level of 40 bits and 64 bits is the same thing; it is just a matter of how the manufacturer decided to label the product. Often, 128-bit cards can be placed in 40/64-bit mode. In addition, when encryption is in use, the access point will issue an encrypted challenge packet to any client attempting to associate with it. The client must use its key to encrypt the correct response in order to authenticate itself and gain network access.

Beyond layer 2, 802.11 WLANs support the same security standards supported by other 802 LANs for access control (such as network operating system logins) and encryption (such as IPSec or application-level encryption). These higher-layer security technologies can be used to create end-to-end secure networks encompassing both wired LAN and WLAN components.

Many corporations postponed rollouts in strategic areas until they were convinced that hackers, corporate spies, and competitors would not intercept wireless data transmissions. In 2001, a consulting firm in Chicago discovered that numerous Wi-Fi transmissions could be intercepted in Chicago's Loop business district by using an empty Pringles can as a directional receiving antenna connected into a Wi-Fi enabled laptop computer.

The good news is that a new WLAN/Wi-Fi-specific security standard has been developed and approved: 802.11i. The launch and availability of this new security standard was a huge step forward for Wi-Fi technology because the one thing that caused the most anxiety and consternation with this technology was its weak security.

20.3.2   Auxiliary Wi-Fi Standards

Wi-Fi has spawned a multitude of related standards efforts to support the growth of this hugely popular technology. Most wireless standards come from the IEEE, whose 802.11 category covers more than a dozen standards for specific wireless areas. But the Internet Engineering Task Force (IETF), Wi-Fi Alliance, and WiMAX industry group also develop related standards.

20.3.2.1   Transport

- **802.11a, 802.11b, 802.11g**: All are IEEE standards for the transmission of wireless signals providing maximum throughput of 54 Mbps in the 5 GHz band, 11 Mbps in the 2.4 GHz band, and 54 Mbps in the 2.4 GHz band, respectively. All are in use.

- **802.11n**: This new IEEE standards effort seeks to develop 100 Mbps maximum throughput, most likely in the 5 GHz band and through the simultaneous use of multiple channels. The standard is slated for completion in 2007.

20.3.2.2 Security

■ **802.1 x**: This respected IEEE authentication standard is used by the 802.11 standards.

■ **802.11 i**: Recently approved by the IEEE, this encryption standard replaces the vulnerable static-key WEP standard. It allows for dynamic shared encryption keys.

■ **WPA2**: The Wi-Fi Alliance's WPA2 standard assures interoperability among 802.11i-based devices (WPA1 was an interim standard issued before the final 802.11i standard, to keep vendors from straying from the draft 802.11i standard).

20.3.2.3 Management, Roaming, and QoS

■ **802.11e**: An IEEE effort to be completed in summer 2005, it defines prioritization levels and provides basic levels of QoS for data, voice, and video traffic.

■ **WME, WSM**: The interim Wireless Media Extensions and Wi-Fi Scheduled Media standards from the Wi-Fi Alliance are based on the draft 802.11e. They are meant to ensure interoperability and consistent deployment across different vendors' prestandard products. WME is available now. WSM is expected in spring 2005.

■ **802.11 f**: This IEEE standard defines communication between APs for layer 2 roaming, but it does not support roaming across different WLAN segments.

■ **802.11 r**: This is an IEEE effort to standardize handoff for fast roaming among APs, including authentication keys, allowing fast roaming that will support voice over wireless in addition to data over wireless. It would also address roaming across segments. The standard is slated to be complete by 2006.

■ **802.11 s**: Another IEEE effort, this standard is designed to wirelessly connect APs for backhaul communication and mesh networking. The standard is expected to be complete by 2006.

■ **CAPWAP**: Standardizing the taxonomy of mechanisms for the control and programming of wireless APs is the goal of this recently completed IETF effort. The acronym stands for Control and Provisioning of Wireless Access Points. CAPWAP is the IETF's version of the proposed IEEE 802.11v.

■ **LWAPP**: An in-progress IETF effort, the Lightweight Access Point Protocol governs how lightweight APs communicate with WLAN system devices and with the controllers that manage the lightweight APs.

20.4   Wi-Fi Integration with Other Wireless Technologies and Macrocellular Networks

Some industry analysts speculate that Wi-Fi could either supplant or complement the deployment to 3G services in the wireless world. Motorola, Nokia, and Ericsson are developing Wi-Fi phones that will allow users to move from Wi-Fi to cellular networks without even noticing. There are major implications to roaming in terms of operations, billing, compatibility, and pricing.

There is nothing stopping Wi-Fi technology from being technically able to integrate with macrocellular networks, especially 3G networks. This type of integration would allow for a truly seamless, wireless access and transport model. This is what is required to make this type of integration to happen:

1. Development and availability of multimode wireless capability for (laptop) NICs, mobile handsets, and/or PDAs. Chipmakers such as Intel would need to modify their Wi-Fi-centric Centrino chipsets to be able to support macrowireless frequencies along with Wi-Fi frequencies. The same applies to PDA manufacturers and mobile handset manufacturers. Everyone's devices will need to interoperate on a number of frequency bands: 850 MHz, 1,900 MHz, 2.4 GHz, and the 5 GHz (UNII band). So ultimately, quad-mode chipsets will be required.

2. Wi-Fi operators, venue owners (i.e., hotels, airports), and wireless carriers would need to work together on RF engineering models to allow for handoffs between the Wi-Fi network and the macrowireless network. A key challenge in this process would be ensuring that a Wi-Fi user's laptop (or other mobile terminal) *first seeks out and uses a Wi-Fi signal* when the end user is in a Wi-Fi-supported venue. This is important because Wi-Fi operators want to ensure access to their system in order to drive revenues to pay for the access points and Internet connections.

Key: This entire effort would be greatly simplified if the Wi-Fi operator and the macrowireless carrier were one and the same. For example, a scenario where SBC Freedom-Link service is offered in an airport and Cingular Wireless is one of the major carriers in that particular wireless market. Both services are offered by SBC in this case.

3. Interconnection agreements would need to be developed and agreed upon between the Wi-Fi network operators and the macrowireless carriers. Disposition of monies obtained from Wi-Fi service that is extended to macrowireless networks is an issue that would need to be addressed.

If this type of interconnection occurs using 3G-based macrowireless networks, the result would be a truly turbo-boosted wireless experience. Although 802.11a (54 mbps) would operate at higher speeds than 3G networks, would the user notice if his or her Internet access were occurring at 400 Kbps speed with a CDMA2000 3X network connection?

WiMAX 802.16 (WiMAX) is a wireless MAN technology that is under development, with launches expected in 2005. WiMAX can deliver 100 Mbps transport over a 30-mile radius.

Like the scenario described in the preceding passage discussing integration with macrowireless networks, the very same integration model could be used with WiMAX when it is available. The key stumbling block is that it will be years (2007?) before WiMAX availability has a large enough footprint to make this concept of Wi-Fi/WiMAX integration a reality. Chipsets are now being planned that integrate Wi-Fi and WiMAX technologies and frequencies.

The interesting thing is that by that time, it is very likely there will be intense competition between WiMAX operators, macrowireless carriers offering 3G services and speeds, and Wi-Fi operators as well. Will they all be able to coexist in some form? Will WiMAX's hype become a reality? Will macrowireless carriers simply crush their WiMAX competition by buying them out? Will the macrowireless carriers out-compete the WiMAX players until WiMAX is a dead technology? These are all questions that can only be answered by watching and waiting, between 2005 and 2008.

20.5 Hype into Reality

These days, almost every single issue of *Network World* magazine, *Telephony* magazine, America's *Wireless* magazine, and *Wireless Review* has a Wi-Fi-related article on the cover. Even *Business Week* ran an April 2004 cover story on Wi-Fi. But it is not all hype.

Since 2001, thousands of Wi-Fi networks have been launched, causing the number of commercial hot spots to rise to over 21,000 across the United States. Starbucks has jumped into Wi-Fi by partnering with T-Mobile Wireless to offer consumers Wi-Fi surfing at more than 2,100 coffee shops.

McDonalds is deploying Wi-Fi at its restaurants, but many analysts question whether the type of people who frequent McDonalds will really want to surf the net while eating a burger and fries. The idea is not necessarily to make money on Wi-Fi service, which goes for three dollars per hour. The plan is to attract new customers and boost sales. McDonalds is also offering a free hour of Wi-Fi access with each Extra Value Meal.

SBC has signed major contracts to supply its FreedomLink Wi-Fi to UPS stores nationwide as well as at all Caribou Coffee shops.

Technology giants are joining the fray as well. Intel is spending $300 million to market its Centrino computer chips, which come equipped for Wi-Fi. According to a source at Intel, the goal is to convert Wi-Fi to an industrial-strength solution that corporations can depend on. In March 2003, Cisco Systems bought Linksys, a Wi-Fi equipment maker, for $500 million. For the first time, this put Cisco into head-to-head competition with Microsoft Corp., which began development of Wi-Fi network gear in 2002. And as more companies join the frenzy, the prices for Wi-Fi equipment will continue to plummet.

A major Wi-Fi trend is for cities to convert their geographical boundaries into one huge hot spot. One trailblazer in this regard is the city of Philadelphia, which is building a citywide hot spot for all its citizens. Individual Wi-Fi cells will be mounted on Philly streetlights, creating a self-organizing and self-healing wireless mesh. Approximately 12 APs will be needed per square mile, and a single worker can install ten units per day. Wireless access is scalable, and connectivity is available as soon as units are installed.

A *Newsweek* cover story in June 2004 listed "10 Hot Wireless Cities." Hermiston, Oregon, was listed as home to the largest Wi-Fi network in the United States, with 35 towers and 75 access points, providing coverage to the entire county. Other cities listed were San Diego, due to the

150 wireless firms based there and the existence of Verizon's 3G network. Las Vegas was listed, as was the Bay Area of California. Lower Manhattan has one of the most heavily used Wi-Fi networks in the world. Washington, D.C., already has 344 commercial hot spots. Austin, Texas, an up-and-coming high-technology city, has 11 access points for every 100,000 residents, including 50 free ones. London boasts 1,110 hot spots, making it the most "Wi-Fi friendly city" in the world.

Other facts that prove Wi-Fi is a rising star include the following:

- There are now 53,779 public Wi-Fi hotspots in 93 countries around the world.

- Over 50,000 of these hot spots charge users for access.

- Over 16,000 of these hotspots are located in hotels.

- Tiny, portable wireless routers are now available, which can be used in hotels to create a user's own private hot spot. These devices plug into the RJ-45 high-speed Internet jack in hotel suites and allow hotel guests to lie on the bed or in a chair, wirelessly conducting business, accessing their corporate network and the Internet.

 Key: The total number of hotspots worldwide is expected to grow 100 percent in 2005! This eclipses the average growth rate of wireless service at its peak in the mid 1990s, which was 60 percent!

- In early 2004, more than 100 Boeing jets were scheduled to be equipped with Wi-Fi. For $25 or so per flight, laptop users can log on to the Net while soaring at 35,000 feet—conducting e-commerce, managing company inventories, or even making voice calls over the Web (Wi-Fi over VoIP). Boeing is so pumped on the new technology that by 2015 it hopes to outfit nearly 4,000 planes with Wi-Fi service via its new Connexion business unit (find out more at www.boeing.com).

- The consumer electronics industry is counting on Wi-Fi as well, to link a host of appliances in the home. Techno hobbyists using Wi-Fi are sending MP3 songs and videos from their computers to their TVs and stereos. This will be even easier with the new generation of Wi-Fi (802.11a) rolls out, raising connection speeds to 54 megabits, equivalent to an hour of MP3 music—per second!

- Dell, Toshiba, and TiVo are building Wi-Fi into computers and digital recording devices. Over 90 percent of new laptops will be Wi-Fi ready in 2005, up from 35 percent by year-end 2003.

- Wi-Fi is getting a boost from the increasing popularity of broadband, which is growing at an annual rate of 30 percent. Thousands of people every day order Wi-Fi routers to support their home network installations. Why? Because Wi-Fi is an inexpensive way to connect several household computers to a single high-speed Internet connection. This is what today's home networking is all about.

- IBM is developing Wi-Fi powered systems to monitor the minute-by-minute operations of distant machines, from potato fryers at restaurants to air conditioners in data centers. Home Wi-Fi gear prices have plummeted. More than 50 companies are in the Wi-Fi chip market alone.

- Intel and MeshNetworks are developing Wi-Fi antennas that can reach for *miles* instead of today's 300 feet. Coming soon: Wi-Fi-ready cell phones and PDAs, and hot spots on trains and buses.

- Prices are dropping fast. An antenna for a laptop now costs just $46, down from $189 four years ago.

- Thousands of enterprising do-it-yourselfers have deployed access points to create their own hot spots. They have even joined together to form networks so that the public can zap e-mail messages and surf the Net for free, no matter where they are. From the streets of Sydney to mountain areas outside Seattle, over 5,000 free hot spots have emerged since 2000. More than 18 million people worldwide have logged on, and the numbers are growing daily.

- Wi-Fi is becoming so pervasive these days that even some nationally known campgrounds (Yogi Bear) are advertising Wi-Fi access.

20.6   Wi-Fi Benefits and the Competitive Advantage of Wireless

The work environment of the twenty-first century is characterized by an increasingly mobile workforce and flatter organizations. Employees are equipped with notebook computers and spend more of their time working in teams that traverse functional, organization, and geographic boundaries. This fact is underscored when mergers occur, and companies suddenly find their business and employee bases stretched across nations or across the world. The productivity of today's workers often occurs in meetings and away from their desks, so users need access to the corporate network far beyond their personal desktops. WLANs fit well in this work environment, giving mobile workers much-needed freedom in the way that they access the network.

WLANs free users from dependence on hard-wired access to a network backbone, giving them anytime, anywhere network access. This freedom to roam also offers numerous user benefits for a variety of work environments, such as the following:

- Immediate bedside access to patient information for doctors and hospital staff.
- Easy, real-time network access for on-site consultants or auditors in corporate offices.
- Improved database access for roving supervisors such as production line managers, warehouse auditors, or construction engineers.
- Simplified network configuration with minimal IT staff involvement for temporary setups such as trade shows or conference rooms.
- Faster access to customer information for service vendors and retailers, resulting in better service and improved customer satisfaction.
- Location-independent access for enterprise network tech support departments, for easier on-site troubleshooting and support.
- Real-time access to study group meetings and research links for students.
- Factory floor workers can access part and process specifications without impractical or impossible wired network connections. Wireless connections with real-time sensing would allow a remote engineer to diagnose and maintain the health of manufacturing equipment, even on an environmentally hostile factory floor (imagine an environment with a multitude of overhead piping, racks, or robotic machinery). Wireless network access in such an environment would be a panacea.
- Warehouse inventories could be carried out and verified quickly and effectively with wireless scanners connected to a main inventory database. Even wireless smart price tags, complete with liquid

crystal display readouts could allow merchants to virtually eliminate discrepancies between stock-point pricing and scanned prices at the checkout lane.

- Eventually, Wi-Fi could even feed information into smart networks in the home or factory to automatically monitor climate controls or industrial supply chains.

The wireless market is expanding rapidly as businesses discover the productivity benefits of going wire-free. According to the research frim of Frost and Sullivan, the wireless LAN industry is expected to grow to $1.6 billion in 2005. To date, WLANs have been primarily implemented in vertical applications such as manufacturing facilities, warehouses, and retail stores. The majority of future wireless LAN growth is expected to be in healthcare facilities, educational institutions, and corporate office spaces. Corporation conference rooms, public areas, and branch offices are likely venues for WLANs.

The lure of Wi-Fi's benefits is proving hard to resist to businesses willing to venture onto the wireless edge. From General Motors to United Parcel Service, companies are using Wi-Fi for mission-critical jobs in factories, trucks, stores, and even hospitals.

WLANs provide a benefit for IT managers as well, allowing them to design, deploy, and enhance networks without regard to the availability of wiring. This saves both effort and dollars. A white paper released by 3Com in 2000 claimed that the benefits of WLAN deployments can add up to as much as $16,000 per user—measured in worker productivity, organizational efficiency, revenue gain, and cost savings—over wired alternatives.

At a high level, WLAN operational advantages include the following:

- Mobility that improves productivity with real-time access to information, for faster and more efficient decision making, regardless of worker location
- Cost-effective network setup for hard-to-wire locations such as older buildings, solid-wall structures, and factories
- Reduced cost of ownership—particularly in dynamic environments requiring frequent structural (i.e., cube/office) modifications—thanks to minimal wiring and installations costs per device and user

The list of possibilities is almost endless and limited only by the imagination of the application designer.

The effort to groom Wi-Fi for business is literally lifting the Internet into the air. A constellation of dependable Wi-Fi hot spots will dramatically extend the range and expanse of the Web, changing its very nature. Wi-Fi connections to the Internet cost only one-fourth as much as the wired infrastructure companies use today.

It is unknown whether the estimated number of Wi-Fi subscriptions—from any provider—will justify the network investment. Bottom line: Can anyone make money in the home networking or wireless world? To make Wi-Fi viable as a business, the job now is to build it into a solid pillar of the networked world.

This chapter could easily have been a book all by itself. In an effort to facilitate the reader's ability to learn more about Wi-Fi technology, please try the following web sites:

www.Wi-Fiplanet.com has a lot of information on Wi-Fi technology itself, the latest industry news on Wi-Fi, and a hot spot search engine.

www.Wi-Finder.com is a hot spot search engine.

See also the listing of telecom and wireless-related trade journals in the appendix.

Test Questions

True or False?

1. _____ There are three operations modes with Wi-Fi: Ad Hoc Mode, Infrastructure Mode, and EZ mode.

2. _____ 802.11a Wi-Fi offers speeds of 5.4 Gbps.

3. _____ The 11 in 802.11b indicates its theoretical data rate (speed).

Multiple Choice

1. The typical coverage for a Wi-Fi hot spot is (no more than):
   a. 1 mile
   b. 10 miles
   c. 300 feet
   d. 30 feet
   e. None of the above

2. Which company formed a brand new business unit, called Connexion, to support its Wi-Fi venture?
   a. McDonalds
   b. Starbucks
   c. SBC
   d. Boeing
   e. T-Mobile

21

802.16
WiMAX

WiMAX stands for Worldwide Interoperability for Microwave Access Forum. Many operators and service providers may be unfamiliar with the details of the IEEE 802.16 standard, but this wireless technology has the potential to revolutionize the broadband wireless access industry.

WiMAX was formed in April 2001, in preparation for the original 802.16 specification published in December of that year. According to the WiMAX Forum, the group's aim is to promote and certify compatibility and interoperability of devices based on the 802.16 specification, and to develop such devices for the marketplace. Members of the organization include Airspan, Alvarion, Analog Devices, Aperto Networks, Ensemble Communications, Fujitsu, Intel, Nokia, OFDM Forum, Proxim, and WiLAN.

WiMAX 802.16 technology is expected to enable multimedia applications with wireless connection and, with a range of up to 30 miles, enable networks to have a wireless last-mile solution. The 802.16 standard, the Air Interface for Fixed Broadband Wireless Access Systems, is also known as the IEEE WirelessMAN air interface. This technology is designed from the ground up to provide wireless last-mile broadband access in the Metropolitan Area Network (MAN), delivering performance comparable to other broadband services such as traditional cable, DSL, or DS1 offerings.

802.16 wireless technology will provide a flexible, cost-effective, standards-based means of filling existing gaps in broadband coverage, and creating new forms of broadband services not envisioned in the wired telecom world. WiMAX is basically the next generation of the wireless local loop (WLL) concept. But the development of WiMAX is taking a much more solid, constructive, growth-oriented approach than previous incarnations of WLL technology took.

21.1 WiMAX: The Wireless MAN Solution

In January 2003, the IEEE approved the 802.16a standard, which covers frequency bands between 2 GHz and 11 GHz. The 802.16 standard is an extension of the IEEE 802.16 standard for 10 to 66 GHz that was published in April 2002. These sub-11-GHz frequency ranges facilitate non-line-of-sight (NLOS) RF connectivity, making the WiMAX standard an ideal technology for last-mile applications. Here is why: Obstacles like trees and buildings are usually present, and base stations may need to be unobtrusively mounted on the roofs of homes or buildings, rather than towers on mountains.

    Key: The most common WiMAX configuration consists of a base station mounted on a building or tower that communicates on a point to multipoint basis with subscriber stations located in businesses and homes. 802.16a has up to 30 miles of range with a typical cell radius of 4 to 6 miles. Non-line-of-sight performance and throughputs are optimal within this typical cell radius.

802.16a also provides an ideal wireless backhaul technology to connect 802.11 wireless LANs and commercial hotspots to the Internet. 802.16a wireless technology gives businesses the ability to flexibly deploy new 802.11 hotspots in locations where traditional wired connections may be unavailable or too time consuming to provision. This capability offers service providers around the globe a flexible new way to stimulate growth of the residential broadband access market segment, especially in areas where cable modem or DSL services simply are not available.

With shared data rates of up to 75 Mbps, a single sector of an 802.16a base station—where sector is defined as a single transmit/receive radio pair at the base station—provides sufficient bandwidth to simultaneously support more than 60 businesses with T1-level connectivity and hundreds of homes with DSL-rate connectivity, using 20 MHz of channel bandwidth. To support a viable business model, service providers need to sustain a mix of high-revenue business customers and a high volume of residential subscribers. 802.16a helps meet this requirement by supporting differentiated service levels, which can include guaranteed T1-level services for business or best effort DSL-speed service for home consumers. The 802.16 specification also includes robust security features and the quality of service (QoS) needed to support services that require low latency, such as voice and video. 802.16 voice service can be either traditional time-division multiplexed (TDM) voice or Voice over IP (VoIP).

Key: The main advantages offered by 802.16 are as follows:

1. The ability to quickly provision service, even in areas that are hard for wired infrastructure to reach
2. The avoidance of steep installation costs
3. The ability to overcome the physical limitations of traditional wired infrastructure
4. The ability to complement and interface with other wireless broadband technologies such as 802.11 Wi-Fi and 3G

21.2 WiMAX Network Architecture

By using a robust modulation scheme, WiMAX delivers high *throughput* at long ranges with a high level of spectral efficiency that is also tolerant of multipath fading. *Dynamic adaptive modulation* allows a WiMAX base station to exchange throughput for range. For example, if the base station cannot establish a robust link to a distant subscriber using the highest order modulation scheme, 64 QAM, the modulation order is reduced to 16 QAM or quadrature phase shift keying (QPSK), which reduces throughput but simultaneously increases effective range.

To accommodate easy cell planning in both licensed and unlicensed spectrum worldwide, WiMAX also supports flexible channel bandwidths —it fosters *scalability*. For example, if an operator is assigned 20 MHz of spectrum, that operator could divide it into two sectors of 10 MHz each, or four sectors of 5 MHz each. By concentrating RF power on increasingly narrow sectors, the WiMAX operator can increase the number of users while maintaining good range and throughput. To scale coverage even further, the operator can reuse the same spectrum in two or more sectors by creating proper isolation (physical separation) between base station antennas to avoid interference. In other words, WiMAX allows for carefully engineered frequency reuse.

In addition to supporting a robust and dynamic modulation scheme, the IEEE 802.16 standard also supports technologies that increase coverage, including mesh topology and smart antenna techniques. As radio technology improves and costs drop, the ability to increase coverage and throughput by using multiple antennas to create transmit and/or receive diversity will greatly enhance coverage in extreme environments.

Privacy and encryption features are also included in the 802.16 standard to support secure transmissions and provide authentication and data encryption.

21.3   WiMAX Standards and Development

Standards are important for the telecom and wireless industry—really any industry—because they facilitate economies of scale that can bring down the cost of equipment, ensure interoperability, and reduce investment risk for WiMAX operators.

Key: Without industrywide standards, equipment manufacturers must provide all the hardware and software building blocks and platforms by themselves. This includes the chip sets, subscriber station gear, WiMAX base stations, and element management software used to provision services and remotely manage the subscriber station. With the 802.16 standard in place, suppliers can amortize their research and development costs over much higher product volume. The WiMAX Forum has facilitated this standards activity.

Standards also specify minimum performance criteria for equipment, enabling a common broadband wireless access baseline (platform) that equipment manufacturers can use as the foundation for ongoing innovations and faster time to market. With its broad industry support, the 802.16 standard lets device manufacturers and solutions vendors do what they do best: achieve overall price/performance improvements and open mass-market opportunities that cannot be equaled by proprietary approaches.

The WiMAX Forum is a nonprofit corporation formed by equipment and component suppliers, including Intel Corporation, to promote the adoption of WiMAX-compliant equipment by operators of broadband wireless access systems. The organization is working to facilitate the deployment of broadband wireless networks based on the IEEE 802.16 standard by helping ensure the compatibility and interoperability of broadband wireless access equipment. This is similar to how the Wi-Fi Alliance promoted the IEEE 802.11 Wi-Fi standard for wireless LANs.

Several industry players are leading WiMAX's implementation. The first of these is Intel, which is making heavy investments into WiMAX as part of a strategy to take the lead in WiMAX the same way it did in Wi-Fi with Centrino. Its research shows that many people use their PDAs, broadband-equipped mobile phones, and laptops to access data networks while mobile, a phenomenon that is causing a significant number of communities to build metro-based broadband access areas to serve them.

Some industry observers believe that WiMAX competes with Wi-Fi. Most informed players argue that WiMAX is really a complementary technology to Wi-Fi, particularly in the metro arena. WiMAX is seen as a broadband wireless alternative to cable or DSL. Industry observers also believe that certified WiMAX products will begin to appear in the enterprise market in mid-2005, although consumer WiMAX products in rural or greenfield areas will provide solutions that are more cost effective than fixed solutions.

In April 2002, the IEEE published its 802.16 standard for broadband wireless access (BWA), also known as WiMAX. 802.16 specifies the details of the air interfaces for wireless metropolitan area networks (MANs). And although some similarities exist between Wi-Fi and WiMAX, in other respects they could not be more different. Spectrum was allocated globally for 802.16 implementation through a 2-year, open-consensus procedure that involved hundreds of engineers from major carriers and vendors around the world. Consequently, 802.16, although still a nascent technology, enjoys global acceptance and what will be a relatively trivial implementation phase once it becomes more widely deployed. Furthermore, the capabilities of the standard are impressive. Whereas Wi-Fi offers megabits of nominal bandwidth over service distances of 300 feet, *WiMAX offers 100 Mbps access over a service radius of several miles*. And because it is orthogonal, it does not require line of sight for connectivity.

Over the last few years 802.16 has undergone a series of modifications, resulting in the existence of various flavors of the original standard including 802.16a and 802.16e.

21.3.1   802.16a

IEEE 802.16a was ratified in January 2003 as an extension to the original 802.16 standard. The 802.16a standard is basically an amendment to the more general 802.16 core standard developed in December 2001 by IEEE Task Group 1. The core 802.16 specification was an air interface standard for broadband wireless access systems using point-to-multipoint infrastructure designs, operating at radio frequencies between 10 GHz and 66 GHz. It targeted an average bandwidth performance of 70 Mbps and peak data rates up to 268 Mbps. It addresses the requirements of both licensed and unlicensed implementations, and supports point-to-multipoint networks as well as mesh topologies within the unlicensed region.

But the 802.16a standard was not complete in many people's minds. It applied only to line-of-sight deployment in licensed spectrum, did not address NLOS transmission, neglected to offer any conformance guidelines and ignored ongoing development of the similar European Hiper-MAN standard. The 802.16a collection of amendments took into account the emergence of licensed and license-exempt broadband wireless networks operating between 2 GHz and 11 GHz, with support for NLOS architectures that could not be supported in higher frequency ranges. Support for NLOS performance was one of primary physical layer (PHY) differences in 802.16a. Therefore, 802.16a was developed with the requirements of lower frequencies in mind. The amended standard also allowed for WiMAX deployment in varying channel capacities to address the different amounts of spectrum that carriers may own from market to market and in different parts of the world.

21.3.2   802.16e

802.16e is under development as an extension of 802.16a. Its potential impact is enormous in that it adds mobility to 802.16a systems—the ability for end users to move while using WiMAX service.

Key: The addition of mobility could support the end-to-end needs of a subscriber in both fixed and mobile environments. However, the addition of mobility has a potentially major impact in an area that not many may be considering. Observers may say that WiMAX represents a real threat to Wi-Fi. But given that WiMAX spectrum has already been allocated around the world, *mobile* WiMAX represents more than a potential threat to Wi-Fi, it represents a huge threat to 3G. Billions of dollars have been spent since 2001 on 3G spectrum, with minimal deployments to date. Now 802.16e-based WiMAX holds the promise of wireless broadband connectivity with global reach. 3G begins to appear as just another option.

802.16e will allow wireless ISPs (WISPs) to enter and take over a market with minimal investment in infrastructure, then offer a complete package of services to subscribers. Wireless broadband, in the form of WiMAX, could also compete favorably with such options as cable modem and DSL (in fact, some analysts refer to WiMAX as wireless DSL). See Table 21-1 for a comparison of the three existing WiMAX standards, illustrating the technology's evolution.

WiMAX is absolutely still in its nascent stage. Members of the WiMAX forum are working together to carefully plot the evolution of 802.16 from a vague standard into a marketable technology. Common ambition and experience deploying Wi-Fi serve as their guide rails.

WiMAX is not Wi-Fi—not yet, anyway—but the next evolution in broadband wireless has set off down Wi-Fi's path. Wi-Fi, based on the 802.11 standard, once claimed only a sparse, polka dot pattern of hot spots around the globe but that pattern is becoming denser by the day— much faster than wireline broadband technologies could. Its success has

Table 21-1

WiMAX Standard Evolution

	802.16	802.16a/RerD	802.16e
Completed	Dec. 2001	802.16a: 802.16 Q3 REVd: 2004	Estimate: 2nd half of 2005
Spectrum	10 to 66 GHz	< 11 GHz	< 6 GHz
Channel Conditions	LOS only	NLOS	NLOS
Bit Rate	32 to 134 Mb/s at 28 MHz channelization	Up to 75 Mb/s at 20 MHz channelization	Up to 15 Mb/s at 5 MHz channelization
Modulation	QPSK, 16 QAM and 64 QAM	OFDM 256, OFDMA, 64 QAM, 16 QAM, QPSK, BPSK	Same as REVd
Mobility	Fixed	Fixed and Portable	Mobility, Regional Roaming
Channel Bandwidths	20, 25, and 28 MHz	Selectable channel bandwidths, 1.25– 20 MHz; up to 16 logical subchannels	Same as REVd
Typical Cell Radius*	1 to 3 miles	3 to 5 miles; max range 30 miles	1 to 3 miles

*Based on tower height, antenna gain, and transmit power (among other parameters)

Source: WiMAX Forum

given proponents of broadband wireless access, whose visions of success have proved mirages before, reason to hope again.

The WiMAX Forum, comprised of more than 100 companies, is gathering a broad array of vendors and service providers from around the world that will contribute to the success of WiMAX. Advancing the technology to market depends largely on the ability of this industry cooperative, which will shape an intentionally vague standard into something more marketable and receptive to mass deployment. The key objective consists of devising specifications for product interoperability across a wide range of equipment vendors and overseeing testing of the products at third-party labs in order to certify them for interoperability.

> *Key*: In the long-term future, the vision is for WiMAX to become a last-mile access technology integrated in laptops and other end-user devices. But in the near term, it probably will have the most viability for backhauling the rapidly increasing volumes of traffic being generated by Wi-Fi hot spots. This would reflect an end-to-end wireless last mile, all the way to the end users' device!

WiMAX has key benefits for carriers and operators. By choosing interoperable, standards-based equipment, the operator lowers the risk of deploying broadband wireless access systems.

- Economies of scale due to standardization will help reduce monetary risk.
- Operators are not locked in to a single vendor, or small group of vendors, because base stations will interoperate with subscriber stations from different manufacturers.

Ultimately, operators will benefit from lower-cost and higher-performance equipment, as equipment manufacturers rapidly create product innovations based on a common, standards-based platform.

21.4 WiMAX Application Options

The WiMAX standard will help the telecom industry provide wireless solutions across multiple broadband segments.

21.4.1   Wireless Carrier Backhaul Network

Long-haul and Internet backbone providers in the United States are required to lease lines to third-party service providers, an arrangement that tends to make wired backhaul relatively affordable. The result is that only about 20 percent of cell base stations are backhauled wirelessly in the United States. With the possible removal of the leasing requirement by the FCC, U.S. wireless providers may look to 802.16-based wireless backhaul as a more cost-effective alternative to leased lines. The robust bandwidth of 802.16a technology makes it an excellent choice for backhaul to support commercial enterprises such as hotspots, as well as point-to-point backhaul applications.

21.4.2   Broadband on Demand

Last-mile broadband wireless access can help spur the deployment of 802.11 hotspots and home/small office wireless LANs, especially in areas not served by cable or DSL or in areas where the local telco may have a long lead time for provisioning broadband service. Broadband Internet connectivity is mission critical for many businesses, to the extent that these organizations may actually relocate to areas where broadband service is available. In today's marketplace, local exchange carriers have been known to take 3 months or more to provision a DS1 circuit for a business customer, if the service is not already available in the building. WiMAX technology enables a service provider to provision service with speed comparable to a wired solution in a matter of days, and at a significantly lower cost.

Key: 802.16a technology also facilitates instantly configurable, on-demand, high-speed connectivity that can support hundreds or even thousands of users for Wi-Fi hotspots for temporary events, such as trade shows. In this context, WiMAX operators could use 802.16a solutions for backhaul to the core network. Current (wireless) technology makes it possible for the service provider to scale-up or scale-down service levels, literally within seconds of a customer request.

On-demand connectivity also benefits businesses such as construction sites that have irregular broadband connectivity requirements. On-demand last-mile broadband service, such as WiMAX, represents a significant new profit opportunity for any carrier willing to deploy it.

21.4.3   Supplementing Gaps in Cable and DSL Coverage

The cost and complexity associated with traditional wired cable and telephone infrastructure have resulted in significant broadband coverage gaps in the United States and across the globe. Early attempts to use wireless technology to fill these coverage gaps have involved a number of proprietary solutions for wireless broadband access that have fragmented the market without providing significant economies of scale.

A surprisingly large number of areas in the United States and throughout the world still do not have access to broadband connectivity. Providing a wireline broadband connection to a currently underserved area through cable or DSL can be a time-consuming, expensive process.

Practical limitations prevent cable and DSL technologies from reaching many potential broadband customers. Traditional DSL can only reach about 18,000 feet (3 miles) from a telco central office, and this limitation means that many urban and suburban locations cannot obtain DSL connectivity. Cable also has its limitations. Many older cable networks have not been equipped to provide a return channel, and equipping these networks to support high-speed broadband can be expensive. The cost of deploying cable is also a significant deterrent to the extension of wired broadband service in areas with low subscriber density. This is because the capital outlay required to implement the cable, which is mostly labor costs for trenching, would take a long time to recoup, if it could be recouped at all. The low subscriber density is the mitigating factor. Even if all potential customers bought into the service, their low numbers would likely not be sufficient to cover the cost of supplying cable to the area(s).

For wireline telcos, WiMAX could prove cheaper and easier to install than DSL, and the evolution toward 802.16e's mobile capability in the future could help telcos compete better with the wireless carriers. Also, WiMAX with mobility could provide a broadband evolution for mobile carriers that have not yet decided about 3G. WiMAX will surely get the attention of wireless carriers because 3G CDMA (EV-DO) operates at lower rates than WiMAX. 802.16e should reach the marketplace sometime in late 2006 or 2007.

> *Key*: The current generation of proprietary wireless systems are relatively expensive for mass deployments because, without a standard, few economies of scale are possible. This scenario also limits procurement leverage of WiMAX carriers. Standardizing a technology leads to more equipment manufacturers getting into the game, which is good for operators because it gives them additional leverage with manufacturers when negotiating.

Cost and deployment inefficiencies will change with the launch of standards-based systems based on WiMAX. The range of 802.16a solutions, the absence of a line-of-sight requirement, the high bandwidth, and the inherent flexibility and low cost of WiMAX-type service helps to overcome the limitations of traditional wired and proprietary wireless technologies.

Service providers seeking to build broadband infrastructure in developing countries should take fiber as far as they can practically take it, then go wireless. WiMAX technology will allow these companies to deploy broadband quicker and cheaper. For any carrier looking to fill the holes in its cable or DSL networks, wireless is the most viable solution today. This is because the cost of wireless components has dropped dramatically since the late 1990s, and WiMAX standardization—like any technology standardization—will facilitate widespread deployment.

21.4.4   Areas Underserved by Broadband

Wireless Internet access based on the WiMAX standard is also a natural choice for underserved rural and outlying areas with low population density. In such areas, local utilities and governments usually work together with a local WISP to deliver Internet access service. Recent statistics show that there are more than 2,500 WISPs who take advantage of license-exempt spectrum to serve over 6,000 markets in the United

States.[1] On an international basis, most deployments are in licensed spectrum and are deployed by local exchange carriers who require customers to purchase voice services in addition to high-speed data. This is because in these areas the wired infrastructure either does not exist or does not offer the quality to support reliable voice, let alone high-speed data. This is essentially a wireless local loop application.

21.4.5 Extendable Wireless Access

As the number of Wi-Fi hotspots proliferates at a blazing pace, users will naturally want to be wirelessly connected, even when they are outside the range of the nearest hotspot. The IEEE 802.16e extension to 802.16a introduces roaming capabilities that will allow users to connect to a WISP even when they roam outside their home or business or go to another city that also has a WISP.

▮▮▮ 21.5 Licensing

The WiMAX Forum has no intention of navigating the jumble of international spectrum policy to bring the licensed frequencies into the WiMAX fold. The Forum has selected a few bands that meet the world's needs, and they have formed a regulatory group that will work toward harmonization. Essentially, the WiMAX Forum is concentrating on two critical bands: the 2.5 GHz or MMDS band and the 3.5 GHz band—by far the most abundant broadband spectrum allocated across international borders. There are other bands under consideration that WiMAX could support, but the Forum and most of its vendor members have chosen to concentrate their energies on those two bands. Because WiMAX is intended to be a specification that enables volume sales through standardization, the objective is to generate a lot of buzz about a few spectral bands rather than little interest in numerous bands. The Forum is getting approached by many governments that want to know how they can work toward harmonization.

---

[1]Source: ISP-Market 2002

Key: As is the case with so many other wireless technologies, the United States is unique when it comes to the potential for broadband wireless. While the rest of the world is jumping on the 3.5 GHz spectrum, regulation has kept those bands off of the table in the United States. Although discussion is happening at the FCC regarding opening up the similar 3.6 GHz band, most of the United States hopes lie in the 2.5 GHz MMDS held by Sprint and Nextel. Neither of these carriers have announced any WiMAX-related plans to date for their MMDS spectrum.

WiMAX is not a new technology—it is a more innovative and commercially viable adaptation of a proven technology that is delivering broadband services around the world today. In fact, wireless broadband access systems from WiMAX Forum members are already deployed in more than 125 countries around the world. These leading equipment providers are on a migration path to WiMAX.

21.6   Proceeding with Caution

For now, WiMAX proponents are hopeful about the prospect for broad market deployment, but they are also trying to keep that hope from mutating into too much hype. It is in the back of everyone's mind that overinvestment, too much marketing hype, and exorbitant product cost were among the factors that killed off a previous generation of broadband wireless service providers, namely LMDS and MMDS. Sprint purchased MMDS spectrum licenses, launched Sprint Broadband service around 2000, only to grandfather the service around 9 months later. This is not to say that many carriers who have purchased LMDS and MMDS licenses will not eventually put them to use again at some point in the future. How that is done is the million dollar question. WiMAX Forum members believe the mistakes of the past have helped them contribute to a detailed, revised standard and a product certification plan that broadband wireless did not have 5 years ago. This time around, Wi-Fi now exists as a last-mile complement that paves the way for WiMAX. Theoretically, WiMAX will be a natural extension of something that already exists: Wi-Fi.

> *Key:* With the advent of a standard and economies of scale driving down equipment prices and improving performance, many analysts predict a virtual explosion in the market for broadband wireless. In-Stat/MDR forecasts the market will grow to more than $1.2 billion by 2007.

Perhaps one of the most exciting aspects of WiMAX—one with the potential to make this a significantly bigger market—is the evolution to true mobility. The WiMAX Forum worked aggressively to incorporate mobile capabilities into the 802.16 standard by the end of 2004. In the 2006 timeframe, it is expected that WiMAX will be incorporated into end-user devices like notebook computers and PDAs along with Wi-Fi and Bluetooth, enabling the delivery of wireless broadband directly to the end user—at home, in the office, and on the move. This potential could allow for true end-to-end wireless communications, a truly ground-breaking moment in the world of wireless communications. In the 2007 timeframe, it is expected that WiMAX will be integrated into 3G phones along with Wi-Fi, providing a simplified network connection for voice and data. The WiMAX Forum is driving a common platform for harmonization between standards that will enable users to remain connected wherever they go.

But for all of the industry hope and hype that has been placed at the feet of WiMAX, the technology remains today in an early, premarket stage.

Market success for WiMAX is not assured. Though analysts believe the WiMAX market will be worth anywhere from $2 billion to $5 billion by 2009, the defining aspects of that anticipated growth remain to be pinpointed. It is very likely that the initial service providers adopting WiMAX in 2005 will be wireless ISPs, many of whom already are demanding precertified products from vendors.

WiMAX has the potential to be the great equalizer in broadband access—to close the gap of the digital divide, which has never been more within reach.

Test Questions

True or False

1. _____ The 802.16e standard is the version of WiMAX that will offer true mobility.

2. _____ Areas underserved by broadband access are not a potential application for WiMAX.

3. _____ WiMAX is sometimes labeled the "wireless LAN" solution.

Multiple Choice

1. Many industry observers speculate that WiMAX will compete with Wi-Fi. To what other wireless technology could WiMAX actually pose a bigger threat?

   a. Bluetooth

   b. 802.20

   c. 3G

   d. None of the above

2. What type of wireless operator stands to benefit greatly with the availability of 802.16e?

   a. 1,900 MHz wireless carriers

   b. Landline-based ISPs

   c. Wireless ISPs (WISPs)

   d. GSM carriers

CHAPTER

22

Home Networking

22.1   Introduction

With the increasing availability of high-speed Internet access services and the growing number of multiple-PC households, interest in home networking is growing at an accelerated pace. People are readily adopting high-speed Internet access and want to share the benefits of a single high-speed Internet link among all family members.

In the late 1980s, the rise of the personal computer fostered by Apple and IBM introduced the rest of the world to personal computing. Computers used to be the tools of the technically inclined, now few people do not have access to a computer. By the end of 2005, industry experts predict over 40 million households will have high-speed Internet connections in North America.

22.2   Home Network Basics

A home network is a group of computers, printers, scanners, game consoles, or other devices connected together inside a home using a hub and/or router. This network enables these devices to communicate and share information with each other and to share access to one high-speed Internet connection.

With a home network, users can share a digital subscriber line (DSL) Internet connection between computers in the home, easily share files and printers amongst PCs, play digital audio and video from the PC to the stereo and television, control home automation systems through the PC, and much more. Users operating a home network are able to perform computing tasks much more conveniently, while simultaneously reducing the overall cost of computing within the home.

A typical home network includes one or more computers (PCs and/or laptops), a modem, network interface cards (NICs), hubs or routers, and network cabling or wireless media. All of these basic components enable the computers and devices to communicate with each other.

22.2.1   Computers

A minimum of one desktop PC or laptop computer is required for a home network. NICs are also required, which enable the PC to interface and connect to a network—in this case a home network. NIC cards plug into desktop computers in the PCI interface and into laptop computers in the PCMCIA interface. The Personal Computer Memory Card International Association (PCMCIA) is an industry group organized in 1989 to promote standards for a credit-card-size memory or I/O device that would fit into a personal computer, usually a notebook or laptop computer. The PCMCIA 2.1 standard was published in 1993 and, as a result, PC users can be assured of standard attachments for any peripheral device that follows the standard. [1]

Key: NICs can be either wired or wireless. In the case of a wired connection, the NIC commonly connects and communicates over Ethernet cable to an Ethernet hub. Ethernet is the most widely installed LAN technology in the world. An Ethernet LAN typically uses Category 5 (CAT5) twisted pair cable to attach computers to hubs or routers, but Ethernet is also used in wireless LANs, which include home networks. The most commonly installed Ethernet systems are called 10BASET and provide transmission speeds up to 10 megabits per second. Devices are connected to the cable and compete for access to the LAN. In the wireless scenario, the NIC connects and communicates over 2.4 GHz wireless radio frequencies to a wireless access point, also known as a *wireless gateway*. The 2.4 GHz spectrum used in wireless home networks is the same spectrum that is used in Wi-Fi access points (hot spots)—the unlicensed ISM band. This connection enables the two-way transmission of data between the PC and other devices on the network.

22.2.2 Network Interface Cards, Hubs, and Routing

Computer NICs can be either internal or external. Internal NICs are housed inside desktop PCs and are inserted into PCMCIA slots in laptops. External NICs can be plugged into a USB port in either a desktop

---

[1]Source: whatis.com)

or laptop computer. With the rising popularity of Ethernet networking, most new computers come with NIC cards built in.

The hub is the central connection point for all devices in a network. PCs and devices are connected to the hub using a wired or wireless connection that plugs into ports on the hub. When one PC sends data to another PC or other device on the network (e.g., printer), all data travels through the hub. The hub takes data from the sending PC and broadcasts it to all PCs on the network. Receiving PCs listen only to data that is addressed to them. See Figure 22-1.

A network hub can have anywhere from 4 to 16 ports, and all networked devices have cables connecting them to the hub. Hubs also range in speed, and the total speed capability of the hub (bandwidth) is usually either 10 or 100 Mbps. Today's hubs are usually 100 Mbps (or even higher, such as one gigabit per second) because component costs and Ethernet technology have become very inexpensive and ubiquitous.

Key: A router looks the same and performs the same functions as a hub with one difference. Instead of forwarding data from the sending PC to all the PCs on the network, the router can determine exactly what device is the intended recipient of data and send the data to only that device. This functionality is much more efficient than a hub and therefore speeds up data transmission over the entire home network. The ideal router type to use in a home network is a wireless router.

22.3 Why Home Networking?

The major telephone companies have a long history of building, maintaining, and operating world-class voice and data networks. For this reason, more and more people are turning to their traditional telephone companies to meet their needs for high-speed Internet service.

At the same time voice, video, and other applications become more popular, DSL users are beginning to want their DSL connectivity distributed throughout the home. They accomplish this by building a home

▩▩▩ ▩▩▩ ▩▩▩
Figure 22-1
Network hub

network. The main drivers behind the popularity of home networks are as follows:

▩ Shared Internet access

▩ File and printer sharing

▩ Telecommuting (using work-supplied laptop at home)

▩ Home control and automation

▩ Multiplayer gaming

▩ Entertainment—audio and video distribution

22.3.1 Home Network Applications

The primary use for a home network is to share a high-speed Internet service between all of the computers within the home. Anyone in the home can then enjoy the benefits of the high-speed connection to e-mail, surf, chat, explore newsgroups, and more, all at the same time. DSL and home networking means Mom can work from home, while Dad does the banking, and the kids play games with the family across the street.

Users on a home network can easily share files and other devices like printers and storage devices. In this manner, files can be shared between users and computers much more conveniently than by using disks or removable drives. Ultimately, home network users benefit by saving the cost of buying additional printers and file storage devices for each computer in the home.

Many users are interested in connecting their company-owned laptops to the Internet, which allows them to telecommute from home.

Before the advent of DSL, doing so meant tying up the home telephone line. Now DSL gives these users Internet access speed at home that is roughly on par with what they would have at the office and frees up their telephone line. Furthermore, with the advent of virtual private networks (VPNs), users can *securely* access corporate e-mail, files, intranets, or the Internet.

The ability to control the environment within the home is also a promising opportunity for DSL users. Home networks can be configured to interface with home control devices to manage lighting, heating, air conditioning, and home security over the Internet. Picture being at work and adjusting the temperature in your home, turning on the lights in your living room, and double-checking your home security system via the Internet. A webcam could also be mounted in and around the home, which could be remotely monitored from any Internet connection.

Capabilities of gaming systems and consoles continue to develop, encouraging Internet gaming with other people across town or across the world. The home network of today can be set up to support all kinds of gaming applications.

The home network can also be used for entertainment activities like enjoying music and movies around the home. This is accomplished by distributing MP3 audio files and video files from a home computer or Internet service to the stereo system or televisions in the home. Another option would be the ability to watch any movie at any time by accessing the movie via your Internet service. Or imagine being able to listen to an enormous selection of customizable, commercial-free music on your stereo system or media player.

22.4   Home Network Architecture

A basic home network is built around a base station, which could be also called a router or residential gateway. This router is the central device that allows two or more computers to share an Internet connection, files, folders, photos, a scanner, and printers. Figure 22-2 shows how the router ties the computers together in a home network.

Computers 1 and 2 in Figure 22-2 can connect to the wireless router or base station in one of two ways: wired or wireless.

Figure 22-2
Home network wireless router function performed by base station

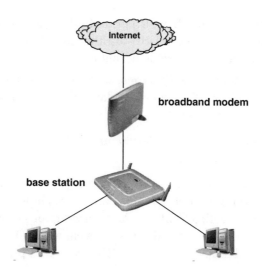

    Key: Even though several means exist to tie devices together, the wireless approach to home networking is the cleanest, least intrusive, most efficient, most cost-effective approach. It also requires the fewest number of components. Wi-Fi technology allows computers in other rooms of the house to connect with each other and the Internet, although the type of structure and distance from the base station could impact the effective range of wireless transmissions in a Wi-Fi-based home network. Use of wireless routers/gateways in home networks is akin to basically building your own Wi-Fi hot spot.

The most versatile routers available today usually have four built-in Ethernet ports available to connect computers within the same room, as well as a built-in wireless access point. A single device can address most home networking connectivity needs. This single device effectively functions as Ethernet hub, router, and Wi-Fi access point—all in one.

A NIC is required on every computer on the network so that it can send and receive data to and from the router. Network adapters could take several forms depending on the actual computer and what ports are available.

Wired adapters include the following:

- Laptops: Newer models have Ethernet built in, so no adapter is required. If not available, a USB-to-Ethernet adapter can be used or a PCMCIA card can provide Ethernet connectivity.

- Desktop PCs: Today new PCs come with built-in Ethernet adapters (NICs), but if not, a PCI Ethernet adapter or USB-to-Ethernet adapter can be used as well.

Wireless adapters, which are the dominant type in use in home networks, include the following:

- Laptops: Newer models have Wi-Fi adapters (NICs) built in, so no adapter is required. Intel developed an entire chipset family to cater to this function, called Centrino. Most new laptops even have the Centrino label on them to advertise this capability. If not available, the most popular option is the PCMCIA card that provides Wi-Fi access.

- PCs: A PCI Ethernet adapter or a USB-to-Ethernet adapter can be used as well.

Wi-Fi equipment comes in various flavors and speeds (802.11b, 802.11a, etc.). The most popular and economical is 802.11b, although some equipment can support both 802.11b and 802.11a, which is faster but has a shorter range. Although it is not necessary to buy the router and adapters from the same vendor, be sure that both the router/base station and your wireless adapters support the same Wi-Fi standard.

Once computers have the *appropriate* network adapter installed, the following actions are required to fully enable the home network:

- Install the network drivers (software) that will support the NIC. The drivers are usually supplied on a CD that comes with the NIC or computer.

- Enable your Wi-Fi adapters to work with the router's built-in access point (wireless base station).

- Turn-on the printer and file-sharing feature.

You should now be able to access your high-speed Internet connection and printer from all computers. This also opens the door for other applications that take advantage of the functionality offered by a home network:

- Play interactive games at high speed

- Backup of files/data to a central computer when desired (and perhaps make backups to a local CD-RW)

- Share MP3 music files

22.5 Home Network Operations

The IP allows for each computer network to have a unique address. Specifically, web sites are assigned a unique IP address (e.g., 64.170.98.10). All computers inside this domain have this address (think of it as the building address), but there is still a requirement to address each individual computer *inside* the domain. A separate, private address space is used inside the domain to communicate with each unique computer inside the building (think of it as the floor number and the cubicle ID), and the router in a home network appropriately translates messages or packets from the internal address space to the external IP address, and vice versa. The router uses a function called *Network Address Translation* (NAT) to appropriately map the internal private addresses to the unique public IP address.

In a home network, an IP address is assigned by the Internet service provider (ISP), and computers in the home network use a private address space. If we want these computers to communicate with other networks in the Internet, the home network also needs a router. Note that for a home network, you need a far simpler and less expensive router than the type a corporation would use, but a router is needed anyway.

22.6 Technology Options for Home Networks

The following paragraphs describe common technologies available for home use, although not all technologies are available in all countries.

22.6.1 Ethernet (IEEE 802.3): 10/100 Mbps

This is the most popular business network because of the low price of network cards, minimal configuration, reliability, privacy, security, and speed. It requires special cabling (CAT5), and it is best used when all computers are in the same room.

22.6.2 Wireless (Wi-Fi): 11 Mbps to 54 Mbps

Various standardized variations are available today, with IEEE 802.11b being by far the most common and least expensive, and IEEE 802.11a being the newer version of Wi-Fi. IEEE 802.11b is a very common technology used in both the home and business. It has become inexpensive and has good speed in most homes (effective range depends on the size and construction materials of the house). It may not be appropriate where security is a major concern—users need to make sure security features are enabled. The 802.11i standard defines security requirements for Wi-Fi networks.

Products using IEEE 802.11a are also available, but are more expensive and the effective range inside a home is not as good as IEEE 802.11b. There is a newer version, which should be standardized by 2005, called IEEE 802.11g, and promising to offer the range of IEEE 802.11b with the speeds of IEEE 802.11a. Some vendors offer solutions that support both b and a and will soon support g as well. Although this equipment may be more expensive, its multimode capability means users can mix and match as needed within the same network, which makes installation easier.

22.6.3 Ultra-Wideband Wireless

Evolving digital homes require connectivity of devices at ranges far beyond the 10-meter-or-less, single-room technology. Consumers are driving the demand for total home networking with a need for computer systems, handheld devices, high-definition televisions, entertainment systems, and other consumer electronics to communicate with each other, transferring high-quality video, photos, text, and audio to all parts of the home. A company known as Pulse~LINK offers a UWB-based approach to home networking, which provides a single solution for total-home networking with wireless LAN capabilities that can be further enhanced through home-electric power lines and cable technologies. Because the chipset automatically selects which networking solution to use based on quantified assessment of all options, the consumer need not be concerned whether the networking is done over wires or wirelessly.

22.6.4 Phone Line (HomePNA™ 2.0): 10 Mbps

This technology, also known as HNPA, supports home networking using the existing copper telephone wires in the home, without interfering with voice or ADSL communications. It is not as fast as Ethernet but, because no new wiring is needed, is a good alternative as long as there is a phone outlet in each room.

22.6.5 Powerline Networking: 10 to 14 Mbps

This technology uses the existing power wires inside the home for carrying data traffic. For home use, this approach provides a good alternative as most rooms in a house have at least one power outlet. However, there is no common standard for this technology. Verification of the standard used by vendors' Powerline adapters is required when purchasing from different sources. An easy solution is to buy all adapters from the same vendor to ensure compatibility. HomePlug® 1.0 (14 Mbps) is one viable option, for example. Products based on this technology are available in many retail computer stores, but note that other implementations of Powerline besides HomePlug are also available. Although data transmission using power lines can be used for broadband purposes, the Powerline technologies mentioned here only address data transmission inside the home or apartment.

22.6.6 Universal Serial Bus (USB)

Some of today's routers (home gateways or portals) come with built-in USB ports. This allows one computer (close to the router) to be connected to the router, without requiring Ethernet, Wi-Fi, or any other technology or adapter—a simple USB cable is all that is needed. Other technologies are still needed for the rest of the computers in the home network, but when available, this single USB port in the router makes connecting one computer very easy.

Most routers, hubs, and switches use Ethernet to interconnect between computers in home networks. So keep in mind that most of your computers that use USB ports might need adapters to connect them in a home network.

Key: The use of these technologies is not exclusive. For example, USB could be used on one computer, Ethernet in the rest of the room, and wireless with a laptop could be used in several other rooms. Different technologies can be mixed and matched to meet the particular home needs.

22.7   Selecting the Right Infrastructure Technology

Do the computers that need sharing reside within the same physical room, or are the computers scattered across several rooms? Do one or more mobile devices (laptop or PDA) also needed to be interconnected? These are a few sample questions that need to be answered when deciding how to approach building a home network.

If the computers that need to be part of the home network are not in the same room, then using HomePNA, HomePlug, and/or Wi-Fi are good alternatives. Wi-Fi is the best option though because it has a widely installed base, is easy to install, and is cheap. If all rooms that have computers that need to be connected to the home network have phone outlets or power outlets, then the easiest alternative is to use either HomePNA or HomePlug. Wireless is still an alternative for those rooms as well—it is a personal or logistical choice. Phone wiring extension kits are inexpensive and fairly easy to use, so extending a phone network can be done.

Key: If at least one mobile device (laptop or PDA) is part of the home network, then at least one wireless base station or access point is needed to support those mobile devices. The mobile devices need a wireless NIC, which can be built in to the laptop (the majority of all laptops sold today have wireless NICs built right into them at the factory). The user does not need to use wireless for all computers, although it is recommended if using wireless for one networked device.

A firewall is a highly recommended feature that helps protect computers inside the home network from malicious attacks from hackers. A software firewall can be installed in each computer in the home network, a hardware firewall (separate box) can be installed that would protect all computers in the home network. Just as a router can have a built-in hub or (Ethernet) switch, some routers also come with built-in firewalls. A router with this feature simplifies network security and generally costs less than buying separate boxes (e.g., router plus firewall).

Setting up a home network could appear to the first timer or novice like an intimidating job. But with careful planning and proper evalua-tion of current and future home network needs, this task does not need to be complicated at all. After doing the preparation suggested here, the potential home networker could take his or her home diagram (showing the computers or other devices and the rooms in which they need to be connected) and visit a local store that sells computer gear. With this home network diagram in hand, most computer stores should be able to assist in selecting the appropriate networking gear as dictated by your budget and networking needs.

▬▬▬ Test Questions

True or False?

1. _____ Home network infrastructures can be a combination of multiple networking technologies such as HomePNA, Wi-Fi, cabled Ethernet, and power-line networking.

2. _____ Wi-Fi is the easiest, most efficient, and most economical home network technology option.

3. _____ A wireless router is also known as a *residential gateway*.

Multiple Choice

1. What is the name of the computer chipset designed by Intel that is specifically designed with wireless (NICs) in mind?
   a. Centrifuge
   b. Greentooth
   c. Centrino
   d. WiMAX
   e. None of the above

A

Test Question Answers

Chapter 1

True/False

1. F

2. T

3. T

4. T

5. F

Multiple Choice

1. C

2. A

Chapter 2

True/False

1. F

2. T

Multiple Choice

1. C

2. C

3. C

4. C

Chapter 3

True/False

1. T

2. F

3. T

4. F

Multiple Choice

1. E

2. C

3. B

Chapter 4

True/False

1. F

2. T

3. F

4. T

5. F

6. F

7. F

Multiple Choice

1. G

2. B

3. C

4. C

5. D

Chapter 5

True/False

1. T

2. T

3. F

4. F

5. T

Multiple Choice

1. D

2. C

Chapter 6

True/False

1. F

2. F

3. F

Multiple Choice
1. D
2. C
3. C
4. C

Chapter 7

True/False
1. T
2. T
3. F
4. T

Multiple Choice
1. C
2. E
3. D

Chapter 8

True/False
1. F
2. F
3. T

Multiple Choice
1. C
2. D
3. A
4. C

Chapter 9

True/False
1. T
2. T

Multiple Choice
1. F
2. C
3. F

Chapter 10

True/False
1. F
2. F
3. T

Multiple Choice
1. C
2. A

Chapter 11

True/False
1. F
2. T
3. F
4. T

Multiple Choice
1. B
2. C

Chapter 12

True/False
1. T
2. T
3. F
4. F
5. F
6. T
7. F

Multiple Choice
1. C
2. B
3. D
4. D

Chapter 13

True/False
1. F
2. T
3. F
4. T

Multiple Choice
1. C
2. D

Chapter 14

True/False
1. T
2. F
3. F

Multiple Choice
1. F
2. D
3. B
4. D

Chapter 15

True/False
1. T
2. F
3. F
4. T

Multiple Choice
1. C
2. D
3. B

Chapter 16

True/False
1. T
2. F
3. T
4. T
5. T

Multiple Choice
1. F
2. C

Chapter 17

True/False
1. T
2. T
3. T

Multiple Choice
1. B

Chapter 18

True/False
1. T
2. F
3. F
4. F
5. T
6. F

Multiple Choice
1. C
2. D

Chapter 19

True/False
1. T
2. F

Multiple Choice
1. C
2. C

Chapter 20

True/False
1. F
2. F
3. T

Multiple Choice
1. C
2. D

Chapter 21

True/False
1. T
2. F
3. F

Multiple Choice
1. C
2. C

Chapter 22

True/False
1. T
2. T
3. T

Multiple Choice
1. C

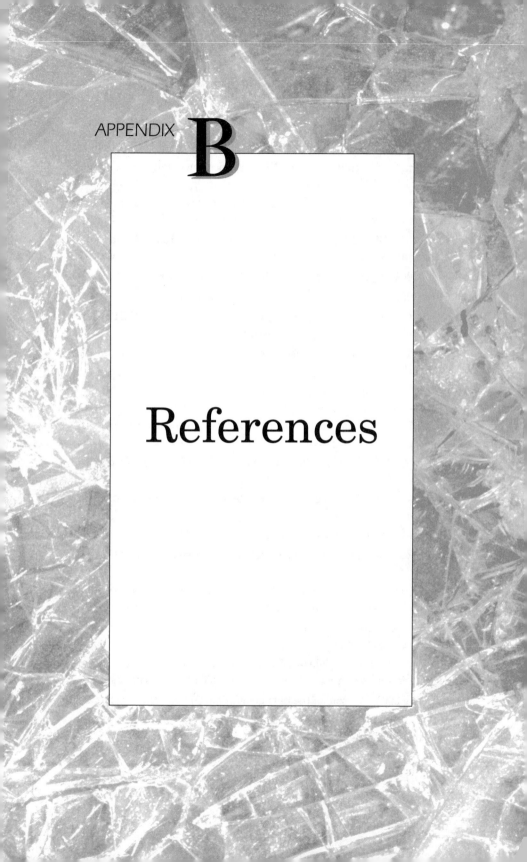

APPENDIX **B**

References

3-G Generation: What is 3G? www.3g-generation.com/what_is.htm.

3GW-CDMA principle. www.3g-generation.com/cdma_principle.htm.

3G overview. www.handytel.com.

Ankeny, Jason. 2004. Call 911: Wireless 411 is in trouble. *Wireless Review*, October.

Balint, Kathryn. Future calling. Copley News Service. www.dailybreeze.com/business/articles/1340326.html.

Beach, Tarre. 2000. Structurally sound? Don't let your towers end up the forgotten element of preventive maintenance. *Wireless Review*, May 15.

Bedell, Paul. 2003. High five for Wi-Fi. Internal newsletter from SBC Pacific Bell Consultant Vendor Sales Group. www.sbc.com/Products_Services/CSG/.

Bedell, Paul. 2001. *Wireless crash course*. New York: McGraw-Hill.

Bedell, Paul. 2000. Interconnection primer. *Wireless Review*, June 15.

Bernal, Jose. 2000. Smart antenna systems. DePaul University TDC 512 Cellular and Wireless Communications Case Study. June.

Bohrer, Becky. 2004. Cell phones cause a howl in the wild. *Chicago Tribune* (AP), April 13.

Boney, Karissa. 1998. Appeasing the locals. *Wireless Review*, January 15.

CDMA Development Group. 2003. The truth about EDGE. White paper from the CDMA Development Group, November 2003. www.cdg.org/resources/white_papers/files/EDGE.pdf#asearch='The%20Truth%20About%20EDGE'.

Colker, David. 2005. Cell phones ring knell on privacy. Technology, *Chicago Tribune,* January 1.

CTIA. November 2004. CTIA semiannual wireless survey. Obtained with written permission from CTIA.

Dell, Stephanie. 2004. Cell phones get juiced. *Telephony*, October 25.

Dvorkin, Larry. 2000. CDMA technology. DePaul University, TDC 512 Case Study, June.

Eisa, Mahir. 2003. UWB technology. TDC 512 Cellular and Wireless Communications Project Paper, Autumn Quarter, DePaul University, Chicago.

Empleo, Philamkelly. 2000. Case study in emergency 911. DePaul University, TDC 512 Case Study, June.

Ericsson Safety Electrical Specialists. EDGE White Paper. White paper from Ericsson. www.ericsson.com/products/white_papers_pdf/edge_wp_technical.pdf.

Fitchard, Kevin. 2005. The soul of the next generation. *Wireless Review*, January.

Fitchard, Kevin. 2004. Consolidation with a vengeance. *Telephony's Wireless and WiFi Weekly*, December.

Note: Internet entries include the source information as listed on references' Web sites.

Flynn, Tom. 2000. CDMA technology: An overview. DePaul University, TDC 512 Case Study, June.

From cell phone to sunflower. 2004. http://news.designtechnica.com/ article6060.html (via CNN.com), December 6.

Gohring, Nancy. 1998. Location, location, location. *Telephony*, May 11.

Gwinn, Eric. 2004. Viruses could become real hang ups for cell phones. *Chicago Tribune*, June 29.

Harter, Betsy. 1999. Ready for inspection. *Wireless Review*, May 1.

Home networking. http://petri.co.il/htm.

Home networking FAQ. www.homenethelp.com/web/faq/start.asp.

HSPDA: The next step for 3G. 3gnewsroom.com. 2004, December 16. www.3gnews room.com/3g_news/dec_04/news_5314.shtml.

Hughlett, Mike. 2004. Spring to take over Nextel. *Chicago Tribune*, December 16.

Intel. IEEE 802.16 and WiMAX: Broadband wireless access for everyone. White paper from Intel, www.intel.com/business/bss/infrastructure/wireless/ 80216_wimax.pdf.

International Engineering Consortium. Smart antenna systems. WebProforum, Online tutorial. http://iec.org/online/tutorials/smart_ant/.

International Engineering Consortium. Wireless intelligent network (WIN). Web ProForum, Online tutorial. http://iec.org/online/tutorials/win/.

Knight, Will. 2005. New hybrid cell phone virus discovered. *New Scientist*, January 13.

Krahenbuhl, Todd. 2004. Software-defined radios. *TDC 593 Wireless System Engineering and Deployment*, Project Paper, Winter quarter, DePaul University, Chicago.

Levy, Stephen. 2005. Something in the air. *Newsweek*, June 24.

Lynx unlicensed microwave systems. www.proxim.com/products/bwa/point/lynx.

Nandhini, V. WiPro migration of GSM nets to GPRS. *Computer World*. www.wipro.com/insights/migrationgsm.htm.

Numerex Technologies. August 2002. Cellemetry data service via cellular. Numerex Technologies. www.nmrx.com/pdf/w_CellemetryNetworkOverview.pdf.

O'Shea, Dan. 2004. Telephony's complete guide to WiMAX. *Telemetry Magazine*, May 31.

Overview of the universal mobile telecommunication system (draft). 2002, July. www.umtsworld.com/technology/overview.htm

Poe, Robert. 2004. Streaming enters the mobile realm. *America's Network Magazine*, December 1.

Pulse-LINK develops first UWB software-defined chipset platform. 2004. *PR Newswire* (United States), May.

Pulse-Link, Inc. Ultra wideband over cable technologies: Enhancing cable television bandwidth without modification to existing infrastructure. White paper from Pulse-Link, Inc. www.pulse-link.net/Pulse-LINK-UWBOC-whitepaper.pdf.

Roberts, Randy. The ABCs of spread spectrum. www.sss-mag.com/ss.html.

Rubin, Daniel. 2005. Ringing up sales with ringtones. *Chicago Tribune*, February 2.

Shah, Niraj. 1998. Microwave path design. *Wireless Review*, September 1.

Siemens. 2004. Mobile communication speed record: One gigabit per second over the air. Siemens.com. Dec 7. www.siemens.com/index.jsp?sdc_p=cz3s3uo1232024pnflmi1218535&sdc_sid=32379314396&sdc_bcpath=1156112.s_3%2C&.

Smith, Clint. 2004. *Wireless Network Performance Handbook*. New York: McGraw-Hill.

Steele, Raymond. What is CDMA? Multiple Access Communications Ltd. *Mobile Communications International Magazine*.

Sullivan, Robert. 2005. The camera phone revolution. *Life* (Chicago Tribune insert), January 28.

UWB and Bluetooth. aetherwire.com.

Van, Jon. 2004. Wireless carriers push 411 access. *Chicago Tribune*, November 12.

Van, Jon. 2003. FCC allows using home number for cell phone. *Chicago Tribune*, November 11.

Van, Jon. 1998. The drive to pinpoint mobile callers. *Chicago Tribune,* June 26.

Vocal Technologies. GPRS white paper. White paper from Vocal Technologies, Ltd., www.vocal.com/white_paper/GPRS_wp1.doc.

What is spread spectrum? www.kmj.com/proxim/pxhist.html.

What is UWB? Palowireless.com, www.palowireless.com/uwb/tutorials.asp.

Whatis.com (online technical dictionary). Technical Definitions of spread spectrum technology; WiMAX; home networking; software defined radios; waveguide; coaxial cable; PCMCIA; PCI.

Warren Publishing. 2004. FCC eyes cognitive radio convergence with satellite communications. *Communications Daily*, May. Warren Publishing, Online newsletter.

Warren Publishing. 2004. Software-defined radio forum says safeguards may be necessary. *Communications Daily*, May. Warren Publishing, Online newsletter.

Young, Deborah. 2000. Ready to roam? *Wireless Review,* May 15.

INDEX